People of the Plow

People of the Plow

An Agricultural History of Ethiopia, 1800–1990

James C. McCann

The University of Wisconsin Press

The University of Wisconsin Press
114 North Murray Street
Madison, Wisconsin 53715

3 Henrietta Street
London WC2E 8LU, England

Printed in the United States of America

Library of Congress Cataloging-in-Publication Data
McCann, James, 1950–
        People of the plow: an agricultural history of Ethiopia,
    1800–1990 / James C. McCann.
        316p.   cm.
        Includes bibliographical references (p. 000) and index.
        ISBN 0-299-14610-3 (cl).—ISBN 0-299-14614-6 (pbk.)
        1. Agriculture—Ethiopia—History.   I. Title.
    S473.E8M33   1995
    306.3'49'096309143—dc20   93-37124

CONTENTS

ILLUSTRATIONS

MAPS

TABLES

TRANSLITERATION

Material presented in this book involves at least three languages of Ethiopia and Sudanese Arabic, though there is more use of Amharic names, place-names, and words than any other. The overriding principle of transliteration I have used is to render words without diacritics into Latin script in a manner which allows the reader to reproduce the pronounciation accurately. Inevitably there are inconsistencies. This system, for instance, creates some problems with vowels, especially in the first and fifth orders. Nonetheless, I have rendered vowels in the following manner:

Order in Ethiopic script	Transcription	Pronunciation
1st order	a	as in "ever"
2nd order	u	as in "crude"
3rd order	i	as in "elite"
4th order	a	as in "father"
5th order	e	as in "bait"
6th order	e	as in "fit"
7th order	o	as in "vote"

In the transcription of consonants, *q* represents the ejective *k*, and the palatized *n* (as in "canyon") is written with *ny*. Double consonants (e.g., as in "Wallo") indicate gemination.

Personal names create the greatest discrepancies in transliteration, because many Ethiopians have chosen to spell their own names in print or in everyday usage in Latin script in ways which do not conform to a single system (e.g., Taddese and Taddesse, Tsehay and Sehay). Where an individual has already used a preferred English spelling, I have used that. In cases where certain spellings are in common or official usage (e.g., Haile Selassie rather than Hayla Sellase), I have accepted those conventions for those individuals. In place-names I have followed the above transcription method (Dabra Zayt rather than Debre Zeit). In verbatim interview texts in Oromofa I have used as much as possible the new Latin transliteration scheme now in use in Ethiopia.

Ethiopian names consist of a given (first) name and a father's name. In the text, the name used after the first occurrence is the given name. For bibliographic purposes I have also alphabetized Ethiopian names according to the first name.

PREFACE

This book has it origins in 1985, when Oxfam America asked me to spend a few weeks in a rural site in Ethiopia (Denki in the Ankober lowlands) to assess the likely effects of their program to distribute oxen, seed, and relief food to a rural population hit hard by two years of drought. I had at my disposal over a year of baseline data gathered by local Ethiopian staff of the International Livestock Centre for Africa (ILCA) and the chance to move freely—or however far my feet or a mule might carry me—around a remote rural site to talk to farmers and observe farm life. After the first week or so of gathering life histories and observing farm activity, it seemed clear to me that what I saw of the landscape and individual farmers' fields and what I heard from farmers about their lives provided a much richer and more complex story than what emerged from the meticulous social science data that my ILCA colleagues had collected in the field. In the end I was able to argue in my report (which became a beginning point for chapter 4) that local responses to the 1982–84 environmental crisis could and should be seen in a longer-term trajectory of social property, patterns in the local environment, and what I knew and had learned about highland farming systems.

Trying to reconcile the social data gathered in the field with farmers' life histories and the stories etched in the landscape itself (abandoned terraces, bits of irrigation, remnants of primary forest, checkerboard distributions of farmers' fields) left me unsatisfied. It occurred to me that my role in the project, though fascinating and, I believe, helpful, had become primarily one of translating the intensely human and subjective stories I gathered from farmers into the "objective" language of social science—opportunity costs, gross margins, social property—most meaningful to my sponsors.

In 1986 I returned for a month to follow the Oxfam project and catch up on another year of farmers' lives. Shortly thereafter I determined to expand these opportunities for applied field study and resolution of short-term problems to try to form a larger more generative book-length work of history. This book is an attempt to describe and explain change in an environmental, social, and technical process—the Ethiopian ox-plow agricultural system—over about 200 years. It has at once the task of telling a story, or a series of stories, and of presenting sufficient data to describe the system as a whole and offering a narrative which links the often disparate pieces of evidence. I therefore have attempted to merge the traditional strengths of history in fashioning a narrative and the social sciences' ambition to construct generative models for explaining, if not forecasting, human action.

In the Prologue I suggest that the narrative descriptions of agriculture taken from nineteenth-century travelers differ substantially in form and substance from twentieth-century empiricism which strives to quantify human action into the empirical precision of the natural sciences and sees agriculture fundamentally as an industrial process. My task in this book has been to reconcile these two descriptive methods into a narrative that incorporates every piece of evidence I can gather of past conditions. Whenever possible, I have attempted to give voice to these sources—travelers, farmers, statistical tables—and let them join my own narrative. The Prologue does not set out with a social science-style theoretical justification, but describes historical sources and how they shape our understanding of agriculture in the past. Unlike literary criticism's New Historicism of the late 1980s in which scholarship uses history to contextualize literary texts, for historians the past itself is the object for study. Sources are merely the lenses we use to view past events and processes, however murky those might prove to be. In this way, this book is almost as much an exploration of the methods and disciplinary bounds of history as an attempt to explain the shape of Ethiopia's agricultural system in the late twentieth century.

I also hope and expect that this book will provide some clarity and insight to those who must make policy decisions in the "fog" of development imperatives and donor expectations. The idea of conjuncture which I explore in this book speaks directly to many of their immediate interests in technology adoption, project design, and social impact.

This book has emerged out of a long journey of over five years, spent in often obscure archives and agricultural school files, in libraries reading arcane accounts of exotic travel, in farmers' homes, at the edge of newly tilled fields and in stimulating conversation with academic colleagues in the field, in the city, and back home. For me, the pleasures of scholarship are often in the travel itself, in whom one meets along the way, and with whom one travels and shares ideas, even if briefly. Looking back to recall and acknowledge those that pushed, nudged, or otherwise accompanied me along the way is a pleasant obligation. The memories of the road itself are not merely intellectual; they are also sensory and personal—the smells, sounds, and physical sensations of sipping spiced tea in a farmer's house or the complex smells of a highland market after a June rain.

The hospitality and insights into rural life offered to me time and again by farmers and small-town folk in my research sites are among my most profound memories of the research for this book. It is therefore difficult to express my appreciation to those Ethiopians who took time to talk to me, take me to their fields, or guide me through a coffee forest to explain their lives and the physical setting because few of them are likely to read the book. I did not offer them direct rewards for their help, nor did they request any. But I hope by passing on their knowledge and understanding of their social and physical world to others I can in some way repay their kindness. Where their help and words contributed directly to my narrative I have acknowledged them by name in the chapters' notes.

This work also owes debts to a great many people and institutions who supported it. In the beginning and throughout, my colleagues and the staff of the African Studies Center at Boston University have provided me with friendship and intellectual support. Joanne Hart and Karen Heybey of the ASC and Yvonne Barr at the Graduate School deserve special mention; they could always find a way to do whatever was needed. A number of international organizations have given me the means to spend time in rural areas in Ethiopia when that was a difficult if not impossible task for a foreign researcher. They also provided field staff colleagues, who became my friends and mentors. I only hope my work for them in small measure repaid what they offered me. These organizations are Oxfam America, Oxfam (United Kingdom), American Jewish World Service, the International Livestock Centre for Africa, the United Nations Environmental Program, and Norwegian Redd Barna. The special individuals who provided me with insights and logistical support are cited in each chapter's notes. Among these, however, I must single out my colleagues from the Highlands Programme at ILCA, whose wise advice and experience I sought before virtually every field visit. These include Abate Tedla, Abiye Astatke, Getachew Asamenaw, Mohamed Saleem, and Samuel Jutzi. Degefe Birru more than any other shared his genius for rapport with farmers, knowledge of Oromofa and his intense curiosity to pursue my questions and shape new ones.

At Addis Ababa University I owe much to the sustained support of the Institute of Ethiopian Studies, its directors, Dr. Taddese Beyene and Bahru Zewde, and the wonderful staff. In the Department of History I have always found lively colleagues ready to challenge my assumptions and share their deep knowledge of Ethiopian history. Shiferaw Bekele in particular deserves special thanks for his talents as a historian, tireless iconoclast, and logistical force majeure. For translations, especially of interviews in Oromofa, I am indebted to Ato Zelalem Banti, Tekalign Wolde Mariam, and Shiferaw Bekele (who arranged it all).

A good deal of the research for this book and a final stage of writing took place in Italy. There I am grateful for the help of Carla Ghezzi, Silvana Palma, and the staff of the Istituto Italo-Africano in Rome. Their cooperation has been invaluable in providing access to their unique collection of publications, documents, and photographs; Dr. Maria Puccioni of the Istituto Agronomico per l'Oltremare in Florence provided access to that greatly underutilized resource on agriculture. For four weeks of writing in 1992 I enjoyed the rural panorama, hospitality, and *cucina* of *la famiglia* Scuderini at Chicco di Grano in Tuscany. None of my work in Italy could have succeeded or been so enjoyable without the diligent help, warm friendship, and hospitality of Irma Taddia of the Università de Bologna (and the forbearance of Alberto Tassinari).

Each of the chapters of the book has evolved in somewhat different circumstances, and I am indebted to many who have offered comments and sharpened my arguments: David Anderson, Sara Berry, Donald Crummey, Dessalegn Rahmato, Stephen Frank, Mickey Glantz, Jane Guyer, Jean Hay, Allan Hoben, Harold Marcus, Tom Spear, Alan Taylor (chief publicist of the Agrarian Institute), and Yaqob Fisseha. Chris Geary taught me the value of imagery as narrative, though she may not have known it at the time. A lively group of graduate students in my seminar on East African environmental history at Boston University in 1991 kept me informed about what I should have read: Thomas Johnson, Kirk Hoppe, Erik Gilbert, Jonathan Reynolds, Shawn Dennard, and Julie Livingston. During this book's gestation Barbara Cooper, and Julie Croston wrote creative, insightful dissertations that provided new ideas about narratives, women's history, and rural history. Tekalign Wolde Mariam, whose own work pushes Ethiopian history on to new ground, has offered many comments and insights from his own field data. Eliza McClennan drew the maps from an eclectic, to say the least, collection of base materials. Leslie Hartwell laid out the text, tables, and figures. Dean Dennis Berkey has provided consistent support for this project and my career.

The final stage of research and writing for this book benefited from important support from the Social Science Research Council, the American Council of Learned Societies, the Fulbright-Hays Program, and the National Endowment for the Humanities. With sabbatical support from Boston University in 1991–92 I had the good fortune to spend that academic year as a Delta Delta Delta fellow at the National Humanities Center where I drafted the major part of the book. It is difficult to imagine a more exciting and supportive intellectual environment. The opportunity to spend 10 months among 39 outstanding scholars of the humanities amid the North Carolina piedmont's dogwoods and loblolly pines shaped many of the ideas presented here. I owe very special thanks to the wonderful staff there and to my colleagues in the Land and People Seminar: Dick and Claudia Bushman, Neal Salisbury, Larry Silver, and Silvia Tomasch. The final shape of this work owes much to their lively inquiry and warm intellectual camaraderie. Though this book would not exist except for the support all of these friends and colleagues, its errors and shortcomings are solely my own.

Emotional support and encouragement from friends and family are far more essential guideposts for leading one to a final goal than is apparent when planning and describing a project in a grant proposal. My friends and mentors Harold Marcus, Charlie McClellan, Steve Howard, and Angela Raven-Roberts have always offered deep and unswerving good cheer and advice. Dorleen Bradley's life and philosophy have long been my quiet and too-often-unacknowledged model. Libby and Martha don't quite know yet what has kept me away from home more than they would like, but I hope they will understand someday soon and know how much I have missed their smiles and laughter when on yet another "trip." More than anyone, my wife and friend, Sandi, knows about this work and how much she has contributed to its final outcome. It is with love that I dedicate this book to her.

Arlington, Massachusetts
September 1994

People of the Plow

Prologue:
Farms, Agriculture, and History

A book on agricultural history, on Ethiopia's agricultural history, might well begin with a narrative prologue in which a farmer or local observer comments poignantly on changes in the farm landscape, providing a perspective from which to analyze historical change. Indeed, those foreign observers who moved their gaze away from the politics of the imperial court or self-interest in trade or hunting reported a richly diverse rural landscape, where cultivation with the ox-drawn plow was almost everywhere apparent. Among the earliest accounts, the Portuguese Jesuit Francisco Alvares in the 1520s argued, "It seems to me in all the world there is not so populous a country, and so abundant in corn, and herds of innumerable cattle."[1] A century later his fellow Jesuit Jerome Lobo in 1626 noted: "The climate is so temperate that at the same time I saw in some places ploughing and sowing, and in others the wheat already sprouting, while in others it was full-grown and mature, in others reaping, threshing, gathering, and again sowing, the land never tiring of continual production of its fruits or failing in this readiness to produce them."[2] Two hundred years later Henry Salt described part of northern Ethiopia's highlands as "so rich in water and pasturage that Europeans could scarcely imagine its beauty," and an agricultural harvest which "must be considered in great measure as owing to the industry of the inhabitants. . . ."[3]

1. Francisco Alvares, *The Prester John of the Indies: A True Relation of the Lands of the Prester John, Being a Narrative of the Portuguese Embassy to Ethiopia in 1520*, trans. C. W. Beckingham and G. W. B. Huntingford, 2 vols., Hakluyt Society, vols. 114-15 (Cambridge, 1961), vol. 1, 131.

2. Donald Lockhart, trans., *The Itinerário of Jerónimo Lobo*, (London, 1984), 245.

3. Henry Salt, *A Voyage to Abyssinia and Travels in the Interior of That Country* (Philadelphia, 1816), 199. 426.

In the last half of the twentieth century, the external accounts of Ethiopia's rural landscape have changed dramatically:

> A culture is dying in Ethiopia. A complete way of life, virtually unassailed for 3000 years, is coming to an end. The Abyssinian high plateau, known to the Greeks as a "cool celestial island," is rapidly turning to dust, merging wearily into the barren and stony deserts that surround it. As it does so, the human populations that it has supported for so long are blowing away too. Having slaughtered their draught oxen and eaten their seed grain, the people are leaving forever their eroded fields and terraces. . . .[4]

Modern images of want, landscapes burned by drought, and tradition-bound farmers unable to respond to the needs of their families, neighbors, or the burgeoning urban population thus contrast sharply with past testimony of fertility and a hardy, productive yeomanry. Yet, empirical studies of criteria such as land size, household nutrition and farm level productivity generally support the idea of a widespread crisis in farm production on the same highlands described as verdant or "salubrious" in over 400 years of travel literature. The obvious conventional wisdom among outsiders is one of a fundamental change in farmers' ability to produce food, though without a clear consensus on causality. Ethiopians, whether farmers or city dwellers, however, also point to the contrast of current crisis with recent memories of past plenty, notwithstanding the knowledge that famine has been an enduring if episodic part of rural life in that region.

The subject of this book is the modern history of Ethiopia's agriculture and the paradox of how the land and farming system which has sustained Africa's historically most productive agricultural system can have fallen into deep fundamental crisis. After all, the agricultural system and the farmers whose ideas and strategies put it into practice have, over the past millennium, evolved distinctive technologies, social institutions, and effective solutions to environmental problems. Highland agriculture has put food on the tables of large complex state systems—Axum, the Solomonic Empire, and an array of small, highly developed kingdoms in the southern highlands for at least two millennia. Why then has there been, since at least the late nineteenth century, fairly consistent evidence of decline in farm productivity? This question is fundamentally historical, requiring an exploration of standard as well as unconventional historical sources which will reconstruct Ethiopia's agricultural and environmental past.

Ethiopia offers an excellent setting for examining the nature of long-term agricultural change in Africa as a whole and for highland environments in particular because its predominant technology and cropping system have been in place for at least two millennia. For that time and certainly the 200 years of this study, its rural economy has fit the historically familiar model of a "preindustrial society charac-

4. Graham Hancock, *Ethiopia: The Challenge of Hunger* (London, 1985), 7.

terized by slow technical change where processes of growth are still dominated by the play between demographic expansion and limited resources."[5] The reality, however, is far more engaging than a simple production equation; the landscapes and contours which the plow has shaped are infinitely varied by environment and by the sets of social institutions, local experience, and episodic events around which farmers organized their economic and social lives. This study will cover regions which differ by altitude, soil, climate, and cultural tradition, but which share a common technology, the plow, a mix of annual cereal crops, and agronomic techniques to adapt these to local conditions.

Ox-plow agriculture first took root in the northern highlands among Cushitic-speaking farmers who domesticated local grasses, but it flourished more dramatically as the economic base for Semitic peoples who made it their trademark. The *marasha* (the plow as a whole or, lit., the plow tip), however, spread more effectively and more permanently than the highland state or Christianity, reaching peoples of the southern and eastern highlands well before Emperor Menilek II's conquering armies of the late nineteenth century. Ox-plow agriculture has continued to spread even in the twentieth century: Somali pastoralists around Hargeisa began using it only within the past 100 years.[6] That the crops planted in the plowed fields—cereals and pulses—vary by local conditions as much as the language, religion, and political traditions of the cultivator is not nearly so surprising as the fact that the ox-plow system continues throughout its range to be recognizable as a discrete technology in the crops it cultivates, in the agronomic techniques used, and in the property institutions it fosters. In fact, the majority of "the people of the plow" described in this book and the local environments they occupy are not of the northern highlands, but peoples who have appropriated the plow and made its productivity fundamental to the economic health of the entire Ethiopian region, as much in 1990 as in 1800.

Method: Agriculture as Ecology

Basic to my approach to this book is the belief that agriculture and its history should be a subset of a much broader ecological history which reconstructs human interaction with the physical environment. William Cronon, whose work has inspired a new generation of writing on environmental history, has posited that environmental history constitutes an "uneasy—and often unacknowledged—alliance" between an argument that human action takes place within social constructions of property, gender, and class and one which places physical setting as a codeterminant.[7] Agricul-

5. Emmanuel LeRoy Ladurie, *The Peasants of Languedoc* (Urbana, 1976), 296.

6. Abdi Ismail Samatar, *The State and Rural Transformation in Northern Somalia*, 1884-1986 (Madison, 1989), 42-43.

7. William Cronon, "A Place for Stories: Nature, History, and Narrative," *Journal of American History* (March 1982), 1349.

tural history therefore should be not only about social practice—property, a gendered division of labor, resource tenure—but also about field systems, food storage, technology, and cropping patterns.

Agriculture per se has not received much attention in the historical literature of Africa. For the most part, African agricultural history in the past two decades has been subsumed under the rubrics of economic, social, or political history and the more general interest in peasant studies. Most historians working on rural issues have placed agriculture within the field of political economy, implying the primacy of politics and the state over climate, crops, and farmers. The emphasis on agriculture as a function of political economy has, understandably for the sake of source materials and documentation, tended to concentrate on the colonial period to the exclusion of longer-term trends rooted in ecology. Even the materialist perspective has recognized this problem. Though choosing to work within the peasant paradigm, Elias Mandala has adeptly pointed out that the antithesis between nature and society in African studies of the 1970s and 1980s was part of the Althusserian preoccupation with relations of production over the analysis of production itself.[8]

Recent popular concern about and scholarly treatment of Africa's famines of the past two decades have generated interest in the agricultural past, but tended to treat agriculture itself as a secondary rather than a primary object for research. The dominant causal interpretation both among scholars—Amartya Sen's entitlement thesis—and in the popular literature—civil war and the "politics of starvation"—focused on political factors, necessarily minimizing the historical evidence of agriculture itself. The issue here is not to deny political causation as a critical factor in famine, but to point out that famine studies have not added much to our understanding of Africa's agricultural past.

Steven Feierman has correctly pointed out the dangers of emphasis on the *longue durée* of Braudelian social history at the expense of understanding rural dwellers' agency in interpreting social practice and managing economic change. Just as Feierman's peasants draw upon past forms of political language and transform it into new political discourse, farmers individually and in the aggregate continually renegotiate the connections among the physical environment, customary practice, experience, and new contexts in agriculture.[9] Agricultural history, however, cannot be simply a chronicle of farmer agency because, to paraphrase Cronon, agricultural history must combine the long-term effects of environmental changes (e.g., climate or soil fertility) with specific and cumulative seasonal adjustments by farmers in crop choices, agronomic practice, and storage strategies, as well as with broader changes in the political domain.

8. Elias Mandala, *Work and Control in a Peasant Economy: A History of the Lower Tchiri Valley in Malawi 1859-1960* (Madison, 1990), 8-11.

9. Steven Feierman, *Peasant Intellectuals: Anthropology and History in Tanzania* (Madison, 1990), 3-5.

African agricultural history here is much less a record of eventful time than the sum of day-to-day and season-to-season adjustments to environmental constraints and shifting conditions in the political and mercantile domains. It is in part a cumulative process, the sum of personal and intergenerational experience. But agricultural history is also *conjunctural* in that it draws together several scales of time, perception, and action which link individual farmers, households, and rural communities with broader and longer-term physical conditions of the environment and short-term events such as drought, war, economic depression, and locust swarms. On a daily basis farmers have had to decide between using stored food as family diet or as seed (capital) or converting it to its exchange value, a point of negotiation in which the domestic domain of gender plays a prominent role. Although seasons are an important part of annual cycles, farmers planting in any given year have nonetheless had to choose which crops to sow and to guess when the annual cycle of rains will permit the final passes of the plow before seeding. Within each generation as well, farm resources of land, livestock, and labor fall under schemes of reallocation through locally specific institutions of inheritance, indebtedness, and the organization of labor within newly formed households.

The central idea of this study is to explore agriculture as environmental history, human manipulation of the natural world as conditioned by a rich variety of social practices. Analysis here does not rest with the political domain, because agriculture seen as a part of peasant economy fundamentally removes farmers from the set of daily, seasonal, and annual choices, which in the aggregate compose agricultural history. Far from acting in concert with nature, farmers have had to outguess the vagaries of the natural world to keep from being steamrolled by its inexorable power.[10] In fact, an important corollary question in this study involves, not the exclusion of politics, but the investigation of its place within the historical process of agriculture and the environment.

My choice in referring to rural producers—male and female—as farmers rather than as peasants is not only a semantic distinction; it also represents my preference to privilege their role as managers of the environment over their relations to the state. Farmers in the Ankober lowlands who planted maize in 1985 and then pulled it up at the green stage did so to pay taxes, not to balance their crop rotation. Yet that decision about the form in which to reallocate their resources was not made by the state itself. In the nineteenth century the imperial state, or its local representations, seems to have had neither the ability nor the desire to intervene in such farm-level decisions. Through the twentieth century, while the power of the central state has grown along with Ethiopia's urban population, urban elites have become increasingly interested in shaping decisions about agriculture through the state apparati they have controlled. Ultimately, their failure to do so constitutes a major theme in this book and in the history of modern agriculture.

10. I am grateful to Dick Bushman for this observation about farm life, in his case drawn from early nineteenth-century southern U.S. farm diaries.

Sources: Agriculture Imagined and Practiced

The historian's basic task in reconstructing the past is to uncover and interpret past signs. Agricultural history and its specific nature in Northeast Africa throw out special challenges in trying to dig into what farmers did in the past, the meaning of those actions for farmers' immediate goals, and the long-term implications of patterns of resource use. Sources have distinctive visions about the nature of the past, what observers valued as history, and what reaches us in the present. Ideally, this study might draw upon historical signs from the agricultural process itself, that is, local records of farms, farmers, and the agricultural community. Such sources would be a product of local institutions recording mundane events and rituals from their daily social and economic lives, agricultural transactions which would allow a precise reconstruction of cropping patterns, labor, market prices, and perhaps even harvest dates. Farm diaries, detailed local tax roles, church baptismal records, local newspapers, and inquisitional transcripts have been the foundation of European and North American rural and environmental history for at least a generation. By contrast, writing on the history of the African environment is comparatively new and rooted in either archaeology or fairly recent colonial documentation.[11]

Ethiopia's well-established literary tradition invites the hope of caches of local documentation which might allow reconstruction of the history of imperial estates, if not smallholder farms as well. A millennium-long tradition of clerical literacy as a part of highland Christian society should not, however, presume its use for documenting the rural economy. The Ethiopic script, derived from Aramaic and Old Phoenician but refined by Ethiopia's clerical literati, was bound up in the affairs of the Church and the imperial court but, unfortunately, not in the economy of production. Rural folk had access to the written word only through the parish clerics and then only to interpret matters of liturgy. Literacy in the Amharic vernacular emerged only in the mid-nineteenth century, and then only in embryonic form used in royal correspondence and a handful of exploratory imperial inventories. Traditional church documentation and royal chronicles have been an important resource in Ethiopian historiography, allowing a finely textured reconstruction of political, religious, and intellectual life of highland Christian society. Those same documents have also been used to chronicle the expansion of highland institutions, particularly the

11. See, for example, Donald Worster, *Dust Bowl: The Southern Plains in the 1930s* (New York, 1979); Donald Worster, *Rivers of Empire: Water Aridity, and the Growth of the American West* (New York, 1985); LeRoy Ladurie, *The Peasants of Languedoc*. A fine example of a farm diary is David Ramsay, *Ramsay's History of a South Carolina Farm from Its First Settlement in 1670 to the Year 1808* (Spartanburg, 1958), cited in Timothy Silver, *A New Face on the Countryside: Indians, Colonists, and Slaves in South Atlantic Forests, 1500-1800* (New York, 1990), 112 n. 17. See also James C. McCann, "Agriculture and African History," *Journal of African History*, 31 (1991); an important exception to the colonial focus is Robert Harms, *Games against Nature* (New York, 1987).

Ethiopian Orthodox church, to new regions, a process which often followed or paralleled the spread of the ox-plow complex. Unfortunately, nowhere have we found anything resembling estate records which might offer a glimpse of the day-to-day workings of farm management, climate history, or the evolution of agronomic practice. The relatively rich set of church documents recently collected by social historians promises to be an excellent source of social history but not of agricultural history in any direct sense. These local church records describe the transfer of offices, property, and income rights rather than the more mundane issues of farming.[12]

State documentation of the agricultural process began only in the post-World War II era and intensified after the 1960s, though it focused largely on the export sector. The lack of indigenous documentation of agriculture in a historically literate society, however, conveys significant meaning. First, it strongly suggests a lack of interest and direct involvement of the state and the elite administrative classes in the actual process of production (see chapter 1), though their interest grew as the twentieth century wore on. Second, the absence of evidence on agriculture in the indigenous written historical record suggests the absence of an indigenous tradition of agricultural history as I have defined it here. The upshot is the need for the historian to be omnivorous in the use of sources drawn from every quarter available and to interpolate historical conditions from contemporary data at times.

My method for exploring the varieties of historical settings for agriculture in Ethiopia has included the use of illustrative examples from different physical environments and historical conjunctures. Three areas receive specific attention: Ankober, the region of a former royal capital of the central highlands; Gera, a region of the heavily forested, humid southwest; and Ada, in Addis Ababa's urban periphery and Ethiopia's most developed and commercially successful area of smallholder agriculture. Further data derive from my research both on Setit-Humera on the Gonder-Eritrea border in the northwest and in the Chercher highlands of Harerge in the east. Understanding the nature of agricultural change in each of these areas has required a different set of source materials and treatment of key themes: the effect of population change on agricultural practice, the role of urbanization on agricultural hinterlands, the effects of annual cultivation on traditions of perennial tree crops, the ascendancy of maize in late-twentieth-century subsistence systems.

12. Two superb histories of the medieval period have relied primarily on hagiographies and Ethiopic documentation: Steven Kaplan, *The Monastic Holy Man in Ethiopia* (Wiesbaden, 1984), 1-10; Taddesse Tamrat, *Church and State in Ethiopia* (Oxford, 1972), 156-205. The research team of Donald Crummey, Shumet Sishagne, and Daniel Ayana have cataloged church documents in the Gondar and Gojjam regions which chronicle land transfers, property sales, wills, and the transfer of political office. See, for example, Donald Crummey, "Three Amharic Documents of Marriage and Inheritance from the Eighteenth and Nineteenth Centuries," *Proceedings of the Eighth International Conference of Ethiopia Studies* (Addis Ababa, 1988), 315-28.

Views from the Farm

Ethiopian farmers as a whole are deeply knowledgeable about agriculture and their physical environment. Historical knowledge about agriculture in the past does exist in the form of individual life histories, information about land holding of ancestors, local political events, and environmental shocks, which help explain patterns of demography and land use. Farmers can recall with accuracy what crops they planted in which fields up to five years into the past but rarely any further. Farmers' historical memories are, however, particularly accurate about key events in the household development cycle: marriage, land and property endowments, settlement and migrations. In my fieldwork I have been able to reconstruct local society and economy using life histories of farmers who could detail, over widely varying lengths of time, individual cropping strategies, livestock ownership, oxen use, and property transfers occurring as much as two generations ago, reaching back in many cases to the period before the 1935–41 Italian occupation. Certain farmers are also amazingly well versed in local, regional, and national political history, including dissident traditions preserved in oral rather than written form.[13]

History for most farmers in the areas of this study, however, is a fragmented phenomenon linking their lives to the larger culture and polity rather than forming a continuous process in which their own daily lives are a part. Local conceptions of *tarik* (lit., "history") related by northern highland farmers conform rather rigorously, in form if not in total content, to those preserved formally by the state. My interviews with farmers were at times focused on specific topics, but also included open-ended monologues by farmers which followed and illuminated their own conceptual categories. The synthesis of this information into a narrative form is my task as a historian, not as an interlocutor for a farmers' history. Farmers' own voices and lives appear in the narrative here because they tell the story effectively. Areas like Gera and Chercher, which did not belong to the political or religious traditions of the north until the late nineteenth century, have different historical traditions, but nonetheless focused on the eventful time conceptions and personalities of political history. In none of the areas covered here is there a coherent and discrete narrative of agriculture in history. For better or worse, that concept is one which I brought with me along with my own questions.

From another perspective, however, the landscape, the local culture, and farmers themselves offer a tremendously rich, irreplaceable repository of knowledge, which ultimately serves as the foundation of local agricultural history. The landscape of any locality is anthropogenic; that is, it is imbued with natural and human-made evidence of the agricultural past which only local knowledge can interpret. Walking with farmers to market or engaging them in conversation along a footpath between inter-

13. For an example of a dissident oral tradition, see James McCann, *From Poverty to Famine in Northeast Ethiopia: A Rural History* (Philadelphia, 1987), 92-104.

view sites stimulates reflections upon certain places and things—abandoned terraces, a new cultivar of sorghum, wild fruit which served as famine food—and
events which can serve as later interview material. Raising questions about the history of a certain variety of wheat, or an irrigated field, or the cost of oxen rental in
1935 also stimulates a wider debate within a local community, triggering collective
memory and placing agriculture alongside local categories of history. A deeper community awareness of my questions and interests invariably produced more focused
and lively interviews and additional data in subsequent sessions. Throughout my
fieldwork, rural walks, market visits, and "small talk" added to my information on
local history and helped me find the most knowledgeable narrators.

If the local landscape serves as a mnemonic device for farmers, it also raises
many questions for the new observer about how the local population has dealt with
environmental or agronomic challenges. Only by working on the landscapes to
which the plow has spread can one truly appreciate the rich variety of settings in
which the plow has succeeded. There is no substitute for work *in situ* for understanding the forces which have shaped local history. Only by working with farmers
in Denki, who live their lives day in and day out under the shadow of the old royal
capital on the western horizon, could I appreciate the context for local history and
agricultural change. Fallowed forest clearings or variations in design of *gotera*
(grain storage containers) in Gera raised specific questions about how maize has so
dominated the region in the last decades.

Clearly the historian in Ethiopia who is able to blend fieldwork with written documentation has distinct advantages over environmental historians of the early North
American frontier. Native Americans who worked the lands and forests of the eighteenth century eastern seaboard have long since disappeared from the scene, along
with their technology and social institutions. We can neither see them cultivate nor
talk to them about their work. Historians of agriculture and environment in North
America and Europe therefore have not had the privilege of observing the technology, social practice, and world view of their subjects nor of discussing their research
questions with those most knowledgeable about the process and about the daily
lives of these earlier peoples.

Views from the Road

Despite the fundamental role of contact with farmers and local environment, this
study has had to rely on the observations of outsiders, European travelers, and more
recently development planners and agricultural specialists (Ethiopian and foreign),
whose records vary widely in sensitivity toward and knowledge about the physical
environment in general and agriculture in particular. Through most of the period of
this study, however, the literature of foreign witnesses constitutes the single most
important source of information about the Ethiopian environment in the past. Given

its centrality to our knowledge of the agricultural past, this body of literature deserves careful scrutiny. The use of travelers as a genre of sources, however, masks the astonishing range of their experience, motives, and abilities. From the first Portuguese Jesuits in the sixteenth century who traveled by foot and by mule, through the twentieth-century observers, who could arrive in Ethiopia by rail or by air and could travel in rural areas, at least in part, by means of the internal combustion engine, the motivations, sensibilities, interests, linguistic ability, and powers of observation of the traveler differed widely.[14] In many cases the writers had deep familiarity with the scenes and settings they describe—such as Alvares, Almeida, or Nathaniel Pearce—while in other cases the observers described landscapes they saw in passing for the first and only time.

Travel literature of the first half of the nineteenth century took the form of personal narratives, which are by their nature egocentric and impressionistic. Unlike the later conventions of anthropological method, however, their form conveys an emotional, sensory, and visceral response as well as a physical description. The form thus allows us to extrapolate a subjectivity from the events and sights described. Travelers to Ankober in 1840, like Major W. Cornwallis Harris (see below), brought a specific experiential vocabulary of Europe and often India along with a firm sense of exotic description anticipated by their potential audience. Like Walter Raleigh's first attempts to describe the New World's oxymoranic "abundant wasteland," travelers to Ankober wrote unabashedly subjective first-person narratives.[15]

The narrative voices of Ethiopia's observers changed dramatically over the course of the 1800–1990 period. Early nineteenth-century travelers wrote consciously subjective, first-person narratives, choosing the literary rather than the scientific idiom. Ethiopia was for them an exotic, isolated kingdom; its agriculture was an organic part of society, polity, and history. Agriculture, like politics, was a measure of civilization as much as an economic or environmental process. Douglas Graham prefaced his detailed description of agriculture at Ankober with a gloss on agriculture's meaning to a nineteenth-century observer: "The different modes of tilling the ground practised among the various nations of the earth, are well worthy of observation and remark, as the progress of agriculture exhibits progress of the population in comfort and civilization and thus forms one of the most important chapters in the history of national manners."[16]

Major Harris' classic 1841 account treated the environment and agriculture of the Shawan kingdom he observed as part and parcel of a metaphor of a decadent Orient:

14. Rita Pankhurst and Richard Pankhurst, "A Select Annotated Bibliography of Travel Books on Ethiopia," *Africana Journal* 9, 2 (1978), 113-32; 9, 3 (1978), 101-33.

15. For discussion of first-person travel narratives, see Mary Campbell, *The Witness and the Other World: Exotic European Travel Writing, 400-1600*, (Ithaca, 1989), 228, 262.

16. Douglas Graham, "Report on the Agricultural and Land Produce of Shoa," *Journal of the Asiatic Society of Bengal* 13 (1844), 253.

Although the majestic fabrics, the pillars of porphyry, and the Corinthian domes of early writers, now exist only in tradition, Aethiopia yet retains the fresh vegetation of a northern soil, the vivifying ardour of a tropical sun, and the cloudless azure of a southern sky. Palaces and fanes, gardens and gushing fountains, have long since departed; but there still remains a fertile country possessing vast capabilities, a salubrious and delightful climate. . . .[17]

For the archaeologist J. Theodore Bent, who wrote in 1895 on the eve of the domination of empiricism, the remains of irrigation works evident at Axum were comparable to "the hills in Greece and Asia Minor" but now merely "a sad instance of Abyssinian deterioration."[18]

For agriculture, such factors as when, where, and how long observers gathered their impressions are critical in the evaluation of a source given the importance of seasonality in cropping, climate, and locale. Agriculture was rarely the single focus of any travel account, and the quality and depth of information could vary even within a single narrative. Johan Krapf's descriptions of the physical environment of Ankober improved in quality and quantity in inverse proportion to how far Negus Sahla Sellase was from Krapf's base at the royal capital. When the *negus* (king) was away Krapf described climate, arrival of visitors, community life; when Sahla Sellase was on the scene the king captured Krapf's attention and Krapf offered primarily political information and no environmental observations at all. Antonio Cecchi's invaluable account of agriculture in Gera owes much to the fact that he was not traveling through but was prisoner for almost a year under suspicion of being a spy for Menilek. Even more to the point, the favor in which Gumiti, the queen *(ghenne)* of Gera, held him on any given day influenced what he was fed, and he could comment authoritatively on a wide variety of local agricultural products and their relative social value.[19] By contrast, the picturesque crater lakes of Ada drew the attention of our best observers so completely that we have virtually no record of nineteenth-century agriculture or the human landscape in an area close to the political center of the emerging empire.

Environmental observation, the essence of agricultural description in European travel sources, is in most cases a single snapshot of a specific place at a certain time of the agricultural cycle and the conditions of a particular year. Such observations thus are inevitably idiosyncratic and synchronic, but with a full corpus of other observations can serve quite usefully as the basis for longitudinal analysis of crops,

17. W. Cornwallis Harris, *The Highlands of Ethiopia*, 2nd ed., 3 vols. (London, 1844), vol. 3, 265.

18. J. Theodore Bent, *The Sacred City of the Ethiopians, Being a Record of Travel and Research in Abyssinia in 1893* (London, 1893), 135.

19. C. W. Isenberg and J. L. Krapf, *The Journals of the Rev. Mssrs. Isenberg and Krapf, Detailing Their Proceedings in the Kingdom of Shoa and Journeys in Other Parts of Abyssinia* (London, 1968); Antonio Cecchi, *Da Zeila alla frontiere del Caffa*, 2 vols. (Roma, 1886-87), vol. 2, 249, 279.

demographic patterns, and agronomic practice, By piecing together several snap-shots of a specific place or process, a mosaic of the agricultural landscape and the agricultural past emerges from disparate shards. For example, the episodes of par-ticularly fine descriptions of the region of Gera's capital in the southwestern forest in 1854 (Massaja), 1881 (Cecchi), 1927 (Cerrulli), and 1990–92 (my own field-work) form a clear and comprehensive picture of transformations in demography, settlement, cropping patterns, and the forest.

Unlike those who reported on social customs or politics, foreign witnesses of the physical environment could report accurately on crops, forests, and soils without need of the linguistic expertise few of them enjoyed. Nevertheless, there are annoy-ing lapses of linguistic precision. Tracing the penetration of maize *(Zea mays)* pre-sents such a puzzle in a cacophony of linguistic imprecision. Nineteenth-century British descriptions of corn refer, of course, to cereal in general—probably wheat, and not maize. Henry Salt's 1814 references to household consumption of *maiz* were to a local name for *taj* (honey wine) and not to the New World grain. "Grano" in modern Italian denotes "wheat," but in nineteenth-century Italian the term could mean most any cereal; granoturco specified "maize." Early translations of Por-tuguese accounts translated Alvares' reference to *milho zaburro* as "maize" (the modern Portuguese translation), though he almost certainly was describing either sorghum or eleusine.[20] Our knowledge of that error, however, derives almost entire-ly from the assumption that New World maize would not have widely penetrated the highlands in the 1520s. In other cases, only context can serve to separate often loose descriptions of agricultural phenomena from accurate ones.

In many ways there is more continuity than we might imagine in the accounts of travelers and post-World War II agricultural science. Robert Chambers' now classic critique of development bias, *Rural Development: Putting the Last First,* offers a valuable primer for evaluating the quality of traveler observations of agri-culture. Chambers identifies several biases built into modern studies of rural economy which skew our understanding of farmers and agriculture. Chambers' term "person biases" suggests the tendency of the rural researcher, as well as and the nineteenth-century traveler to come into social and professional contact more often with local elite, males, and with those who choose to be observed. These biases also include the roadside bias, in which observers travel by established roads and between administrative centers, thus avoiding precisely those areas in which most Ethiopian farmers located their fields and homes. That most travel-ers fed themselves on either imperial largesse at the royal capital or through *dergo* (the rural hospitality that local farmers were forced to extend) necessarily meant that many farmers deliberately avoided contact with foreign freeloaders whenev-er possible.[21]

20. Alvares, *Prester John,* vol. 1, 87-88; 136 n. 1.
21. Robert Chambers, *Rural Development: Putting the Last First* (New York, 1983), 13-27.

Most important from the perspective of agriculture is what Chambers calls dry-season biases, the tendency of observers in tropical environments to concentrate their movements in dry months of the year. Observers from temperate climates thus often miss the dramatic effects of tropical bimodal seasonality, especially the "hungry season" just prior to the first harvests after the rains. In Ethiopia the effects can be especially serious, since travelers invariably miss the primary sowing seasons of midsummer and, if they journey between January and May, may entirely miss seeing some crops in the field, or base judgments on the more erratic short rain *(belg)* cultivation. Rural disease, seasonal hunger, negotiation over seasonal debt, and final plot preparation all take place after the rains soften the highland soils for sowing but make them difficult to traverse. The bias is as pronounced for modern research as it was for Victorian travelers: most, but not all, of my field visits have had to be worked around the academic year, placing me in the field just before the rains or at their onset, a quite different schedule from that of most travelers.

The Lens of Empiricism

Nineteenth- and early-twentieth-century travelers observed Ethiopia's environment and agriculture within a naturalist and, in some cases, literary genre. Their estimates of crop yield, catalogs of cultigens and cultivars, and observations on meteorology and soils were part of an early positivist tradition encouraged by the growth of national geographic societies in Britain, France, Germany, and Italy.[22] These groups supported scientific and geographic inquiry but also maintained a healthy appreciation for the role of personality in their published accounts. In the twentieth century, however, the first-person, self-consciously subjective narrative of travel accounts gave way to the metaphor of empiricism. In this new form, agriculture was no longer part of a natural setting or a measure of civilization, but an object for quantification and abstraction.

With few exceptions, the dominant idiom for describing agriculture in the post-1945 years has been one in which scientific method rather than personal narrative has dominated. The loss of the narrative voice of travelers has not meant, however, an absence of subjectivity, but the submersion of motive and person within a social science and natural science idiom. Twentieth-century empiricists have described agriculture less as a measure of civilization than as an industrial process in which its constituent elements are separable from social context and divisible as inputs and

22. For a valuable compilation of geographic society publications on Ethiopia, see Harold G. Marcus, ed., *The Modern History of Ethiopia and the Horn of Africa: A Select and Annotated Bibliography* (Stanford, 1972). For an excellent review of nineteenth-century travel reports on crop yield and varieties, see Donald Crummey, "Ethiopian Plow Agriculture in the Nineteenth Century," *Journal of Ethiopian Studies* 16 (1983), 21-23.

outputs. The preamble to a 1975 socioeconomic study of Ada District defines their assumptions of the farm family:

> . . . the rural household was viewed as a production-consumption unit with all farm output measured in relation to a known area of land for which the family was responsible. With this conceptualization in mind, the research design could follow a fairly straight-forward tracing of the transformation of measurable inputs through a production process to measurable outputs that could be related to known levels of land, labor, and capital.[23]

If the subjective first-person account has given way to the disembodied voice of data, the individual has also given way to the institution. Studies of and evidence about agriculture by the 1960s and 1970s emerged out of organizations dedicated in one form or another to changing agriculture to an engine of national economic growth, to eliminating famine, or to responding to specific needs in world markets. The meta-narrative has derived from institutional interests in specific aspects of agriculture: crops, fertilizer use, engineering, and even the sociology of land tenure. Such descriptions of Ethiopian agriculture have rarely been published for a public or even scholarly audience, but rather for a narrow consumption as building blocks of economic analysis which require data in a statistical form or at least in the language of neoclassical economics.

The first coherent corpus of "empiricist" materials to study Ethiopia's agriculture was from the Italian Società Geografica's scientific field station at Let Marefia, established near Ankober in 1877, on land granted by then *Negus* Menilek. Though mainly a base for expeditions to the south, Let Marefia engaged in agricultural experimentation on its own 90 hectare grant. Later Italian experimental studies of Eritrea were carried out as early as 1893 under the auspices of the Ufficio Agricolo Sperimentale di Asmara, which sponsored agricultural shows and crop research. Agricultural research also took place under the auspices of the Istituto Coloniale Italiano, geographic societies, and, beginning officially in 1938, under the auspices of the Regio Istituto Agronomico per l'Africa Italiana in Florence. In the post-1941 period that institution's research continued, reorganized as the Istituto Agricolo Coloniale and later as the Istituto Agronomico per l'Oltremare.[24] The Cen-

23. Warren Vincent, "Economic Conditions in Ada Wereda, Ethiopia during the Cropping Season of 1975-76," unpublished paper, Michigan State University, 1977.

24. Isaia Baldrati, *Mostra agricola della colonia Eritrea: Catalogo illustrativo* (Firenze, 1903); Ufficio Studi del Ministero Africa Italiana, "Lineameni della legislazione per l'impero," *Gli Annali dell'Africa Italiana* 2, 3 (1939), 145-50. The best survey of the development of Italian research on Africa is Carla Ghezzi, "Fonti di documentazione e di ricerca per la conoscenza dell'Africa: Dall'Istituto coloniale italiano all'Istituto italo-africano," *Studi Piacentini* 7 (1990), 167-92.

tro di Documentazione of this latter institute, together with the library of the Istituto Italo-Africano, houses a superb collection of published and unpublished agricultural reports, surveys, and field research from the pre-1960 period.

Italian research on agriculture in Ethiopia was pragmatic: As early as 1890 policy makers had in mind a "colonizzazione agricola" for the highlands of Eritrea and Ethiopia, which would provide exportable foodstuffs and support a further "demographic colonization" of Italy's surplus population in Ethiopia's highlands. Such were the schemes for "demographic colonization" of the highlands. These included the ONC (Opera Nazionale per l'Combattenti) plan to settle veterans and their families at Holeta and Ada (Bishoftu)—a dismal failure—or the abortive projects to transplant entire Italian peasant villages to Harerge (Ente Puglia) or Gonder (Ente Romagna d'Etiopia). After 1935, the major goal was the "paneficazione" of Ethiopia's agriculture to provide for the Italian population of the cities and agricultural autarky for the colonies as a whole.[25]

In the end, Italian occupation and investment had little effect on agriculture itself beyond the momentum provided to the general commercial economy through impressive investments in road building and urban infrastructure. The occupation did produce, however, a corpus of excellent crop research, catalogs, local studies, and national surveys: Raffaele Ciferri and his collaborators, for example, compiled volumes of careful crop research on the speciation of highland cereals and a comprehensive national survey of cropping for 1938. Emilio Conforti's route report *Impressioni agrarie su alcuni itinerari dell'altopiano etiopico* is a tour de force, the most detailed and comprehensive survey of local agriculture available.[26] These materials provide roadside glimpses of smallholder agriculture which serve as an

25. A good portion of the agricultural documentation relates to development plans which never came close to implementation. For a summary of Italian schemes of agricultural colonization, see Istituto Agricolo Coloniale, *Main Features of Italy's Action in Ethiopia 1936-1941* (Florence, 1946); and Ufficio Studi del Ministero Africa Italiana, "La valorizzazione agraria e la colonizzazione," *Gli Annali dell'Africa Italiana* 2, 3 (1939), 179-316. For autarky policy, see Giuseppe Tassinari to Mussolini, 23 March 1937, Istituto Agronomico per l'Oltremare (hereafter IAO), fasc. 1990. Italian agricultural policy, if not their implementation, is clearly laid out in articles in the journal *L'Agricoltura Coloniale*. See, for example, A. de Benedictis, "L'autarchia alimentare dell'impero, problemi e prime realizzazioni," *L'Agricoltura Coloniale* 32 (1938), 1-12. A superb account of Italian land and agricultural policy in Ethiopia is Haile M. Larebo, "The Myth and Reality of Empire-Building: Italian Land Policy and Practice in Ethiopia (1935-41)," Ph.D. thesis, University of London, 1990.

26. For numerous Raffaele Ciferri reports see, for example, Raffaele Ciferri and Enrico Bartolozzi, *La produzione cerealicoltura dell'Africa Orientale Italiana nel 1938* (Firenze, 1940); and Emilio Conforti, *Impressioni agrarie su alcuni itinerari dell'altopiano etiopico* (Firenze, 1941).

invaluable benchmark between the first solid reports from the 1840s and the formal development literature of the post-1960s.

Comprehensive data on highland agriculture in the 1941–89 period are more sparse than one might imagine. Sources within the post-war development arena are conceptually flawed in that they are almost universally synchronic and provide virtually no historical data. Ironically, however, with the passage of time many have become historical documents; such surveys provide a different kind of agricultural snapshot of a place and time, a perspective similar to nineteenth-century travelers. Their coverage is, however, geographically skewed. A significant portion of national and international investment was concentrated in capital-intensive cash-crop production and large-scale agribusiness projects using mechanization in the Awash Valley, Arsi, and Walayta (see chapter 6) which fall out of the purview of this study and make up only a few percentage points of Ethiopia's total agricultural production.

A final important source for data on agriculture and rural economy, though not often history, is the genre of fugitive records circulated by bi-lateral aid projects, non-governmental organizations, and a new generation of farming systems research carried out by the Institute of Agricultural Research and the International Livestock Centre for Africa's Highlands Programme.[27] The empirical studies of modern agricultural science are particularly important as a means of unwrapping the relationship between human action and the environment. These studies include technical studies which identify the physical properties of plants, soils, agronomy, and climate. Empiricism in agriculture also extends to surveys of local and regional agricultural production, which provide insights into wider trends beyond the farm level or the individual descriptions of travel accounts. Having a 1937 estimate of national cropping patterns to compare with similar surveys from 1963 and 1983 helps to establish long-term trends less visible at the local level.

For the historian the most important innovation in recent agricultural research has been the rise of farming systems research, a collaborative approach which developed in the late 1970s among agricultural economists and scientists. Farming systems analysis has focused research on understanding the relationship between technical, environmental, and human elements of agriculture, gathering data on smallholder farms rather than from single-sector test plots. Though synchronic in conception, the results of this approach are particularly valuable for this study, since such research can establish empirically verifiable evidence for phenomena observed in historical documents. The farming systems approach, for example, has been able to establish

27. For a survey of agricultural development in Ethiopia see chapter 8 and John M. Cohen, *Integrated Rural Development: The Ethiopian Experience and the Debate* (Uppsala, 1987), 39-46. The best comprehensive bibliography of empirical research on agriculture is I. Haque, Desta Beyene, and Marcos Sahlu, *Bibliography on Soils, Fertilizers, Plant Nutrition and General Agronomy in Ethiopia* (Addis Ababa, 1985).

in quantitative terms the relationship between seasonal nutritional needs of livestock, especially oxen, and farmers' crop choices. For instance, teff is the most palatable and digestible forage for oxen, which partially explains why it is favored by farmers despite its high labor requirements. Maize is vulnerable to drought and does not store well, but its growth/labor cycle suits the seasonal demands of coffee producers. Cows can indeed substitute for oxen but may consequently suffer lower milk production and calving rates. Many of the results of this research are available through technical periodicals, working papers, and unpublished reports of non-governmental agencies.[28] Ironically, however, empirical studies serve primarily to tell us what farmers already know and act upon.

Narrative and Perspective in Agricultural History

In agricultural history, as with environmental history as a whole, the choices of narrative style facing the historian are particularly complex. The actors in the narrative include non-human elements of the natural world—animals, insects, and plants—but also more purely physical players such as soil, temperature, rainfall, and wind. Agricultural history has special qualities because it involves, in part, tracing the purposeful and systematic human manipulation of certain natural elements to minimize the effects of other, more disruptive, elements. Should narrative within such a history portray humans as battling the forces of nature, or can the forces of nature themselves join the story as protagonists? The historian's choices about constructing the narration rest upon deep and fundamental assumptions about ecological causation and agency. Modern environmentalists often blame peasants for the loss of forested land, whereas populists and Marxists see farmers as either positive agents of environmental stability or helpless victims of larger forces.

William Cronon has argued powerfully that an effective narrative's goal is that it "hides the discontinuities, ellipses, and contradictory experiences that would undermine the meaning of its story." His argument is a challenge to reconstructing environmental history whose principal task is to explore the relationship between people and nature in which culture, gender, race, and class affect the outcomes of historical conjuncture.[29] In agricultural history people and their cultural and tech-

28. Guido Gryseels and Frank Anderson, *Research on Farm and Livesock Productivity in the Central Ethiopian Highlands: Initial Results* (Addis Ababa, 1986). ILCA from 1991 has conducted research on cow traction. Preliminary data indicate that nutrition is the critical element in preventing loss of milk production and calving in cross-bred cows used for plowing.

29. Cronon, "A Place for Stories," 1349-50.

nological inventions may well take center stage, but cannot account for the plot in and of themselves. If the plow and its associated set of institutions and social corollaries provide continuity in the history of Ethiopia's agriculture, it is the sharp variety in the landscape and climate which provide its local shape.

This book examines agriculture from two distinct vantage points. The first is from a distance and attempts to describe changes in the physical and demographic landscape. How different did the forests and fields of Gera or the Denki River valley below Ankober look in 1990 from the way they looked in 1900 or 1840. My goal is to re-create in the mind's eye of the reader, and in my own, a panorama which accounts for changes over time in crops, land use, and population settlement patterns. Second, the book's analysis and narrative will also depict changes at the level of the farm and farmers, women and men, who made seasonal decisions about their labor, their diet, and their relations with their neighbors. At the farm level, however, the description of change is less a visual one than one evoked by the voices of the farmers themselves, expressed in interviews or in their life histories.

People of the Plow is a story about the variety of human adaptations of an agricultural system across a varied physical landscape. But our sources do not provide an Ethiopian Thoreau to narrate the changes from a local perspective. The farm voices available to the modern historian are, ultimately, those of the present, not the past; the observers are foreign to the elements they describe and modern agricultural science provides only hints of farmers' rationale for their daily and seasonal acts. This book is therefore an attempt to tell a story for its own sake, but it also has a didactic purpose in demonstrating the essential role of history in understanding environmental transformation and human crises.

Part I

The Plow and the Land

1

The Salubrious Highlands:
A Historical Setting

> . . . there still remains a fertile country
> possessing vast capabilities,
> a salubrious and delightful climate.
> —W. Cornwallis Harris,
> *The Highlands of Ethiopia (1844)*

Africa's highlands—those areas above 1500 meters elevation—constitute only 4 percent of the total land mass but contain the highest population density of any agroclimatic zone, with a livestock density four times the continent's average. In all, Africa's highland zones contain almost 20 percent of the continent's rural human and ruminant livestock population. The Ethiopian region has 490,000 square kilometers of highlands, 42 percent of its total land mass and almost half the continent's highland area. Moreover, Ethiopia's highlands account for over 80 percent of its human and livestock population and 90 percent of its arable land.[1] The highlands' physical properties have bounded Ethiopia's agricultural history, but, in turn, the actions of humans as cultivators, users of fuel, and managers of land have also brought long-term, cumulative changes to the highlands themselves.

The tremendous variety of environmental elements—soils, climate, topography, vegetation—which characterize the highlands has resulted from the evolution of the highland land mass in the period before human habitation. The environmental prehistory of Ethiopia's highlands divides into two epochs of rather unequal length. In the first period, geological forces across a wide region of instability in Northeast Africa acted upon the Precambrian land of marine sediments and limestone. Some

1. Gryseels and Anderson, *Research*, 1-2; Daniel Gamachu, *Environment and Development in Ethiopia* (Geneva, 1988), 5.

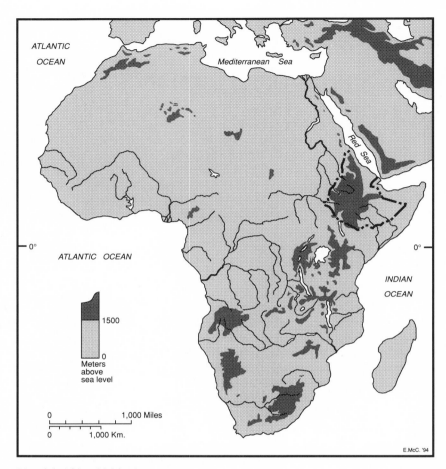

Map 1.1. Africa: highland zones.

65 million years ago tensions within the earth's crust pushed and cracked the surface into elevations of more than 3000 meters, bringing lava to the surface from both fissures and volcanoes, covering most of Ethiopia's older crystalline and sedimentary rock layers and forming the highlands and mountain ranges. Some 10 million to 20 million years ago, these forces resulted also in the formation of the East African Rift Valley, which divides Ethiopia's highlands into two plateaus on either side of the system of southern lakes and the Awash River valley (see maps). Basaltic lava flows thus blanketed the plateaus and the Rift Valley itself and formed the highlands.[2]

2. Daniel Gamachu, *Environment*, 8-9.

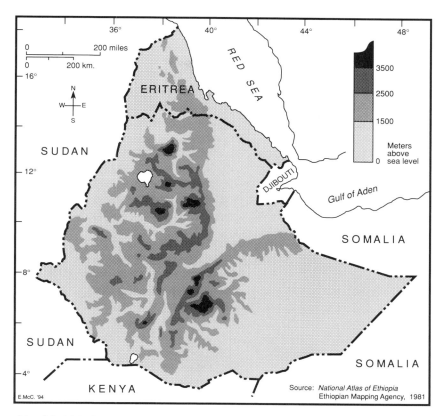

Map 1.2. Ethiopia: highland zones.

Across the Ethiopian region, the highland plateau pushed up by volcanic action presents widely differing shapes and configurations. In Gojjam and the Lake Tana basin undulating plains offer relatively soft shifts between elevations; Johan Krapf preferred a maritime image to describe the rugged highlands in Lasta: "a raging and stormy sea, presenting numerous hills of waves."[3] Along the eastern escarpment, there is a precipitous drop to the lowlands and the flat lowland plains which lead to the Red Sea. In the west, the descent to the Nile basin is less steep.

Almost as soon as volcanic action built the elevated plateaus, other natural forces began to shape them further through the erosion of their top layers by the action of water and ice. Glaciers in the higher areas sculpted upland landscapes while runoff from new river systems decomposed the volcanic material, simulta-

3. Isenberg and Krapf, *Journals*, 464.

neously creating Ethiopia's highland soils and deep river gorges such as the Abbay (Blue Nile) and Takkaze in the north and the shallower Wabe Shebele Valley in the southeast. Other important river systems draining the highlands include the Awash, the Gash, the Gibe, the Juba, and the Didessa. To travelers descending into the mile-deep Abbay gorge, the deeply eroded shear walls reveal the pre-volcanic limestone and sandstone in Ethiopia's geological past. Temperatures at the top of the gorge may be cooler than at the river's edge by as much as 50 degrees Fahrenheit. In the lowland plains below 1500 meters, which surround Ethiopia's central highlands, alluvial valleys, salt plains, and dry lakes show the geological connections between the transformation of the highlands and adjacent lowlands below the escarpment.

Ethiopia's soils result from the decomposition of the rich volcanic material which overlays the sedimentary base, and they therefore differ from the lateritic soils found widely in Africa. The highland soils range from red clay loam to black "cotton" soils. In general, higher altitudes tend to have the reddish brown latosols with excellent permeability, though they are highly acidic and therefore lacking in phosphorous. At lower highland altitudes and in the valleys, black cotton vertisols predominate, composing 24 percent of all cropped land in Ethiopia. Floods, wind, and other natural forces, however, have combined to offer a wide range of variants and admixtures of soil types, blending colors, qualities, and varying clay contents. Overall, the darker the soil is the greater its clay content and the more prone it is to waterlogging when wet and cracking when dry.[4] The peoples who evolved agricultural systems in this setting had to learn to read the meanings of the variation on a field-by-field basis.

Water's corrosive power is evident in the deep gorges which dissect the plateaus, confounding communications but shaping political and economic relations in human history. Ethiopia's highlands have made a profound contribution to the economy and landscape of Northeast Africa, since 84 percent of the water reaching Egypt's Nile Delta flows from Ethiopia's river systems. Moreover, the decomposition of the basaltic blanket covering the highlands has made up virtually all of the silt reaching the Nile Valley, forming the Nile Delta over geological epochs and supporting the flood plain agriculture at the heart of Egypt's predynastic and dynastic civil society.

The single greatest determinant of environmental conditions and of the terms of human habitation in Ethiopia's highlands is elevation, which affects temperature,

4. H. P. Huffnagel, *Agriculture in Ethiopia* (Rome, 1961), 43-50; Asnakew Woldeab, "Physical Properties of Ethiopian Vertisols," in S.C. Jutzi et al., eds, *Management of Vertisols in Sub-Saharan Africa* (Addis Ababa, 1988), 111-23. The classic work on Ethiopian soils is H. F. Murphy, *A Report on the Fertility Status and Other Data on Some Soils of Ethiopia* (Alemaya, 1968).

Figure 1.1. A highland landscape (Lasta, Wallo) 1868.

Figure 1.2. A highland Landscape (Lasta, Wallo) 1972.

rainfall, vegetation, and, highland farmers would argue, human temperament. The agriculturally viable highlands themselves range between 1500 and 3000 meters, with subtle, incremental shifts in characteristics of ground cover, crops, and temperature as altitude rises and falls. The close connections between altitude and vegetation are such that a practiced eye can guess within 50 meters or so the elevation of a site by its natural vegetation, temperature, and a few quick glances at crops in the field. Virtually all languages spoken on the highlands distinguish three general gradations: *daga,* cool highlands about 2500 meters, *wayna daga,* midrange areas between 1800 and 2400 meters, and qolla, hot lowlands below 1800 meters. Inhabitants of the upper elevations also identify *wurch,* the coldest elevations above the tree line. The highlands thus include all three zones, the elevation in each zone creating one of the few consistent elements on the wide variation of highland environments.[5]

Climate: The Seasons

Of the factors which shape agricultural practice, climate is the most capricious. Farmers routinely manage soil fertility, innovate or adapt new technology, experiment with new crops and adjust to changes in political-social systems. The length of growing season, rainfall amounts and distribution, and temperature, however, are not amenable to any human action save migration to a different climatic zone. Ethiopia's agricultural calendar thus is largely a product of a distinctive climatic regime and the patterns of variation within it.

Like Africa as a whole, Ethiopia's climate is predominantly bimodal, with seasonality and variation in rainfall rather than temperature being the limiting factors. In the past decade or so, however, studies of poverty in agricultural communities in non-Western settings and climatological impact research have come to appreciate the importance of interannual variation in climate, particularly rainfall. Recent African droughts have provided strong evidence of these effects, as have microlevel community studies which have demonstrated the critical nature of seasonality, intra-annual variations in climate and agricultural cycles of labor, and food availability.[6]

5. For a study of effects of elevation, see Mesfin Wolde Mariam, *Suffering under God's Environment: A Vertical Study of the Predicament of Peasants in North-Central Ethiopia* (Bern, 1991), 20-28.

6. For Ethiopian and African evidence, see Workineh Degefu, "Some Aspects of Meteorological Drought in Ethiopia," in Michael Glantz, ed., *Drought and Hunger in Africa: Denying Famine a Future* (Cambridge, 1987), 27; Michael Glantz, "Drought, Famine, and the Seasons in Sub-Saharan Africa," in R. Huss-Ashmore and S. Katz, eds., *Anthropological Perspectives on the African Famine* (New York, 1987), 2; for social effects of seasonality, see Robert Chambers, Richard Longhurst, and Arnold Pacey, eds, *Seasonal Dimensions to Rural Poverty* (London, 1981).

Ethiopia's seasons result from three influences on air-mass circulation which regulate rainfall in the Ethiopian region: the Inter Tropical Convergence Zone (ITCZ), the northeast trade winds, and the southwest monsoons. From October through January the ITCZ's southern position brings an extensive dry season to much of the Ethiopian land mass. From March to June the northward movement of the ITCZ encourages the movement of moisture-laden monsoonal air from the southwest. By mid-June—in many Christian areas traditionally on St. Mikael's day—these conditions bring the primary rainy season to the highlands, which can last through mid-September. A Tigrinya proverb conveys the association between farm life and the physical world of climate: Dehri Dabra Tabor yelbon keremti dehri derho nego leyti (After the feast of Dabra Tabor [St. Mikael's Day] comes the rains, after the cock crows comes the dawn.)[7] In a deeper sense, this proverb indicates the association of the factors in time but asserts neither causality nor necessary timing. Farmers know cocks often crow well before dawn and most would consider a neighbor who plants on St. Mikael's Day as more foolish than pious if the rains have not actually begun.

The beginning of the rains brings a chain reaction of physical responses to the highland environment. Rainfall not only adds moisture to the soil but also causes a chemical reaction which releases nitrogen and allows seed germination. Grasses and legumes on open grassland or fallowed fields also return; within two weeks of the rains, brown landscapes turn green and grasses that are palatable to livestock return. Rivers and torrents swell, their force redistributing topsoil to enrich bottom lands and build flood plains, and carry off the rest to the Nile basin. Rainfall moistens and softens the soil to allow plowing for July sowing, but also stimulates the regeneration of rhizomes and seeds of the natural vegetation, which competes with crops. By September the noxious weeds—St. John's wort and Adey Abeba *(Bidens)*—which are dormant in the dry season, cover the highlands with brilliant yellow and purple flowers. Ironically, this season of vegetative renewal also marks the "hungry season," when women's joy at being able to collect clean rainwater within their own compound is mitigated by potential debt and the specter of the year's lowest stocks of food.

Annual rainfall averages range from between 500 and 1500 millimeters per year, but the rhythms of the seasons and long-term climatic patterns display considerable variation across the landscape from north to south and among altitude zones. The primary features of this variation are spatial—there is a major decline in the annual average from southwest to northeast and with variations in relation to altitude (rainfall averages drop to 500 millimeters or less in the lowlands). Not only do average amounts of rainfall steadily decline toward the northeast and down the slopes, but also rainfall shadow effects and topographic idiosyncrasies create a patchwork

7. Carlo Conti Rossini, *Proverbi tradizioni e conzoni tigrine* (Roma, 1942), 54. This translation and all other translations are my own unless otherwise attributed.

Figure 1.3. The main rains, Blue Nile Falls, September, 1973.

Figure 1.4. The dry season, Blue Nile Falls, May, 1974.

quilt of microclimates throughout the highlands. Addis Ababa thus receives 1200 millimeters per annum with 55 percent of its rain occurring during the main rainy season. Gonder receives 94 percent of its annual rainfall between April and September (60 percent in July and August alone); Asmara (Eritrea) in the far north, receives 74 percent of its rainfall during the main rains.[8] In other areas such as the Dabra Berhan highlands at 2900 meters, frosts and flooding push the growing season to after September or into early spring.

The bimodal character of rainfall sometimes breaks with high pressure systems in the Indian Ocean in mid-February, which in some years and in some parts of the country (primarily the east-central region) brings a period of agriculturally significant small rains to the south, east, and central parts of the country. In Wallo, Arsi, Wallaga, Bale, and parts of Shawa, spring rains allow a small crop of barley, pulses, or maize, which adds to food stocks and mitigates debt for resource-poor farms in those years when *belg* (spring) rains are agriculturally significant. In frost prone areas it is the only time in which cereals (usually barley) can grow. *Belg* production is directly dependent on the amount and reliability of February-to-April rains which soften and moisten the soil for short-maturing crops. Though in the mid-1980s *belg* rains were more reliable than the main rains, the historical pattern is the opposite. Nor could farmers in all highland zones build their agriculture around these rains. In the far southeast highlands of Harerge spring rains constitute the main growing season. In recent years while *belg* production amounted to only 3–5 percent of total national food production, it accounted for almost half the annual cereal crop in the key areas of Shawa and Wallo at the heart of the historical ox-plow complex.[9]

Signs of the Climatic Past

The importance of famine to recent politics in the Horn of Africa has generated a nascent historical literature on famine and drought, prompting climatologists to look to the historical record. Climate history as a means of understanding agricultural change in highland Ethiopia presents multiple methodological problems to the historian. Historical studies of Ethiopia's climate have often confused the records of famine with drought or other environmental shocks in royal chronicles, church

8. See Workineh Degefu, "Some Aspects," 25-26. Also Mesfin Wolde Mariam, *An Introductory Geography of Ethiopia* (Addis Ababa, 1972), 62-63; Frederick Simoons, *Northwest Ethiopia: Peoples and Economy* (Madison, 1960), 7; Daniel Gamachu, *Environment*, 14-16.

9. Relief and Rehabilitation Commission, *1987 Belg Production: Early Warning System Belg Synoptic Report* (Addis Ababa, 1987). The most comprehensive work on *belg* production is that of Alemayehu Lirenso. See his "Socio-economic Constraints to the Production of 'Belg' Crops in Ethiopia." (Social Science Research Council Working Paper No. 4), New York, 1991.

documents, and the travel literature.[10] Lists of such events have been attractive because Ethiopian Orthodox Christian hagiographers and imperial scribes often noted unusual natural phenomena—eclipses, droughts, locusts and so on—as markers of time or signs of divine judgment, resulting in a largely anecdotal record of putative climate events. Such records are spotty, impressionistic, and tend by definition to report only those events of political importance to elite interests, that is, end points such as famine rather than precise data on causation. For example, an often quoted letter from Ras Walda Sellase to English traveler Henry Salt [1811] offered a bizarre blend of personal greetings and descriptions of environmental calamities in:

> Peace be to you and may the peace of God be with you. How are you really, really my friend Henry Salt? Locusts have devoured my country. There is a drought. I had intended to make a [military] campaign, but I remained. The rainy season began. So I will spend the rainy season in my country. And now how are you, my friend Salt? I am glad you arrived safely. How are you really, really Henry Salt? In my country, however, there is suffering; locusts have devoured it. Peace be to you and may the peace of God be with you.[11]

This type of evidence places certain events in time, but rarely in space with sufficient detail to serve as a basis of climate or environmental history. Historians of Ethiopia have yet to use or collect a critical mass of proxy data such as lake levels, Blue Nile floods, or pollen studies which might corroborate events or patterns of climate change. Thus, despite the advantages of a documentary tradition, the evidence on Ethiopia's climate and its social and historical impact remains sketchy at best.

10. For treatments of history and drought, see Bahru Zewde, "A Historical Outline of Famine in Ethiopia," in Abdul Mejid Hussein, ed., *Rehab: Drought and Famine in Ethiopia* (London, 1976); Workineh Degefu, "Some Aspects," 28-31, relies on a mixed list of drought, locust, and famine episodes to conclude that "there is no pattern for Ethiopian drought episodes." A more substantial study of the social effects of drought, in this case the 1889-92 famine, is Richard Pankhurst, "The Great Ethiopian Famine of 1888-92: A New Assessment," *Journal of the History of Medicine and Allied Sciences* 21 (1966), 95-124, 271-94. A geographic study of the origins of vulnerability to famine (rather than drought) is Mesfin Wolde Mariam, *Rural Vulnerability to Famine in Ethiopia, 1958-77* (Addis Ababa, 1984). Most lists of environmental crises are based upon Adrian Wood, "A Preliminary Chronology of Ethiopian Droughts," in D. Dalby, R. J. Harrison, and F. Bezzaz, eds., *Drought in Africa-2*, African Environment Special Report 6 (London, 1977), 68-73.

11. Sven Rubenson, ed., *Acta Aethiopica, vol. 1: Correspondence and Treaties 1800-1854* (Evanston and Addis Ababa: 1987), 7. Compare these sources with those cited in Emmanuel LeRoy Ladurie, *Times of Feast, Times of Famine: A History of Climate since the Year 1000* (Garden City, 1971).

Intriguing evidence of long-term climate change comes from the peculiar debate between successive generations of travelers about the presence of snow in the northern highlands. Jerónimo Lobo, the Portuguese Jesuit writing in the early seventeenth century, argued that assertions that snow from Ethiopia's mountains was the source of the Nile waters were "ridiculous fantasy" but allowed that snow existed "on the mountains of Semen in Tigré and on those of Namora in the kingdom of Goiama [Gojjam]." A century and a half later James Bruce argued that "snow was never seen in this country nor do they have a word for it." Salt in 1814 argued he could "plainly distinguish the snow lying in large masses" in the Simen mountains. Two decades later Rüppell, however, also reported seeing snow on the Simen mountains on several different occasions. Interestingly, the debate among travelers was less an evaluation of the evidence of climate change than challenges to one another's veracity. In fact, the differences between accounts before and after 1850 may well point to regional climate change when the Little Ice Age yielded to warmer temperatures until after 1840.[12] Though warmer temperatures might have reduced rainfall and increased evaporation of soil moisture in tropical climates such as Ethiopia's, the evidence is too vague to support a conclusion.

Perhaps more compelling evidence of long-term climate and environmental change and its agricultural impact has been in the shift in Ethiopia's political center steadily southward since the decline of the Axum Empire at the end of the first millennium A.D. The migration of populations along a north-south axis, the seeming desiccation of the Eritrean and Tigray highlands, and the recent frequency of drought in those regions suggest a connection with the broader patterns of desertification of the Sahelian zone farther west. Unfortunately the cause and effect in this historical pattern are unclear, since a corollary question might ask whether intensive cultivation and deforestation had induced both climate change and later migration.[13]

The greater problem with understanding the factor of seasonality in a historical context is, not the episodic appearance of drought, but the variation of seasons from year to year and the patterns of variation over time. In the actual rainfall regime faced by farmers the questions have been whether rains which began last year on 12 June will do so again this year and whether the level will be sufficient

12. Lockhart, *Itinerario*, 240 n. 1; James Bruce, *Travels to Discover the Source of the Nile; in the Years 1768, 1769, 1770, 1771, 1772, and 1777* (Edinburgh, 1804), vol. 5, 256; Eduard Rüppell, *Reise in Abyssinien* 2 vols. (Frankfurt, 1838-40), vol. 1, 356, 402, vol 2, 14, 243, 244, 249, 251; Salt, *A Voyage*, 310; LeRoy Ladurie, *Times*, 312-13.

13. A key corollary to the effect of human action on climate is the "Albedo effect," which holds that removal of vegetation exposes soil to direct solar heat, raises surface temperatures, and affects rainfall -"drought follows the plow." See National Meteorological Services Agency, "Climatic and Drought Conditions in Ethiopia," paper presented to the Scientific Roundtable on the Climatic Situation and Drought in Africa, Addis Ababa, 1984, 13.

to begin the full cycle of events necessary to plant and harvest a crop. The uncertainty of levels from one year to the next underlies farm-level decisions historically. Farmers may well know the features of seasonality but they cannot predict the more subtle annual variations of the season's onset and duration, and thus must build risk-aversion strategies around these uncertainties. In most areas historically prone to drought, year-to-year variability is a strong feature; climatic data show rather conclusively that the lower the annual total rainfall, the greater the degree of variability.[14]

Seasonal aspects of climate are therefore critical prerequisites for the utilization of human, technical, and capital resources within historical agricultural systems. A small spatial or temporal variation in critical climate variables can produce a substantial effect on the physical limits of production. In the case of Africa, and Ethiopia in particular, it is the intra-annual distribution of rainfall that is one of the most relevant climatic variables of food production. The timing of rainfall and its particular relationship to the constellation of labor, cropping patterns, and capital requirements of a specific farming-social system are a critical but neglected aspect of both the historical and contemporary development of rural society and economy. It is, after all, the patterns of seasonal rainfall which trigger the social and economic processes of labor, the renewal of resources (food, seed, forage), and the shortage or abundance of harvests.[15]

The historical relationship between patterns of climate and human adaptation raises questions critical to the field of climate history as well as to agricultural history. The relationship between socioeconomic activity and seasonal rhythms of climate is not direct but mediated by environmental factors such as forage availability, crop characteristics, soil type, and socioeconomic institutions of labor exchange, kinship obligations, property distribution. Jane Guyer argues persua-

14. Eugene Rasmusson, "Global Climatic Change and Variability: Effects of Drought and Desertification in Africa," in Glantz, *Drought and Hunger;* 8 and Glantz, "Drought, Famine, and the Seasons," 8. Rainfall data derive from Tsegaye Wodajo, *Agrometeorological Activities in Ethiopia: Prospects for Improving Agroclimatic/Crop Condition Assessment* (Columbia, Mo., 1984).

15. For example, in Iceland, where temperature is the key climate variable, a difference of one degree in annual temperature can decrease the length of the growing season by 27 percent. See Reid Bryson and Christine Paddock, "On the Climates of History," in Robert Rotberg and Theodore Rabb, eds., *Climate and History in Interdisciplinary History* (Princeton, 1981), 8-9. Rasmusson, "Global Climatic Change," 6-7; Sharon Nicholson, "The Methodology of Historical Climate Reconstruction and Its Application to Africa," *Journal of African History* 20 (1979), 38. Glantz's notion of a meteorological drought being far less serious than an agricultural drought is an important case in point. He defines agricultural drought as "the lack of adequate soil moisture to sustain crop growth and production." Glantz, "Drought, Famine, and the Seasons," 15.

sively: "Through living in a particular social and ecological environment over time, people develop a known repertoire of strategies which become part of customary knowledge, socially maintained as a resource and predictably mobilized where the need arises. . . "[16] Case studies included in part 2 of this book detail local variation across the ox-plow complex in which climate, economic culture, and historical experience interact.

People, Agriculture, and Changing Landscapes

Ethiopia's highlands offer a distinctive and, in many ways, ideal setting for human habitation and the evolution of agricultural societies. For highland peoples and for visitors to Ethiopia, the highlands' climate, their seasonal verdancy, and striking topography appear in sharp contrast with the hot, semi-arid lowlands surrounding them. Before commercial air travel in the 1950s, travelers to Ethiopia necessarily experienced the highlands in juxtaposition to the hot and topographically monotonous journey from Sudan or the Red Sea coast. Subjective views of the environment have not been peculiar to Europeans. Ethiopia's highlanders have their own views, though they share with outsiders an anthrocentric view of environmental-human interaction, especially the effects of elevation. They associate altitude zones with temperature, disease risks, cuisine, musical styles, and even personality— lowlanders being hot-headed and highlanders being cool and deliberate.[17]

The Ethiopian highlands on which human action has taken place (i.e. during the second epoch of Ethiopia's prehistory) thus were composed of rolling plateaus broken by shear escarpments and mountain ranges exceeding 4000 meters. Average elevation in the highlands ranges between 1800 and 3000 meters. Vegetation regimes and the appearance of highland landscapes vary by the effects of geology and hydrography and more profoundly by the weight of human activity which has disturbed vegetative cover and the soil beneath it. Grassland and pasture undisturbed by the plow probably historically enjoyed a net balance of seasonal soil loss and new soil formation. Current estimates of soil loss place the removal by water and wind at 1,493 million metric tons per year—a staggering rate and probably the cumulative effect of human actions of cultivation and deforestation—which exacerbated the natural effects of seasonal torrents on highland soils. Recent alarmist estimates place the annual loss of soil productivity at 2–3 percent. Annual loss rates on cultivated zones range from estimates of 2.3 mil-

16. Jane Guyer, "Synchronizing Seasonalities: From Seasonal Income to Daily Diet in a Partially Commercialized Rural Economy (Southern Cameroon)," in David Sahn, ed., *Causes and Implications of Seasonal Variation in Household Food Security* (Washington, D.C., 1987), 1.

17. See, for example, Donald Levine, *Wax and Gold: Tradition and Innovation in Ethiopian Culture* (Chicago, 1965), 77-78.

limeters in dry, warm zones in the lowlands to 4.9 millimeters in the cold highland zones at about 2600 meters.[18]

Vegetation and forest across the highlands have historically varied from open grasslands or scattered wooded savannah in the central highlands, to moist evergreen montane forests in the southwest with dense forests of tall broadleaf hardwoods. The hardwood forests visible today in the southwest may have originally covered as much as 21 percent of the Ethiopian region. Today they account for less territory than that but still constitute 65 percent of Ethiopia's total forest resources. The hardwood forest soils are deep compared with the brown-black soils of the northern plateau, but concentrate their nutrients in the upper layers, making them more vulnerable to erosion and requiring use of forest fallow cultivation. In the less-well-watered regions to the north and east dry evergreen montane forests occur with shorter softwood *podocarpus (zigba)* and junipers *(tid)*.

Interaction of the highlands' vegetative cover with the ox-plow complex has brought significant changes. Under the influence of the plow, virtually all the northern highlands' dry evergreen forests and grasslands and a large part of the moist evergreen forests have changed to open farmlands and pasture. The *juniperus* forests of the central and northern highlands perhaps once covered as much as 15 percent of Ethiopia, but were the first lands converted to cereal agriculture; less than 1 percent of the original forest area there remains.[19] Acacia woodlands along the eastern and western escarpments have more recently succumbed to short-fallow cultivation.

The history of the highlands' vegetative cover has also been subject to a deep-seated mythology. Christopher Clapham has pointed out that the commonly repeated "fact" that Ethiopia's forest cover has diminished from 40 percent to 4 percent within the twentieth century is both apocryphal and false. These figures, which suggest an alarming deforestation, have been reported as unchanged in development literature since at least 1960. Evidence presented in chapters 4 and 5 indicates that forest loss is neither unidirectional nor permanent.[20]

18. Daniel Gamachu, *Environment*, 11-12. For loss of soil productivity, see Hans Hurni, "Degradation and Conservation of the Soil Resource in the Ethiopian Highlands," paper presented at the First International Workshop on African Mountains and Highlands, Addis Ababa, 1986.

19. Daniel Gamachu, *Environment*, 11-13; see also United Nations Food and Agriculture Organization, *Vegetation and Natural Resources of Ethiopia and Their Significance for Land Use Planning* (Rome, 1982); Hurni, "Degradation and Conservation."

20. Christopher Clapham, *Transformation and Continuity in Revolutionary Ethiopia* (Cambridge, 1988), xi-xii. Clapham points out that the highlands contain 21 varieties of endemic montane nonforest birds and few montane forest birds. For two citations of the apocryphal 40 percent to 4 percent, see United Nations Development Program and World Bank, *Ethiopia: Issues and Options in the Energy Sector,* Report no. 4741-ET, ii; (New York, 1983) and Relief and Rehabilitation Commission, *Combating the Effects of Cyclical Drought in Ethiopia* (Addis Ababa, 1985), 14-16. The first reference to the 40 percent and 4 percent figures I have found is in Huffnagel, *Agriculture*, 395, 405-6, who estimates a 1.4 percent annual loss.

If those figures on deforestation are misleading, there is nonetheless strong evidence that the highlands' agricultural economy evolved since at least the sixteenth century in an open environment without heavy forests in place. In the first instance, human effects on forest and vegetation are by no means a modern phenomenon: the first charcoal making on the highlands appeared 2500 years ago, and dung has been a major source of fuel in the nonforested highland regions for at least 400 years.[21] The Portuguese Jesuit Almeida reported the conditions in the north in the 1620s:

> Generally speaking there is not much woodland in Ethiopia. In some parts, especially in Gojam, there are forests of trees of various kinds, like wild cedars, zigbas, which closely resemble cedars, and other wild unproductive trees. . . . The whole of this country is well supplied with thorn bushes and the trees are so tall that where there are many of them together they seem to be groves of pines rather than thorn bushes; they are used for firewood.[22]

By the nineteenth century little had changed: Salt in 1811 noted that groves of trees were "rarely met with in Abyssinia," and Walter Plowden in the 1850s observed that "of large timber in the territories of the Christians, there is little." Forests in the highlands have never been continuous, being interrupted by waterlogged highland plains, which impeded forest growth in as much as 24 percent of the total highland area and produced open, seasonally flooded pasture.[23]

Human habitation has continually altered the vegetative cover, not simply removed it. For example, evergreen scrub, the thorn bushes in Almeida's description, formerly found only in shallow soils and rock crevices has expanded with the shrinkage of the original forest, because the erosion associated with cereal cultivation has expanded its habitat.[24]

In the opening epoch of highland environmental history, the highlands themselves formed and took their initial geological contours from natural forces. In the second phase with increasing expansion of human habitations, people began to manage and shape the soil and vegetation of those contours with agricultural tools and evolving economic cultures. From sometime before the first millennium A.D. the ox plow became the dominant tool, and the combination of livestock and an annual cereal crop base became the highlands' economic context. Under these conditions, the ability of farms over time to have a cumulative effect on the land was a

21. Manoel de Almeida, *Some Records of Ethiopia, 1593-1646* trans. and ed. C. F. Beckingham and G.W.B. Huntingford (London, 1954), 48; Alvares, *Prester John* , vol. 2, 510; Tewolde Berhan Gebre Egziabher, "Historical and Socio-economic Basis of the Ethiopian Ecological Crisis," unpublished typescript, Asmara University, 1989, 11.

22.Almeida, *Some Records,* 48.

23. Salt, *A Voyage,* 201; Walter Plowden, *Travels in Abyssinia and the Galla Country* (London, 1868), 32.

24. Tewolde, "Historical and Socio-economic Basis," 3.

fairly direct, if not always historically recoverable demographic equation. Cultivation expanded with population growth and concentration, but contracted when numbers of people on the land declined through movement or slowed growth. Yet, the people which wrought changes in landscape were themselves subject to short- and long-term changes in the physical environment that arrested or altered agriculture and land use. The mix of the geological and climatic factors which lay beneath highland agriculture predetermined little if anything in its historical development. For the 1800–1990 period covered in this book, however, human actions and natural forces worked in tandem, if at different paces, to shape the movement of agriculture through time and across the highlands' localized environments. The technology and economic culture described in chapter 2 further set the patterns, if not the events, and the ultimate shape of modern agricultural history.

2

The Ox-Plow Complex:
An Ecological Revolution

Agricultural Genesis

The rich black and red clay soils and open woodland of the north central highlands provided the first home for the Ethiopian ox plow. The first direct evidence of a plow with its typical long curved beam comes from two sites in Eritrea, where images from cave paintings depict humpless oxen pulling long-beam plows. While rock art is difficult to date, those paintings can be tentatively attributed to the first millennium B.C. From the same period, a small figurine from the pre-Axumite period (end of the first millennium B.C.) shows a yoked humpless ox.[1]

Early evidence of the plow in Ethiopia appears bound up with, and confused by the interaction of Cushitic-speaking residents of the highlands and Semitic-speaking migrants from South Arabia. Though early scholarship tended to ascribe a "civilizing" mission to South Arabian migrants, which included a plow-based agricultural system, the best recent evidence supports the idea that plow agriculture preceded South Arabian influence. Joseph Michels' proposed settlement model based on the Axumite surface evidence from Yeha, which he dates from 700 B.C. to 400 B.C., argues for "dry farm" settlements which predated and then coexisted

1. The Italian archaeologist Radolfo Fattovich suggested these dates to me. The first evidence of humped cattle in Ethiopia comes later, in the first centiury A.D. See Antonio Mordini, "Un riparo sotto roccia con pitture rupestri nell'Ambà Focadà," *Rassegna di Studi Etiopici*, 19 (1941), 59. Michael Goe, "The Ethiopian Maresha: Clarifying Design and Development," *Northeast African Studies* 11 (1989), 71-73; Steven A. Brandt, "New Perspectives on the Origins of Food Production in Ethiopia," in J. D. Clark and Steven Brandt, eds., *From Hunters to Farmers* (Berkeley, 1984), 178.

with early South Arabian forms of irrigated agriculture and terracing. In fact, Michels argues, the collapse of the South Arabian center at Abba Penteleon result- ed from the local environmental limitations on specialized, intensive agriculture, that is, irrigation and terracing. Though the environs of the city of Axum itself show evidence of intensive agriculture, the dominant form of cultivation in the empire at its zenith (100 B.C.–A.D. 1000) was dryland agriculture, most likely based on the ox plow. Michels points to the archaeological evidence for six phases of develop- ment in and around the site of Axum in which, except for the 400–100 B.C. peri- od, dryland cultivation of the rich red clay and loamy soils of the northern plateau was the dominant form of production.[2]

Highland agriculture did not evolve piecemeal but as a system which has wed- ded a primary tool (the plow), the use of oxen, and a diverse repertoire of annual crops. This formula has proved itself a powerful environmental adaptation by its persistence on the northern highlands and also in its spread across a diverse ethnic and physical highland landscape. The spread of plow agriculture across its modern domain is difficult to trace, though the prominence of local cultigens (teff and eleu- sine) in the early forms of plow agriculture suggests it probably remained within the 1800–2200 meter range until the arrival of Near Eastern cereals (barley and wheat) allowed an expansion further up the slopes to 2900 meters at the upper limit.

Plow agriculture is oldest in the northern highlands, but today encompasses most of the Ethiopian highlands, save the areas occupied by pockets of hand-hoe horti- cultural economies in the south and southwest.[3] Its spread to new areas and new peo- ples has not, however, necessarily paralleled the expansion of the Amhara-Tigrayan ("Abyssinian") state or the national market economy. From a millennium long per- spective, the single greatest path for the expansion of plow agriculture has been the engagement of the economic cultures of highland Semitic-speaking populations and Oromo-speakers between the 16th and early twentieth centuries. In the first instance, Oromo who arrived in the highlands of eastern Wallo and Tigray, northern Shawa, and as far north as the Lake Tana basin shifted within one or two generations from pastoralism to plow agriculture. In the densely forested Gibe basin of the southwest highlands, Oromo pastoralists who arrived in the mid to late eighteenth century were already established in their own sophisticated ox-plow, cereal-based system by the time of our first detailed ethnographic descriptions in the mid-nineteenth century.[4] In Harerge, along the ridge of the southeastern highland plateau, pastoral Oromo shift-

---

2. Joseph Michels, "The Axumite Kingdom: A Settlement Archaeology Perspective," in *Proceedings of the Ninth International Congress of Ethiopian Studies* (Moscow, 1988), 176-78.

3. For descriptions of these horticultural economies, see Hermann Amborn, "Agricultur- al Intensification in the Burji-Konso Cluster of South-western Ethiopia," *Azania* 24 (1989), 71-83; C. R. Hallpike "Konso Agriculture," *Journal of Ethiopian Studies* 8 (1970), 31-43.

4. For these descriptions, see Cecchi, *Da Zeila*; Guglielmo Massaja, *I miei trentacinque anni di missioni nell'alta Etiopia*, 12 vols. (Milano, 1886); Mohammed Hassen, *The Oromo of Ethiopia: A History 1570-1860* (Cambridge, 1990), 114-32.

Figure 2.1. Humpless oxen and long beam plow at Amba Focada (Eritrea), c. 500–1000 B.C.

ed to agriculture more slowly, learning from Harari town dwellers and finally settling into plow agriculture in the Egyptian period (1875–85). Plow cultivation on the Somali plateau around Hargeisa began less than a century ago under the influence of colonial rule and urban market growth.[5] By the twentieth century, northern highlanders retained the cultural mantle of being the consummate people of the plow, even though the majority of Ethiopia's plow agriculturalists in fact lived within the southern highland sweep from Wallaga in the west to Harar in the southeast.

In addition to evidence of spontaneous movement, the plow also has had a strong historical association with the expansion of political economy of the northern highlands' centralized state system. In periods of expansion such as under Emperor Zara Yaqob (1434–68), the imperial government's reach extended beyond the northern highlands to include tributary states to the south and east. The plow's own hegemony expanded with military presence, and also through more subtle cultural influences and diet. The key to the state's expansion was imperial control over the assignment of patrimonies *(gult)* on the frontier or their reshuffling in dissident core areas as a means of cementing ties of loyalty and channeling resources to the

5. See Muhammad Hassan, "The Relation between Harar and the Surrounding Oromo between 1800-1887," B.A. thesis, Haile Sellassie I University, 1973, introduction and 30; interview with Yusef Chela, Lelissa (Harerge), 4 July 1987. For Hargeisa agriculture, see Samatar, *The State and Rural Transformation,* 42, 55, 61.

imperial court. In periods of imperial expansion military conquest produced new *gult* grants which expanded imperial revenues and increased the pool of loyalists who owed their economic and political base to imperial indulgence. That expansion also brought with it Ethiopian Orthodox Christianity, conventions on social property, forms of taxation, and an overarching political culture. The distinguishing hallmark of these expansionary states was the center's ability to expand its influence during periods of economic and political coherence and its countervailing decline during periods of political retrenchment. In other periods, notably the Zamana Mesafint (1769–1855), the center exercised no clear power; while its political influence withered, the domination of plow agriculture went unchallenged.

Incorporation of new areas particularly along the frontier between agricultural and pastoral highlands, involved military conquest and also the expansion of Christianity, language, social institutions, and plow agriculture. In the long run, it was the successful expansion of these latter factors rather than military conquest which resulted in incorporation into the highland political economy. Being an Amhara or Tigray—Habasha—was not a tribal identity dependent on lineage but a social category which implied adherence to religion, knowledge of language, and, more subtly, being part of a system of cultivation.[6] The initial military conquest settled northern soldiers on new lands and provided patrimonies for new churches to serve them. The spread of the plow, however, proved a much more effective indicator of an area's political and economic assimilation to the highland state; to wit, the imperial system often conquered, but rarely effectively ruled, areas in which environmental conditions prevented adoption of the highlands' agricultural system.

By the opening of the nineteenth century virtually all the highlands above 9 degrees latitude—the upper two thirds of the Ethiopian region—were fully incorporated into an annual cropping, ox-plow rural economy. The expansion of imperial state authority and cultural traditions associated with the northern highlands was not, however, organically linked to farm-level decisions about adoption or use of the plow or a particular cropping system. In fact, neither the imperial state nor local gult-holders retained any degree of control over farm-level decisions about production, that is, what crops to grow, what tools to use, or the forced adaptation of the "specialized" techniques of terracing or irrigation. This characteristic of the highland political economy stood in marked contrast with European feudal relations where landlords took a direct role in production, introducing new field systems or technologies.[7] Innovation and incorporation of new forms of agriculture, especially in new cultivars or adaptation to new settlement areas took place at the farm level and therefore appear rarely in the "eventful" recordings of historical sources.

6. Donald Donham, "Old Abyssinia and the New Ethiopian Empire: Themes in Social History," in Donald Donham and Wendy James, eds., *The Southern Marches of Imperial Ethiopia: Essays in History and Social Anthropology* (Cambridge, 1986), 10-12.

7. Donham, "Old Abyssinia," 13-14.

Figure 2.2. A northern setting for agriculture: Axum, c. 1810.

Figure 2.3. A southern setting for agriculture: a white sorghum field, c. 1850.

43

If the expanding state did not require use of the plow, it did demand tax payments in the annual crops of the plow complex. The rural subjects of Axum and its successor highland states provided a soldiery as well as agricultural products consumed by non-farming bureaucratic classes in the church and political hierarchy. Teff, the highest prestige food throughout the highland system, requires intensive seedbed preparation possible only with the ox plow. Barley, wheat, pulses and almost all annual crops save maize (a late addition) and sorghum also require use of the plow.

Until the post-World War II period, markets also played a remarkably weak role as an engine of change in the farm economy, largely because of the dearth of urban centers and predominance of political clientship as the chief means of food distribution. Towns and population concentrations did exist across the highlands, but the population's demand for food was met by granaries filled with tax and tribute payments rather than market purchases.[8] Rural markets, of which there were many, served as points of exchange between economic zones and long-distance trade rather than as points through which farmers could expect to acquire capital. Transport costs of grain—a high volume-to low-price commodity—limited access to grain markets in Eritrea and Sudan to only a few areas. Therefore the spread of the ox-plow complex derived less from political expansion of the northern imperial state than from the characteristics of its own technology and its transformative powers in social relations on the farm and in the rural community.

Microagronomic Economies

The thoroughness of the ox-plow complex's movement across the highland landscape should not obscure the power of localized environments to alter its forms. This localization has taken place down to the smallest environmental and political units, sometimes only a few kilometers wide. In agricultural terms the variations within these "microagronomic" economies consist of adjustments to design of the plow, agronomic field systems for managing water and soil, crop mixes, and crop-livestock integration. The idea of microagronomic economies helps account for diversity within the ubiquitousness and power of the ox-plow system itself. Thus it is possible to see horses plowing in a system overwhelmingly dominated by oxen, the use of broadbeds or terracing only in a small river valley that is isolated from areas where such practices are unknown, or entirely different responses to new inputs.

8. For urbanization, see Donald Crummey, "Some Precursors of Addis Ababa: Towns in Christian Ethiopia in the Eighteenth and Nineteenth Centuries," in *Proceedings of the International Symposium on the Centenary of Addis Ababa* (Addis Ababa, 1987); also see Frederick Gamst, "Peasantries and Elites without Urbanism: The Civilization of Ethiopia," *Comparative Studies in Society and History* 12 (1970), 373-92; Donham, "Old Abyssinia," 15.

These microagronomic economies are products not only of environmental conditions—topography, soils, microclimates—but also of cumulative and/or conjunctural effects of historical factors such as demography, the presence of a royal court, and specific economic shocks. The following sections on the plow and oxen, field systems, and crops describe both the overarching principles of the ox-plow complex as a whole and the infinite variation within it. Part 2 offers case studies which indicate the interaction between systems and the ox-plow complex's localized manifestations of historical conjuncture.

The Plow

The highlands' farming system centers on the *marasha,* or plow, a simple but brilliantly adaptable tool, which when drawn by a yoke of oxen breaks but does not invert the soil. This scratch plow, or ard, stands at the center of the agricultural system because of its simplicity and its efficiency in human labor. The Ethiopian version consists of eight basic parts, all available locally: the beam *(mofer),* the plowshare *(marasha* lit. "that which plows"), the sheath *(wogal),* the stilt *(erf),* two wooden ears (sg. *diggir)* inserted into the plowshare's sheath, a yoke *(qanbar),* and a leather strap *(mangacha)* which adjusts plowing depth.[9] None of these basic parts has changed over the course of the plow's recorded history, though some materials have adjusted to new conditions: blacksmiths fashion twentieth-century plowshares from leaf spring steel of automobile carcasses rather than locally smelted iron, and the wood used for the beam has also changed from the favorite *wayra (Olea africana)* to the more widely available eucalyptus.

The finished product's parts are replaceable and adjustable to specific needs of a field or task; the plow is light enough (12–20 kg) for the farmer to carry long distances from field to field. Farmers adjust the angle of pull to the desired plowing depth by varying the share and ear length, adjusting the angle between share and beam, or exerting downward pressure on the handle during plowing. The angle of pull may vary also; in Dabra Berhan in northern Shawa it varies from 15 to 25 degrees. Farmers thus regularly adjust their plows for the conditions of each of their fields and whether they are making a first pass (a shallow angle) or a seed-covering pass (deep angle). Farmers take the raw beam to a local artisan, who, through charring, burying, and shaping it under rocks, produces a *mofer* which conforms to exact local needs.

The *marasha* design allows it to break the soil's surface at a fairly shallow angle, an effect substantially different from heavy steel moldboard plows of European origin. The clods of soil pushed to the side on the first pass can be broken up

9. Michael Goe, "Tillage with the Traditional Maresha in the Ethiopian Highlands," *Tools and Tillage* 6 (1990), 131.

by repeated passes, the number depending on soil conditions, prior field use, and the crop to be sown. Teff requires up to eight passes, though more commonly five, and horsebeans may need only one. A final pass covers broadcast seed with a layer of soil. In many cases farmers may simply shorten the angle of the plowshare to cut more shallowly. In Dabra Berhan the soil wedge moved on short fallow plots may be as much as 150–200 square centimeters, whereas hard long-fallow plots may allow only 43–150 square centimeters with furrow depths varying from 10 to 15 cm.[10]

The plow's simplicity of design has also allowed its adaptation for specific conditions across the highlands. Farmers over time have adjusted or reengineered the *marasha* for local field conditions and for plowing depth, soil conditions, and local variations in oxen. Farmers in Gojjam and Gonder, for example, use longer handles than farmers of northern Shawa, whereas Eritrean beams are shorter than those of other areas by as much as one meter and are the heaviest of all regions. Farmers in Ada use short beams that have a large diameter which are less likely to break in the area's heavy clay vertisol. The short beam also is easier to lift in damp soil, which cakes on the share once the rains have begun. As a whole, these differences reduce or increase the weight of the soil wedge, as well as allow for adjustments to the strength, size, and condition of the farmers' oxen. Though the plow's weight may vary between 13 and 20 kilograms by region and individual design, in every case the plow's simple, efficient design allows the cultivator to carry it from field to field and home again on a daily basis. Recent attempts to introduce heavier steel moldboard plows have ignored the essential element of portability which allowed farmers to cultivate highly fragmented land holdings at different altitudes and long distances apart.

Yoke design also has adapted historically to local conditions and farmer preference. In rocky areas in northern Shawa, wide yoke beams permit cultivation around large rocks and stumps or on slopes. Longer yokes also add to the ease of turning the oxen on slopes or along stone-fenced borders. Elsewhere, such as in the Mecha area of western Shawa, farmers prefer shorter yoke beams, which, they argue, concentrate draft power for heavy soils, and yoke beams with special extrusions to protect the oxen's humps.[11]

Ox-plow farms across the highlands also historically used supplementary hand tools for preparing the soil and the harvest. These tools varied by region in name and design but generally included an *akafa* or *doma* (short-handled hoe), a *machid* (sickle), and a *magelabecha* (a wooden spade for winnowing). Heavy soils in virgin or fallowed fields may have required use of a *makotkocha* (a long-handled hoe) or in Harerge a stone-weighted digging stick *(dongora)* to break clods or the soil's surface. Food processing, performed almost exclusively by women, required a set

10. Goe, "Tillage," 148.

11. Goe, "Tillage," 141-42; personal observation in Tagulet.

of grinding and pulverizing tools complemented by basketry shaped into sieves, containers, and mats for finer winnowing.[12]

The revolutionary qualities of the ox-plow system reside partly in its adaptability but more fundamentally in its clear savings to labor over hand tillage. Field studies from Burkina Faso which compared hand tillage to an ox-drawn plow suggest dramatic advantages to the plow. Compared with hand tillage ox plows required 31 percent less labor time per hectare with 16.7 percent higher yields for sorghum; also, the plow was shown to increase yields more than 200 percent on fertilized plots whereas fertilizer had less than half that effect on hand-tilled plots. Beyond yields, the same study indicated substantially larger areas cultivated by ox-plow households than those cultivated by hoe households. An adult can cultivate annually only about 1.25 hectares with hand-hoe technology; ox-plow farms in highland Ethiopia routinely cultivate two to four times that amount.[13]

We have few, if any, historical records of the transition to the plow in the northern highlands; indeed for the early Cushitic population of the northern highlands it may be better to consider the plow an autochthonous technology.[14] Most Oromo groups of the north and central highlands appear to have adopted the plow directly rather than after an intermediate hand-hoe stage, reflecting a choice of agriculture over pastoralism rather than a specific choice of one form of tillage over another. In the case of the southern and southwestern highlands, however, the plow supplanted farming systems based on hand-hoe tillage of perennial crops and tubers. The transition therefore involved a shift in cropping systems from perennial to annual cultigens and from root crops to cereals. In one such case, Amon Orent argues that the kingdom of Kaffa shifted to plow agriculture in the seventeenth century not as a producer-based response to increase overall food production, but as a result of the royal court's preference for the prestige value of teff and cereals over *qocho* (ensete), yams, and taro, spurring elites to require tribute in cereals. Cereals were better for tax collectors since they could be stored, divided, and moved. In other areas, Gurage, Gedeo, Malle, Aro, Walayta, where ensete dominated traditional

12. For descriptions of supplementary equipment, see Oklahoma State University, *Ethiopian Farm Tools*, pamphlet (Alamaya, 1960); Huffnagel, *Agriculture*, 156-61; Graham, "Report," 266-7.

13. Vincent Barrett, et al., "Animal Traction in Eastern Upper Volta: A Technical, Economic, and Institutional Analysis," MSU International Development Paper No. 4, 1982, 63, 11-12, 6. For highland Ethiopia the constraint on farm size is land availability rather than technical capacity: a single yoke of oxen can plow 0.25 hectare per day. For historical comparative value of ox-plow cultivation over hand tillage in the eighteenth-century Kaffa region see Amon Orent, "From the Hoe to the Plow: A Study in Ecological Adaptation," in *Proceedings of the Fifth International Conference on Ethiopian Studies* (Chicago, 1979), 187-94.

14. Evidence on origins of the plow is not at all conclusive, though assumptions of its Egyptian or South Arabian origin have been effectively refuted. See Goe, "Maresha," 2-3.

diets, the adoption of plow agriculture may have come with changing elite tastes rather than with the attractions of greater productivity.[15]

Oxen must have been an intrinsic part of plow cultivation from its earliest use, indicating a long-standing and sophisticated practice of specialized animal husbandry which produced, trained, and maintained oxen as traction animals. The earliest evidence of the plow—the rock paintings of Amba Focada—depicts humpless oxen yoked to a typical *marasha*. Most highland farmers use a pair of zebu *(Bos indicus)* oxen, a humped species which arrived in the region from two separate migrations of Indian stock.[16] Highland farms also historically kept ancillary livestock: sheep or goats for food, savings, and small-scale exchange; donkeys for marketing; chickens for food and exchange. Specialization in mule husbandry in certain areas such as Lasta also reflected sophisticated management. Mules were, however, few in number and used as prestigious riding animals rather than as a part of agricultural production.

Specialized breeding of cattle from an early stage is evident in the finely tuned speciation of zebu varieties: Boran have twice the water intake capacity of other zebu and also make good plow animals; sanga are adapted to diseases of lower zones, highland zebu are small, hardy and well adapted to confined pasture. A dwarf breed still visible around Jimma, by contrast, suggests a type adapted for the forest rather than one bred for the plow. Skills in husbandry resided in individual households and in rural communities rather than with specialists. In the early seventeenth century, Almeida described considerable diversity and sophistication in household herds which provided oxen for food as well as work: "There are very big oxen and cows and the horns of the gueches which are big oxen, show that they do not work but are used for slaughter."[17]

A major weakness in household economy over time, however, has been the number of oxen any given farm could produce and sustain. To maintain a yoke of working oxen through time a single household probably required 10 head of cattle (2 oxen, 2 young "apprentice" bulls, a stud bull, and 4 or 5 milk cows). In highland areas of the longest settlement (northeast and central areas) pasturage available to each farm was inadequate to sustain such a herd in many local economies. The roles of regional specialization, middle distance trade, and interhousehold exchange were also critical to maintaining adequate supplies of oxen across the highlands. As new ox-plow communities matured over time, pasturage inexorably gave way to new

15. Orent, "From Hoe to the Plow," 187-93. We have little empirical evidence on the adoption of plow culture in the societies of Gamo Gofa, Kaffa, and Ilubabor, many of which adopted the plow within the last century. The best study of these southern regions is remarkably silent on the issue. See Donham and James, *Southern Marches.*

16. M. Alberro and S. Haile Mariam, "The Indigenous Cattle of Ethiopia-Part I," *World Animal Review* 41 (1982), 3-4. Abyssinian zebu are divided into four varieties.

17. Almeida, *Some Records,* 49-51.

cultivation, reducing livestock carrying capacity (see chapters 3 and 6 on "closed" ox-plow economies). Epizootic diseases, such as rinderpest, anthrax, blackleg and bovine pleural pneumonia periodically thinned herds in particular areas, and endemic problems such as liverflukes, intestinal worms, ticks, and lungworm reduced productivity at lower elevations, where open pasturage was more available.[18]

Farms thus depended on a steady, if historically unnewsworthy, percolation of oxen around the highlands through market purchases, exchanges between relatives, and raiding. Alvares in the sixteenth century observed the practice of *ribi*, a livestock propagation and pasturing agreement. That practice continues as a means of dealing with shrinking pasturage in densely settled highland zones.[19] Recent studies suggest that rates of calving and longevity in highland areas are lower than in lowlands, supporting localized evidence that highland zones have historically been net importers of cattle and oxen in particular. Across the highlands there has likely been a net movement of oxen from the south toward the northeast, that is, in the direction of the most mature ox-plow zones with the least pasturage. Historically, the movement of salt, mined in Tigray and along the eastern escarpment of northern Wallo, into long- and middle-distance trade networks drew oxen and cattle out of surplus areas like Gojjam, the Lake Tana basin, and southeastern Wallo into the oxen-poor areas to the north and east. The rise in trade in manufactured goods from Eritrea beginning in the late nineteenth century also helped sustain the northeast's agricultural base by providing exchange goods which continued to draw oxen and cattle into the northeast's rural economy.[20] By the second half of the twentieth century the crisis in oxen supply has become more generalized, affecting farms on the southern periphery as well.

For the highland rural economy oxen were the consummate form of capital and often the scarcest economic resource. For many highland areas—especially those of the longest settlement and highest population density—the breeding, purchase, borrowing, and maintaining of oxen determined household patterns of land use, debt, labor allocation, and cropping strategies. If "rights in persons" (i.e. slavery) was an economic engine of many West African rural economies suffering from a shortage of labor, oxen or "rights in oxen" dominated rural resource strategies in the Ethiopian highlands. Through networks of social cooperation, rental agreements, and labor exchange, households that controlled oxen could almost always obtain either labor or land on favorable terms.

For Ethiopia, the *prima facie* historical evidence indicates the overwhelming dominance of oxen as the draft animal of choice. No other animal has supplanted it over the history of the use of the plow. Why? Horses replaced oxen in Eng-

18. Gryseels and Anderson, *Research*, 14.

19. Alvares, *Prester John*, vol. 1, 92-93; the *ribi* relationship remains a common exchange between highland and lowland households where pasture is in short supply.

20. See McCann, *Poverty to Famine*, 73-79.

lish smallholder agriculture in the fourteenth century; in the New World horses, mules, and oxen have all served as traction animals. On Ethiopia's highlands no other animal fits the needs of the smallholder farm so well as the highland zebu ox. Questions to farmers or livestock specialists about the use of other traction animals usually generate anecdotal responses ("Once I saw. . .") referring to the use of donkeys, horses, camels, or even humans yoked to a plow. Draft power that allows adequate plowing depth, however, is a function of body weight distributed over the front legs. Highland oxen are consistently heavier than any other domestic animal, except large mules, which are scarce and cost 2–10 times the cost of an ox.

Horses and cows on rare and localized occasions substitute for oxen. In Tagulet (just across the Jamma River from Enewari where horses are used) oxen-poor farmers will yoke a cow with an ox, but the cow's smaller size limits the plowing depth, and cows fed on unimproved forage may suffer a loss in milk production and fertility. Highland horses are light and require special yokes to prevent choking. Where and when horses have been used, they are faster than oxen but can manage only a few centimeters of plow depth and cannot plow before the rains or after muddy conditions.[21] In fact, in particular highland microagronomic economies—in Agawmeder (Gojjam) and Jirru (northern Shawa)—horses have substituted for oxen and, at times, have even dominated localized traction systems. Those specialized exceptions, however, are rare and usually conjunctural (i.e. they occur when oxen prices soar). In contrast with both cows and horses, oxen thrive at all elevations and retain an additional appreciation value as both labor and food.[22]

Crops

The ox-plow complex's crop repertoire has historically been almost as stable as the technology of the plow itself. Although prestige at the Axumite court derived from cultural traditions of literacy, religion, and commerce from the Red Sea-Mediterranean world, the members of Axum's elite fed themselves with the produce of a home-grown agricultural system. In fact, there is better information on what Axumites and highland populations ate—the early highland crop repertoire—than on the plow itself. The evidence of both language and plant biology indicates that the

21. In 1992 I interviewed farmers at Enewari and Injabara who plowed with horses. In each case farmer life histories indicated that they used horses only at early stages in their household capital cycle or during a crisis. Those with sufficient resources always used oxen, though several mentioned the greater speed of horse plowing.

22. Michael Goe and Robert McDowell, *Animal Traction: Guidelines for Utilization* (Ithaca, 1980), 25. See also Barrett, et. al. "Traction," 93.

Ethiopian highlands were a center of secondary dispersal for a wide variety of crops, especially annual grain crops; Ethiopia is especially rich in cultivars of barley, wheat, sorghum, teff, and eleusine (finger millet). Vavilov also identified pulses (chick-peas, lentils, and peas) as part of the highlands' crop patrimony and listed in all 24 grains and oil plants for which Ethiopia was a world center. Christopher Ehret has used linguistic reconstruction of the Afro-Asiatic language family to argue that cultivation of teff and eleusine took place on the highlands as much as 7000 years ago; highland farmers later added Near Eastern cereal crops (wheat and barley) during the last tropical wet phase in the third millennium B.C. The same evidence also associates livestock domestication with these crops. Today, Ethiopia's highland agriculture has the world's most important pools for genes of durum wheat, barley, sorghum, linseed, eleusine, chick peas, cow peas, niger seed, and Ethiopian rape seed.[23]

Ethiopia's ox-plow system innovated, adapted, and evolved an array of cultigens synchronized to conditions of climate, soils, and an impressive—and highly stratified—cuisine. Farmers cultivated crops such as teff, eleusine, chick-peas, and lentils, for which the highlands were a primary center of origin, and barley, wheat, sorghum, and a wide range of oil seeds for which Ethiopia has been a secondary center of dispersal. Douglas Graham, who reckoned in 1844 that Shawa might make an "inexhaustible granary for all the fruits of the earth," counted 43 species of "grain and useful products," including 12 cultivars of sorghum. Théophile Lefebvre noted 12 varieties of barley and 7 of wheat, and Italian agronomists Raffaele Ciferri and Guido Giglioli in the 1930s found 4 dominant wheat cultivars.

Crop regimes varied considerably by region, reflecting altitude, soil, and local economic tradition. Alvares, who gave us our earliest account, remarked in the sixteenth century on the variety of crops in Lasta (Angote) and the effects of elevation on cropping:

> The kingdom of Angote is almost all of it valleys and mountains, and the tilled land has little wheat and little barley, yet it gives much millet, tafo [teff], daguça [eleusine], chickpeas, peas, lentils, beans. . . .

A bit further along his route in southern Wallo, he described a new scene:

> On Monday we travelled over flat country between mountains which were very populous and much cultivated, for a distance of two leagues. We ascend-

23. N. L. Vavilov, *The Origin, Variation, Immunity and Breeding of Cultivated Plants* (Waltham, 1951), 37-39; Huffnagel, *Agriculture*, 177-80; Brandt, "New Perspectives," 174; Christopher Ehret, "On the Antiquity of Agriculture in Ethiopia," *Journal of African History* 20 (1979), 176-77; also see Tewolde "Historical and Socio-economic Basis," 4.

ed a very high mountain without cliffs or rocks, or jungle; all used for cultivation. . . . They [local people] told me that here on this hill we divided the
country from millet from that of wheat, and that further on we should not find
more millet, but wheat and barley.[24]

Vegetables, herbs, and a rich variety of spices appeared in homestead gardens
(gwaro) as condiments to complement basic farm diets of cereals and pulses, which
only rarely included meat. Oil seeds—niger seed, linseed, safflower, and small
amounts of sesame—also appeared here and there in regions of middle to low altitude, though less often in main crop rotations. Fruits such as peaches, bananas, citrons, limes, and oranges appeared sporadically in markets, though rarely on ox-plow
farms themselves. Most outside observers mixed their wonder at the highlands' agricultural potential with disappointment that highlanders were lamentable farmers,
contrasting the land itself with the agricultural system and the people's work habits.
Portuguese accounts noted the wide variety of fruits in evidence but lamented the
poor cultivation of them. Alvares commented on a Tigrayan monastery's gardens:

> In the narrow valleys which belong to this monastery there are orange trees,
> lemon trees, citron trees, pear trees, and fig trees of all kinds, both of Portu-
> gal and India, peach trees, cabbages, coriander, cress, worm-wood, myrtle,
> and other sweet smelling and medicinal herbs, all badly utilized because they
> are not good working men; and the earth produces them like wild plants, and
> it would produce whatever was planted and sown in it.[25]

Lobo a century later offered the common travelers' complaint about the lack of fruit
in the travel diet: "The land does not produce any more fruit than it does because of
lack of cultivation rather than an inability to produce them; for the fact that it has
some varieties of fruit is a clear indication that it could produce the others if they
were planted."[26]

The restricted cultivation of tree crops was less an issue of taste, climate, or agronomic skill than of the incompatibility of such perennial crops within the annual
cropping regime (see below). Perennial or multiyear crops such as coffee, *chat
(Catha edulis),* yams, and ensete existed almost exclusively on the ox-plow com-

24. Alvares, *Prester John*, vol.1, 230, 251-52. For cultivars, see T. Lefebvre, A. Petit, and
Vignaud Quartin-Dillon, Vignaud, *Voyage en Abyssinie exécuté pendant les années 1839,
1840, 1841, 1842, 1843:* (Paris, 1845-48), vol. 1, 245; vol. 2, 92, Graham, "Report," 269-70,
mistakenly lists sorghum varieties as *zea mays* (maize). See chapter 3 herein for confusion
over reports on maize; R. Ciferri and R. Giglioli., *I cereali dell'Africa Orientale Italiana, vol.
1: I frumenti dell'Africa Orientale studiati su materiali originali.* (Firenze, 1939), 234.

25. Alvares, *Prester John*, vol. 1, 75.

26. Lockhart, *Itinerário*, 162.

plex's southern periphery, where horticultural traditions and more consistent soil moisture allowed local farms to retain perennial tree crops and provide diet supplements, nonalcoholic stimulants, or market income. Cotton was a crop important in the lowlands to the west and east. Graham described Ethiopia's cotton varieties as "more luxuriant than any species I ever saw in India"; the eastern Ifat variety that he saw below Ankober produced for five years, after which the plants were stumped and oversown with grain. Shoots from the old stumps then regenerated, producing two more cotton harvests.[27] Cotton's characteristics as a perennial crop with a lowland range, however, conflicted with ox-plow farming, so it was rarely integrated into highland ox-plow farms.

Coffee, until the early twentieth century, was not a cultivated crop but a natural forest product gathered in the southwestern moist montane forest in the range of 1800–2200 meters. In the early nineteenth century Christian aversion to it as "savouring too strongly of the abhorred Mahomedan" restricted its market in the Christian highlands; transport costs prevented its large-scale international export until the completion of the Franco-Ethiopian railway in 1917. Coffee cultivated as part of ox-plow agriculture began with Ethiopia's connection to world commodity markets and the growth of Addis Ababa. Coffee on small plantations and on smallholder farms began around Harar under Emperor Menilek's cousin Ras Makonnan, at Lake Tana, and in the newly conquered Kaffa region, where it supplemented collection of wild varieties which grew "like a weed over the rich surface of the country."[28] Coffee's need for shade, rich soils, and consistent soil moisture restricted its range to the southwest and eastern highlands, but by the 1930s its specific labor needs fairly quickly produced a distinctive niche in the southwestern version of the ox-plow economy (see chapter 5).

Similarly, the cultivation of the perennial tuber ensete *(Ensete ventricosum)* or the "false banana," illustrates the long-term interaction between the plow culture and a competing, and perhaps older, horticultural complex. Ensete is a large, fibrous plant with distinctive long banana leaves, whose root (corm) and pseudostems, rather than its fruit, provide its food value, when prepared as a starchy edible paste made into a bread *(qocho)*. Ensete shows distinctive signs of long-standing adaptation in the highland environment, perhaps even its innovation as a cultigen.[29] Although it grows wild elsewhere in East Africa, only in Ethiopia is it a cultivated food crop. In contrast with the ox-plow complex's dedication to broadcast sowing, cultivators propagate ensete vegetatively; in fact its seeds are virtually sterile. Ensete appears in household compounds throughout the central and southern high-

27. Graham, "Report," 272-74.

28. Graham, "Report," 270-72; Huffnagel, *Agriculture*, 204-26.

29. Ehret, "Antiquity," 175-76 suggests that coffee's cultivation may date back 7000 years and may have been an Omotic or Eastern Cushitic innovation. Also see Brandt, "New Perspectives," 185-88.

Figure 2.4. Ensete cultivation and preparation, c. 1860.

lands, and, curiously, in isolated areas of the Simen mountains. It appears as a primary food crop primarily in the south among the Gurage, Gedeo, and Kambata peoples; in the southwest and in Walayta it coexists with plow cultivated cereals.

Seventeenth-century Portuguese accounts, and later James Bruce, described ensete cultivation on the northern highlands especially around Gonder and in Gojjam. Almeida described it as "a tree peculiar to this country [i.e., northern Ethiopia] so like the Indian fig that they can be distinguished only from very near," but Lobo offers great detail in describing its preparation and vegetative propagation as a food crop: "When cooked it resembles the flesh of our turnips, so that they have come to call this plant 'tree of the poor' even though wealthy people avail themselves of it as a delicacy, or 'tree against hunger,' since anyone who has one of these trees is not in fear of hunger."[30] Almeida noted that it grew together thickly and that when propagated, "500, 700, and sometimes a thousand grow from the same one."[31]

By the mid-nineteenth century, however, memories of ensete as a dietary base had vanished in the north.[32] Although Lobo had observed its propagation and consumption in Damot (Gojjam) in 1640, according to Charles Beke's detailed 1840s'

30. Bruce, *Travels*, vol 3, 584; Lockhart, *Itinerário*, 245-46. Almeida, *Some Records*, 47-48.

31. Almeida, *Some Records*, 47-48.

32. Simoons, *Northwest Ethiopia*, 89-99 is the most comprehensive source; also see Almeida, *Some Records*, 47-48; Lockhart, *Itinerário*, 245-46.

narration Gojjam was a dedicated cereal-consuming zone. Almeida in the 1640s described ensete as "the sustenance of most of the people of Nareâ [in the Gibe basin]." Antonio Cecchi's 1880 narrative depicts that region clearly dominated by the cereal-pulse complex, with ensete present but in a minor role.[33] The historical evidence, though far from conclusive, suggests a successive recession of an older horticultural tradition in the face of annual cereals, a tradition which appears to have continued through the twentieth century.

Root crops played a minor role in northern highland farming systems until the introduction of the Irish potato and the more recent spread of sweet potatoes from the southeast to the northeast and southwest. Other root crops, however, have long been a part of cropping systems in the south and west of Ethiopia. These include *godare* (taro), *boye* (yam), and in some areas a small potato-like tuber *YaOromo dinich (Coleus edulis)*. Though each has its own peculiar characteristics, in general these root crops often compensate for gaps in the annual food calendar, particularly in the months of April through June. Root crops are never stored for any time above ground; rather, farmers by harvesting them as needed and when individual plants become ready.[34]

Above all crops, teff *(Eragrostis teff)* drew the greatest attention of travelers and achieved pride of place across elite diets in the highlands. It was unique to Ethiopia, grew at the elevations travelers most traversed, appeared larger than life in Ethiopians' own conception of their cuisine, and was fed to honored guests as the primary ingredient of *enjera,* Ethiopia's distinctive thin, fermented, batter "bread." Teff cultivation ranges between 1700 and 2200 meters elevation and is easily the most labor-intensive crop in the highland system. Indeed, its domestication from the wild *eragrostis* genus took place in Ethiopia, though wild *eragrostis* remains have appeared in Egyptian pyramid bricks and as wild pasture grass in both hemispheres. In its domestic form teff requires intensely prepared seedbeds and heavy labor in weeding and is even adapted to flood cultivation similarly to lowland rice. As perhaps the world's smallest grain, teff also requires great care in harvest and threshing: Almeida described it as "a seed so fine that a grain of mustard might be equal to ten of Tef."[35]

Teff, however, richly rewards farmers' efforts, since it is the highest prestige cereal food across the ox-plow landscape, yielding the highest exchange value and the longest storage period; it also yields the best building straw and most digestible cattle fodder, and is somewhat drought resistant. Ethiopians eat teff only as *enjera,* and most highly value it in its *manya,* or whitest, variety. Red *(qay)* or black *(tiqur)*

33. Almeida, *Some Records,* 47-48; Cecchi, *Da Zeila,* vol. 2, 221.

34. Field notes from Gera field trip, January 1990, November 1991. Farmers in Gera argued that the new variety of taro did not store well in the ground but offered a higher yield and therefore was preferable.

35. Almeida, *Some Records,* 45.

varieties or mixed-seed *(sergegnya)* harvests are more common. The color valuation of teff in fact is not a pure distinction of cultivars but the percentage of admixture in a given field. In recent times teff's high price and prestige have made it the only crop amenable to ox-plow farm commercialization and specialization (see chapter 6).

Through most of Ethiopian history, however, teff has never dominated either farm diets or hectarage. Barley and sorghum—occupying higher- and lower-altitude zones, respectively—have been more widely cultivated and have a richer range of cultivars in highland agriculture. Maize was a late New World arrival which first appears in a travel account in 1809 but now has become Ethiopia's, and all Africa's, dominant crop in total production in recent decades. In fact, the evolution of the ox-plow system's repertoire offers an insightful measure of dynamism and stability in the system, that is, not so much the importation of new crops as the transformations in the crop mix, which reflect subtle farm-level choices written across an agricultural landscape. Chapter 3 examines these trends in depth.

Field Systems:
Microagronomic Economies

Beyond subtly adjusting crops to local landscapes and attempting to minimize crop loss, farmers have used the tools, and agronomic strategies accumulated over time as customary knowledge to manage fertility and soil moisture and to minimize effects of climatic intrusions: frost, waterlogging, and drought. These practices have responded to an almost infinite variety of conditions, localized down to shades of difference between microclimates, valleys, and even soil conditions and the slope of individual plots. Nevertheless, the techniques have drawn upon a known, if not fixed, repertoire of solutions to potential problems. The historical sources thus reveal a sometimes surprising historical stability in agronomic practice, revealing not so much conservatism as the depth of experience and risk aversion built into local agriculture.

Soil management, especially the maintenance of soil fertility has been the fundamental task of small farmers on the highlands. At its most basic level, alternate plantings of cereals and pulses in a short-fallow rotation have helped achieve a balance of available nitrogen, though most Ethiopian soils are relatively rich in that substance. Local rotation practice, however, has varied considerably by elevation, soil conditions, and the exigencies of individual farms. In Dabra Berhan, above 2500 meters, the pattern is two seasons of cereals followed by a pulse and then a period of long fallow (15 years). In Ada at 1800 meters, a nine-year rotation on black vertisol alternates six years of teff with single plantings of chick-peas, barley or wheat, and white sorghum on reddish, sandy soil. Ada's nine-year cycle includes alternating cultivars of teff over five or six seasons followed by single years of a cereal and two years of a pulse. Northern migrants to Gera's forest soils left pulses out of their short-fallow crop rotations altogether. In long-fallow areas, farmers observe the natural succession of grasses as "indicator" plants, which signal readiness for a new rotation.[36]

Population pressure in the post-World War II period has virtually eliminated fallow in many highland areas, including Ada and most areas of the north-central highlands. The historical evidence for changes in rotation schemes is sketchy, since few historical travelers remained long enough in one setting to record crop succession. Graham observed central highland rotation schemes for two years in the 1840s:

> In all the districts of Shoa, a regular system of cropping has been established, and these rotations of crops are scarcely ever departed from, founded on the principle of preventing the soil from becoming impoverished. . . . In the valleys, teff, jewarree [sorghum], cotton, oil and wheat follow in succession. On the high country, barley and wheat in alternate seasons, and in the cold moors of the table land, the ground is left fallow for one year to recover itself before a fresh crop can be taken from the exhausted material.[37]

Rotation and fallowing sustains fertility by fixing nitrogen through planting pulses and natural regeneration by plowing organic matter back into fields. Rotation also breaks the reproduction cycle of crop-specific pests. Across the highlands farmers have used manure to various degrees, but primarily for their household garden (gwaro) crops; the practice of stubble feeding—allowing animals to graze postharvest crop residue in the field—produces limited deposits of manure. In densely settled, dry, evergreen montane regions such as northern Shawa and modern Ada, women have carefully collected and dried manure as household fuel rather than fertilizer (see chapters 4 and 6). Its ubiquitous presence was witnessed by Graham, who lodged a specific complaint: "The unseemly dunghill, which in other countries is carried far away to improve the soil and means of the proprietor, is here suffered to accumulate and rot adjoining the entrance to the dwelling."[38] In long-fallow systems farmers might fire grasses and plow in the ashes.

In the northern Shawa highlands farmers have practiced soil burning (gaye) to increase fertility on long-fallow plots. After initial plowings both men and women gather clods and stubble into mounds half a meter high and filled with cattle dung, which they then ignite. The mounds may burn for a few days to over a week, creating small plumes of smoke across a wide landscape (see figure 4.8). Gaye in the

---

36. Huffnagel, *Agriculture*, 143-44; Goe, "Tillage," 143; my own interviews from five regions included data on rotations and field management.

37. Graham, "Report," 262-63.

38. Graham, "Report," 263, 269; my own field observation.

short run increases available nitrogen and phosphorus, raises soil pH, and increases crop yield in the first two to three years, with a steep drop off in productivity afterward. After two to three years of enhanced cereal crops, farms then leave the *gaye* plot to fallow for an additional 8–15 years.[39]

The process of plowing and the skill of individual farmers has been central to the success of the individual farm and the system as a whole. Outside observers have had little understanding of the use of the scratch plow, referring to the *marasha's* action on the soil as "haphazard" and as "feeble scratching . . . carried on in every direction where ever the animals can find room to turn."[40] Plowing with the light, single-tine *marasha,* however, requires skill, stamina, and knowledge of the conditions of each field; each plot receives individual treatment depending on what is to be sown, the slope, drainage conditions, and previous usage. Pulses, lentils, chick-peas, field peas, and horsebeans—may require only two passes of the *marasha.* Cereals are smaller with fragile seed coats and require several more plowings. In the case of teff, the final soil clumps may be broken by driving goats or sheep repeatedly across the surface to break down clods and compact the soil before a final seed covering pass.[41] Though stones in some areas can cover as much as 20–25 percent of the soil surface, they are usually left in place, because farmers claim they retain soil moisture and prevent soil runoff in heavy rain.

Despite outsiders' impressions of a random cultivation practice, ox-plow farmers control drainage and soil erosion through contour plowing and the use of drainage furrows. On plots with a slope of 10 degrees or more, farmers plow along the contour with subsequent intersecting furrows either on or just off the contour. Where flooding is likely, farmers plow drainage furrows at three to seven meter intervals after seed germination.[42]

Adaptations to specific conditions, however, have been quite localized historically. In the early sixteenth century Alvares noted severe conditions of seasonal flooding in Warailu in southern Wallo:

> We went to sleep in a great field of grass, where the mosquitoes were near killing us. These fields are not used except for pasture, as they are rather marshy, and the people do not know how to drain off the water at the base of the mountain in drainage channels. . . . From here we took our road through many large valleys and yet they had very poor fields of wheat and barley;

39. See Goe, "Tillage," 143-45; and T. M. M. Roorda, "Soil Burning in Ethiopia: Some Effects on Soil Fertility and Physics," in *Management of Vertisols in Sub-Saharan Africa* (Addis Ababa, 1988), 124-25.

40. Augustus Wylde, *Modern Abyssinia* (London, 1901), 258; Graham, "Report," 266.

41. Goe, "Tillage," 145-48.

42. Goe, "Tillage," 148-49; and Huffnagel, *Agriculture,* 142.

some were yellow, as though dying from the water, and others were dying of drought, and so we puzzled at the way these crops perished.[43]

By contrast, a few hours' walk south across the Jamma River at Enewari, farmers faced with the same conditions have for generations employed broadbed drainage furrows, hand-built by women and young children on the area's flood-prone vertisols.[44] It is curious why such variation in agronomic practice exists between two areas which have geographic proximity, common cultural profiles, and traditions of contact through marriage and trade. The most plausible explanation may be demographic: Enewari's population density, the availability of only flood-prone land, and the former status as a royal "kitchen" (see chapters 4 and 6) have forced the intensification of land use as a solution.

Across the highland landscape dominated by dryland, rain-fed agriculture, historical sources remarked on the scattered evidence of what John Sutton has called specialized field systems, that is, irrigation and terracing.[45] Highland irrigation has historically consisted of small-scale use of gravity-fed rivulets captured from small streams or springs. Alvares in the sixteenth century reported spring-fed irrigation in southern Tigray church fields, at Yeha near Axum, and around Warailu in southern Wallo. A wide range of nineteenth- and twentieth-century observers have also described the irrigation systems around the royal capital at Ankober (see chapter 4) and "in every situation where a supply of water can be obtained without much trouble."[46] In contrast with their disdain for most Ethiopian farm practices, foreign observers admired the labor devoted to irrigation. Henry Salt described both his admiration for highland farmers and the presence of irrigation in Tigray:

The productiveness of the soil must be considered in great measure as owing to the industry of the inhabitants and their skill in irrigating the land, the effects of which, where a constant supply of water can be procured, prove highly beneficial. The common mode practiced here consists in digging small channels from the higher parts of the stream, and conducting them across the plain, which is thus divided into square compartments according to the general practice adopted in India.[47]

Augustus Wylde noted with some sympathy the costs of such specialized cultivation: "The amount of labour expended on the system is often very great, and one

43. Alvares, *Prester John*, vol. 1, 253.
44. Oxfam/ILCA joint project; personal communication from Samuel Jutzi, project agronomist.
45. John Sutton, "Towards a History of Cultivating the Fields," *Azania* 24 (1989), 98-103.
46. Graham, "Report," 263; see chapter 3 herein.
47. Salt, A Voyage, 199. Salt was reporting on the 9th of March, that is, in the dry season.

cannot help but admiring the natives for their ingenuity and the hard work that has to be done every year to keep the small water courses in order." In most cases the water sustained year-round cultivation of annual cereals, household vegetables, and spices rather than perennial crops, though in some cases lowland fruit trees required such sustained moisture to supply highland markets.[48]

Wherever highland smallholders have engaged in irrigation, they have almost always built terraces also. In some areas, such as Chercher in Harerge, bench terraces permit moisture retention for perennial crops such as coffee and chat. In other areas, such as in northeastern Shawa, terraces appear to have stabilized soil on steep hillsides brought under year-long cultivation of annual crops. In some cases, but not all, irrigation brings water from upslope springs or rivulets to terraced crops.[49] Terraces vary widely in their width and design; they also appear to vary widely in the forms of labor used to construct and maintain them. In Konso, terraces were individual property prior to the 1975 land reform but were built by collective labor; on the northern highlands, though the evidence is weak, individual farm households have built many terraces slowly over several years on individual plots rather than as collective community enterprises. David Buxton, traveling in northeastern Shawa in the late 1940s described terrace building still practiced there:

> Many of the slopes are unobtrusively traced, both in the gorges and on the plateau, and it is not uncommon to see new terraces in process of formation. A wall ("gur") is built on the lower side of the area selected for improvement, advantage being taken of any natural flattening of the hillside. The "gur" has a vertical face on the downward side, but is sloped on the up hill side and therefore triangular in section. The filling up of the empty space behind the wall, to the new terrace level may be partly accomplished by hand but tends to be left to nature. During the rains, especially on the steeper slopes, there is a continuous downward movement of soil, which is arrested by the terrace walls and the pockets above them gradually fill up.
>
> In the course of centuries an immense amount of simple terracing of this kind has been done, the effect being to accentuate the existing irregularities

48. Wylde, *Modern Abyssinia*, 262, 306. Also see chapter 4 herein; Antonio Cecchi and Giovanni Chiarini, "Relazione dei signori G. Chiarini and A. Cecchi sui mercati principali dello Scioa," *Bollettino della Società Geografica Italiana*, 16 (1879), 445-55; Bent, *Sacred City*, 14.

49. The primary purpose of terraces is under some debate. Hans Hurni (personal communication), whose primary work has been in the Simen mountains, argues that highland terraces were built for moisture retention, not to prevent erosion, whereas David Buxton, traveling in the late 1940s, argues that terracing prevented erosion. David Buxton, "The Shoan Plateau and Its People: An Essay in Local Geography," *Geographical Journal* 114 (1949). For irrigation of terraces see Huffnagel, *Agriculture*, 147-50.

of the steeper slopes. The landscape has certainly been greatly modified in this way. . . . [50]

Although many travelers have described such specialized field systems, these forms concentrated in areas of dense settlement and have not been a regular part of the landscape: most accounts note them in Tigray and Lasta in the north, Harar, Ankober, Yajju, and Konso in the center and south. Even the evidence of Axum suggests that irrigated agriculture concentrated only in specialized locations near large towns. Frederick Simoons' survey of the Gonder region found only small, isolated cases of irrigation, with little to none in the lowlands.[51] Across the highland landscape, however, there is evidence that specialized agriculture appears to have declined within the period of this study. Chapter 3 discusses this evidence and its implications for agricultural history.

Protecting the Harvest

After seedbed preparation, selecting the crop, and sowing, farm activity shifts to protection, harvest, and storage. If timing in field preparation is critical to potential yield, the application of labor in protecting crops maximizes the yield. The intensity of summer rains on sloped plots requires men to plow drainage furrows; women and children weed cereal crops, especially teff; young men use slings and whips to frighten away birds from sorghum during the day and guard against baboons at dusk and porcupines at night. The labor process following the initial sowing lent a gender specificity to the agricultural process, with labor and collected food moving between male- and female-dominated spheres of control.

As a source of affliction to agriculture, the desert locust *(Schistocerca gregaria)* commanded both the power of biblical imagery and the physical ability to denude fields of crops. Jerónimo Lobo argued that locusts were God's plague on Abyssinians for their "willful perfidy." His seventeenth-century description of a locust invasion mirrors farmers' accounts I have in the 1980s:

At the time when the seeds sown have sprouted, this plague [locusts] appears, is seen to be a darkish yellow as when it is beginning to rise, and since they fly high in the air, one cannot see them but can observe the indications of how numerous they are by the change in the sunlight. With this evidence that the locusts are coming, all the people begin to set up a cry, imploring God's mercy; for since the roots are tender, wherever they light they destroy all of it down to the roots. And since they are so numerous that they occupy all the land for many leagues around, they take everything irremediably. . . .

50. Buxton, "Shoan Plateau," 167-68.
51. Michels, "Axumite Kingdom," 176-77; Simoons, *Northwest Ethiopia,* 74-75.

Ethiopian couplets see locusts akin to unwelcome human guests: *"Hayyal anbeta ekelka beliu ab deguolka yiseffer.* (The locust eats your crops then settles in your yard.)"[52]

Locust swarms have been a part of the agricultural environment throughout the period of this study, appearing so often in the record of travelers as to distort their impact somewhat on rural life. European and Ethiopian sources recorded at least 19 major invasions of swarms (mature flying locusts) and hoppers (immature wingless locusts) between 1812 and 1930 and perhaps many more localized ones have gone unreported.[53] In fact, Ethiopia's highlands receive the highest rate of invasion of any region in the distribution area of the desert locust. Breeding in the lowlands, locusts swarm twice a year, in the fall and spring, though not in all places in all years. Behavior of the swarms depends almost entirely on climate—and thus on the highland's capricious microzones—because wind, temperature, and rainfall determine both their breeding and flight paths.

Despite locusts' dramatic and tragic local effect, endemic pests have probably more directly affected farm strategies for crop protection. Army worms *(deyri* or *temch),* stalk borers *(gind korkur),* grasshoppers *(fenta),* and birds have accounted for fairly consistent losses, but ones for which farmers can reduce effects through the selection of cultivars, rotation, and vigilance. In areas of low population density there can be heavy loss to wild animals that consume large amounts of grain (especially maize and sorghum) before it reaches storage. In many areas, wild pigs *(sg. asama),* warthogs *(sg. karkaro),* porcupines *(sg. jart),* and baboons *(sg. zinjero)* damage and consume considerably more grain than is lost in storage. Some farmers in the Gera lowlands have estimated maize loss to warthogs and wild pigs at 50 percent. In other areas baboons have accounted for similar levels of damage. In one of the few agricultural themes in religious literature, an eighteenth-century illustrated manuscript depicts Saint Takla Haymanot's miracle of ordering animals and birds not to eat the crops of his monastic order.[54]

Harvest time for crops sown in the main rains can vary by crop and region from late November to early January. Collection of mature crops draws men and sometimes women into the fields to cut grain with a *machid* (sickle) or to pull dried pulses by hand to be stacked carefully for drying and to avoid the effects of a freak

---

52. Lockhart, *Itinerário*, 192-93; Conti Rossini, *Proverbi*, 31.

53. Richard Pankhurst, *Economic History of Ethiopia, 1800-1935* (Addis Ababa, 1968), 220; A. Chiaromonte, "Il problema delle cavallette nell'A.O.I.," in Sindicato Nazionale Fascista Technici Agricoli, *Agricoltura e impero* (Roma, 1937), 71; Huffnagel, *Agriculture,* 169-71. Locusts are grasshoppers with a capacity for gregarious behavior in either their immature or their winged stages.

54. Lockhart, *Itinerário*, plate 5 (originally from the British Library, OR 721, fol. 167 v). Donald Crummey's survey of church manuscripts has turned up only a handful of agricultural images in religious painting.

harvest season downpour. Ironically, those small grains which are resistant to storage loss and wild animals, such as teff and millet, are the most vulnerable to mold damage prior to threshing. In southern areas with heavy rainfall, postharvest loss of teff has been substantial. A northern Ethiopian couplet states: "YaHadar zenab bimeta / yishalal anbeta (It is better to have locusts / Than rain in November [i.e., during or after harvest])."[55]

Shortly after harvest, farmers turn cattle into the open fields of teff, wheat, or barley stubble or of sorghum stalks, since by harvest season livestock forage is in short supply. Despite its poor storability (see below), maize left on the stalk is resistant to rain damage and therefore is far less sensitive to the timing of its harvest and transfer from field to storage. This feature may partly account for its rapid spread in traditional coffee-producing regions, where labor conflicts between coffee and annual crops at harvest are particularly acute. Maize stalks can be left in the field while the coffee harvest proceeds.

Threshing of cereals, one of the few cooperative farm labor activities in the ox-plow communities, has been fairly uniform in its methods across the highlands and across time. Common dry-season scenes in the countryside have always included men preparing the open-air threshing floor which were "well leveled and consolidated with white earth and cow dung." Graham's account of Shawan threshing of the king's demesne crops in 1844 mirrors that seen on a smaller scale across the ox-plow landscape in the late twentieth century:

> The threshing out of the wheat, barley, and jewarree, is performed in the field by the tramp of muzzled oxen and other produce is also cleaned in the open air by means of long crooked sticks wielded by the arms of sturdy peasants, in as short a time as possible; beer and bread being prepared in great quantities; hundreds assemble on the spot; the process commences with an uproarious song of exaltation, and a most animated scene of noises, labour and confusion ensues, until the grain is entirely separated from the straw.[56]

If other threshing scenes have contained fewer participants, the process shows remarkable historical stability. Timing, again, is critical but remarkably variable, depending on the availability of local labor and oxen, the arrival of the harvest, and, historically, the presence of a tax collector to assess the result before taking the grain to storage and the straw (valuable oxen fodder) to stall. In other areas farmers prefer to thresh grain as they need it. There are few hard and fast traditions across the ox-plow complex.

55. Told to me January 1990 by Tigrayans who had just lost a poorly stacked harvest to mold; their new settlement in Gera is more susceptible to December rains than their native Tigray is.

56. Graham, "Report," 267.

Figure 2.5. Oromo granary, 1880.

Storage: The Social Life of Granaries

The storage of crops from one annual harvest to the next is a critical if poorly under-stood process. Food in the form of stored grain within rural economic culture is a form of social property which individuals, households, and communities manage in the same socioeconomic environment as they do land, livestock, and labor. Food, however, has attributes which are very different from those of other forms of rural

Figure 2.6. Maize granary (Gera), 1990.

social property: it is a highly perishable good (unlike land), is very divisible (unlike livestock), and appears and is consumed within a fixed annual cycle (unlike labor). Moreover, to farm households food represents, at one time or another, income, potential capital, and sustenance. Historically, claims on food by individuals within households and communities reflected the same overall terms of exchange as access to labor, livestock (especially animal traction), and even land. The sharing and exchange of food for other goods and services reflect broader rights and obligations in property as well as relationships of gender, age, and status. The storage of food within the annual cropping system is an attempt to reconcile the rigid seasonality of production with the consistent daily demands of consumption and income to meet annual obligations to the state and investment needs for the farm enterprise. The technical and social means by which farm households have managed stored food therefore provide a valuable index of changing social and property relations within the ox-plow complex.

Managing stored food has also been an acutely gendered process. Unlike in much of West Africa, women in Ethiopia's ox-plow households are structurally distant from the primary act of cultivation (see below), retaining sovereignty over the conversion of grain from produced, stored wealth and property to household consumption. Women's authority over food processing shifts when grain is transferred from the threshing floor to storage, where food as social property moves through gender-specific domestic domains. From the field to the granary food moves from the male sphere to a mixed status. Traditionally, a large exterior granary or pit has been opened in the presence of a male household head, yet, once allocated for household consumption, grain resides purely in the woman's domain. Decisions about converting stored grain to capital equipment or taxes are male but are negotiated with women's prerogatives to convert it to food directly or through exchange for another form of food (e. g., exchanging sorghum for horsebeans).[57]

Estimates of farm-level food losses of up to 30 percent are common in the community of agricultural specialists in Ethiopia as well as in the small body of specialized food-storage literature. More recent work on rural economies has challenged the conventional wisdom of large-scale losses and argued that such estimates are not empirically based. Farmers do not, in fact, express storage loss in terms of annual percentages since that is not how they experience the losses. Rather, they indicate the amount of time a particular grain can be stored—the threshold of storage—before they dispose of it by consuming it or taking it to market. Farmers, however, do not often take the risk of keeping vulnerable grains for a year to find out how much they might lose.[58]

The agents which cause damage to grain in storage have been well known and fairly universal across environments and cropping systems. Weevil damage is a direct function of temperature and humidity conditions, which affect the weevil's reproduction. In lowland regions warm weather leads to quick generation of weevils in stored grain and damage in as little as two months. Rodents also constitute a threat to stored food, though of much less concern than weevils, according to the farmers interviewed. Rats are most destructive of large grains and pulses and less destructive of small grains such as teff and millet. Termites (Amharic: *mist*) constitute a particular problem to grain stored indoors in sacks. Mold is a problem only in areas of high humidity and most significant to postharvest losses in the field before threshing.

Food storage was a part of the historical farming system. Farmers built granaries just prior to harvest, when they could make an accurate assessment of their storage needs by crop. The size and construction of new *gotera,* and those from past seasons were, in fact, excellent indicators of farmer perceptions of past yields, current crops in the field, and expectations for the future. It was men who designed, built, and repaired *gotera,* a reflection of their experience with house construction and domination in the domain of major crop production. Within houses women managed the grain; that which was to be consumed within the current month was stored in a wide variety of small containers as shelled grain (maize and sorghum), flour (hand or mill ground), or fermented batter. Long-term, in-house storage took place in a large mud-plastered jar called *dibignet.* These vessels, built by women using basketry techniques, were best suited for small grains such as teff

57. The arena of negotiation over stored food and its usage is unexplored for Ethiopia. See Audrey Richards, *Land, Labour, and Diet in Northern Rhodesia* (London, 1939), 188-93.

and *dagusa,* which resist weevil damage. Sacks or goat skins could also serve for small amounts of grain.

Seed stocks appear to have been selected carefully and stored within the house in considerable secrecy. Farmers stored ears of maize and sorghum heads selected for size and uniformity, hanging them from rafters of their houses, where smoke dried the kernels and discouraged weevils. Other seed stocks were stored separately in the house in earthen jars and periodically checked for damage. In recent food crises seed stocks appear to have been well preserved in both their variety and amount.

Perhaps as important as the adoption of specific technologies, storage strategies have also been based on local assessments of the political climate and potential threats to food stocks from neighbors, the state, or outsiders. Communal storage of grain in Ethiopia's highland farming systems is rare both historically and in recent practice. Food is property, and consistent evidence from smallholder agriculture in the north and in sedentary Oromo agricultural systems in Harerge and Gera indicates that the dominance of individual property versus collective ownership is characteristic of the ox-plow social system. Food, like oxen labor and social ties, flows primarily vertically from resource-rich households to resource-poor ones, rather than through horizontal channels of kinship or neighborhood. Since farmers in these farming systems individually manage resources like oxen, labor (except cooperative labor—*dabo*—for threshing), and land, it is not surprising that there is no evidence of collective storage by kinship, community, or other horizontally oriented institutions.

The historical evidence strongly supports this principle. Political climates have also historically conditioned storage strategies. In the nineteenth and early twentieth centuries underground pits were a primary means of storing sorghum in the lowland areas of Harerge, Wallo, and Tigray, but not in any areas above 2000

58. A 1964 food storage study by the Imperial Ethiopian College of Agricultural and Mechanical Arts estimated storage losses at 25 percent to 50 percent, with an occasional 100 percent loss in underground pits. That report also estimated merchant losses of up to 25 percent. Wesley Hobbs and Leonard F. Miller, *A Proposed Grain Storage Programme for Ethiopia* (Dire Dawa, 1965), 36. These figures were used primarily for lowland regions and for maize and sorghum crops. Interviews with Abate Tedla and Abiye Astatke (International Livestock Centre for Africa), 3 January 1990 and Dr. Taddese Gebre Medhin (Institute of Agricultural Research), 4 January 1990. During my interviews, agricultural specialists at ILCA also mentioned the figure of 20-30 percent for annual farm-level losses. Robert Chambers has challenged the "myth" of 30 percent farm-level loss. Chambers, *Rural Development,* 58. Chambers himself cites studies from India and Bangladesh that indicate losses of less than 10 percent. For empirical study of losses see Yamane Kidane and Yilma Habteyes, "Food Losses in Traditional Storage Facilities in Three Areas of Ethiopia," *Towards a Food and Nutritional Strategy for Ethiopia: The Proceedings of the National Workshop on Food Strategies for Ethiopia* (Addis Ababa, 1989).

meters or where there is high moisture content in the soil. Wylde reported such pits in eastern Tigray.[59] William Coffin, an English traveler of the early nineteenth century described their construction: "It is a common custom, in all parts of Abyssinia, for the inhabitants of the villages to have gudguads, large pits under-ground, plastered within with cow-dung and mud, and having the mouth very narrow, some of which are made to hold forty or fifty churns of corn, between three and four hundred English bushels."[60]

Underground pits common in the northeast have not only been effective at lowering storage temperature and thus slowing weevil reproduction; such pits have also the added advantage of secrecy since they can be covered by earth and overplanted with annual crops. Coffin witnessed, and his contemporary Nathaniel Pearce even participated in, small-scale raids on stored grain in Oromo agricultural villages in eastern Tigray. Farmers in that area plowed over storage pits or hid them under dung heaps, though often to little avail. Raiding parties, he observed, were well acquainted with pit storage:

> After destroying a village or finding it deserted by its inhabitants, they form into different parties, and keeping in a close body they begin to sing their own warlike songs, stamping and going on in a regular pace. . . . In this manner they continue until they find the ground hollow under their feet, when they lay their shields in a circle round the spot, and every one sets-to with both hands, as eager as hyenas after their prey; they soon claw out all the earth, break in the rafters, and then begin to fill their skins or bags.[61]

Areas of eastern Tigray and northeastern Wallo, where pits are still most common, are also those areas with the greatest political instability over the past few generations. In the more tranquil Shawa region, Graham observed grain stored "within the walls of the domicile in wicker baskets and large earthen jars; for a detached barn is nowhere to be found in Abyssinia." Frederick Simoons claims that storage pits were common in the Gonder region, but declined in use in the twentieth century with greater political stability.[62] Strategies can also change within a shorter time frame. In one case, a farmer in southern Wallo who in the 1960s had 10–15 gra-

59. Wylde, *Modern Abyssinia*, 261. In no instances, for example, had Wolloye farmers who had resettled in the moist environment of Gera even attempted to use pits because of their assessment of rainfall and humidity conditions, though they argued that pits could store large amounts (several tons) of sorghum for more than a year.

60. Nathaniel Pearce, *The Life and Adventures of Nathaniel Pearce*, 2 vols. (London, 1831), vol. 1, 206-7. Coffin authored two chapters in Pearce's narrative. This account implies communal storage at the village level; such forms are rare in the more recent examples.

61. Pearce, *Life and Adventures*, vol. 1, 207-8.

62. Simoons, *Northwest Ethiopia*, 81.

naries outdoors in his compound, had by the mid-1980s shifted his grain storage to a hidden room inside his house, reflecting the value of secrecy over visible wealth in a socialist political climate.[63]

State granaries in the imperial north were property of the state but were built and managed by individual farmers on behalf of the state rather than as large, collective granaries. They differed from farm-level storage only in ownership, relative size, and source of the grain. Grain from these stores could be distributed to cement political ties, supply local feast cycles, or feed the destitute in times of stress (e. g., the 1889–92 famine), but these occasions of food distribution were the result of political decisions by the owners, not of any collective claims on community resources. Even church holdings distributed in times of need were not communal stores, but "property" of the tabot (the ark representing the church's founding saint).

The Ox-Plow Revolution:
Society, Property, and the Plow

In important ways, the variety of crops, the staggering array of cultivars, and the range of geographic and ethnographic settings mask a more fundamental simplicity in the history of ox-plow cultivation. Crops cultivated in the ox-plow system have overwhelmingly been cereals and pulses planted in rotation and, equally important, were *annual* cultigens. The annual regime of seedbed preparation, harvesting, threshing, and storage not surprisingly has produced common rhythms of farmers attempting, however imperfectly, to synchronize their labors with nature. Annual crops depend on the seasonality of soil moisture; rain in annual cropping systems in fact forms a far more rigid, socially defining regime than it does in horticultural, perennial cultivation. These confines have also tended over time to focus and narrow the array of the social institutions of property, marriage, gender division of labor, annual ritual cycles, and the strategies of environmental resource management of plow-based populations across the highlands. In its spread across the highlands, ox-plow cultivation can be said to have brought with it what Carolyn Merchant has called an ecological revolution, a fundamental alteration in the use of and consciousness about nature, involving new social institutions of property, gender, and organization of the physical world.[64]

63. Interview with ILCA forage agronomist Abate Tedla, 3 January 1990; Graham, "Report," 267.

64. Carolyn Merchant, *Ecological Revolutions: Nature, Gender, and Science in New England* (Chapel Hill, 1989), 4-5. She identifies ecological revolutions in Marxist terms as "major transformations in human relations with nonhuman nature. They arise from changes, tensions, and contradictions that develop between a society's mode of production and its ecology, and between its mode of production and reproduction."

Scholarship on Ethiopia's history has long argued that the historical expansion of northern highland cultural hegemony rested on the spread of three components: land tenure, the Amharic language, and Ethiopian Orthodox Christianity.[65] In fact, these features have been secondary to a more profound social transformation brought on by the spread of ox-plow agriculture itself, which necessarily brought with it the "ecological revolution" that expanded across the highlands and over time transformed the physical landscape and social organization of populations on it.

The expansion of ox-plow agriculture across the highland landscape, slowly in its first two millennia, then with rapid spurts in the sixteenth century (the Oromo expansion) and again in the twentieth century (Menilek's empire and modern central state growth), was not, in the final analysis, part of the baggage of political expansion from the center. The conversion of horticultural societies, pastoralists and agrarian systems, which were based on perennial crops, to the ox-plow complex based on annual crops and integrated livestock management, implied more than a simple change of economic base or political authority. The use of the plow meant a transformation in environmental management: a livestock system which produced, trained, and maintained a stable supply of oxen, a reorganization of seasonal labor to fit a new set of crops, and a resource tenure system in which fixed multiannual resources—perennial crops—lost pride of place to annual crop fields. For most areas, the principal transition was multileveled: from digging stick or hoe to the plow, from ensete to cereals, from livestock as milk producers to livestock as labor. For most, the change was not piecemeal but, over time, total. By the modern period, northern highland farmers, who had once integrated ensete with annual crops, had abandoned that perennial cultivation of tree crops except in concentrated, specialized environments.[66] Farther south, perennial cultigens which previously coexisted with annual cropping, have tended to decline in importance or disappear altogether in the face of requirements of cereal-pulse cultivation. Annual ritual cycles in the ox-plow complex focus more tightly around the timing of rainfall than the cycles in pastoralism or perennial crop systems do. In its most obvious manifestation, the Oromo celebration of *butta,* an eight-year ritual cycle of the gada system, has disappeared in most areas in favor of annual-cycle agricultural festivals.[67]

65. See Donham, "Old Abyssinia," 11.

66. Merid Wolde Aregay, "Land Tenure and Agricultural Productivity, 1500-1850," in *Proceedings of the Third Annual Seminar of the Department of History* (Addis Ababa, 1986), 116-17; for early Portuguese reports on ensete and fruit see Almeida, *Some Records,* 47-47, and Lockhart, *Itinerário,* 245-46. Most fruit cultivation takes place in lower altitudes unsuited for agriculture.

67. See chapter 5. In Gera, *butta* was last celebrated two or three generations ago; in Wallaga more recently.

Ox-plow technology and the annual cycle of cropping which has accompanied it are at least as old as, and probably older than, the distinctive set of institutions of social property, land tenure, and gender relations which characterize "Abyssinian" (i.e., core northern highland) society. Though variation has existed across the highlands and through time, the ox-plow system in almost all cases has traveled with the specific baggage of gender division of labor, social property, and land tenure. Jack Goody has argued, for example, that the plow separated Ethiopian society from African models through its labor efficiency, producing land scarcity, endogamy, bilateral property rights, and stratified rural culture. Donald Donham has revised this argument by pointing out that the core highland political economy rested on an elite class's rights over income from an agricultural peasantry, not on title to the land itself.[68] *Gult,* the bundle of hereditary or temporary income rights held by elites over the rural population, provided a point of articulation between classes but did not involve European feudal-style control over the land itself. Tenancy, though widely visible especially on the periphery of the state, also provided for income to a landlord, but not for domination of farm-level choices of crops or agronomic strategies. Thus the agricultural process in the highland political economy was largely autonomous of state control (see chapter 7).

Annual cropping was a powerful transformative innovation. At any given time the systems of placing land in the hands of farm households have varied widely across the ox-plow landscape. Yet, the domination of annual cropping has simplified the process of land reallocation by making farm plots *tabula rasa* with each annual cycle in contrast with the complicated negotiations over rights that were embedded in tree or perennial crops fixed on land for more than one season. Little in the way of capital investment is immovable on an annual basis. The plow's light, mobile design has enhanced this tendency. Even specialized agricultural systems such as irrigation are short-lived and in their design require annual labor rather than permanent structures to sustain them. The system has thus sustained a very fluid land allocation and property system with little abiding interest in long-term capital or labor investments on the land itself. Farmer strategies and customary knowledge about subtle environmental variation have allowed ox-plow households to move easily to new areas opened by political expansion. Along the periphery of the expanding ox-plow complex, local prop-

68. Jack Goody, "Class and Marriage in Africa and Eurasia," *American Journal of Sociology* 76 (1971), 601; Donham, "Old Abyssinia," 15-16.

69. The use of *rest* by settlers on newly incorporated lands in the early twentieth century likely paralleled earlier processes of adoption. See, for example, Charles McClellan, *State Transformation and National Integration: Gedeo and the Ethiopian Empire, 1895-1935* (East Lansing, 1988), 87, 96. For the process in northern Shawa during an earlier period, see chapter 4. *Rest* was a metaphor readily adopted, since it implied ownership rather than tenancy.

erty systems appear to have adapted variants of the northern land *(rest)* system encouraged by new northern landlords or by the annual cropping regime. Thus within a generation in most newly conquered areas, farmers, especially northern soldier settlers, added appropriated concepts of *rest* to claims over land original-ly claimed for the emperor.[69]

Gender, Class, and Debt

Historically, the division of land encouraged by the annual cropping regime also appears strongly as a theme in family inheritance, gender relations, and the house-hold development cycle across the ox-plow highlands. Highland ox-plow farms, in contrast with northern European manorial estates, have historically been fragile, short-lived enterprises which held land, housing, and property only within a single generation. Ambilineal descent (i.e., equal property rights accorded to all children), broke up a household's holding in land, livestock, and equipment on the death or old age of the parents or through dissolution by divorce. As a social ideal, children could expect to establish a new, single-generation estate at marriage through inher-itance, borrowing, and strategies of accumulation characteristic of young house-holds in the growth stage. In other words, highland ox-plow farms have historically been the product of marriage, which "capitalizes" the household's initial formation and the subsequent accumulation of land, labor resources in children and relatives, equipment, and livestock.

In its most basic form, the allocation of core highland agricultural land has revolved around the concept of *rest,* a cognatic inheritance system that allowed individuals to claim land use rights through male or female ancestors who were linked with an original settler of a particular region. The potential claimants consti-tute what Allan Hoben has called a descent corporation for a particular land divi-sion.[70] Rights in the land have resided in the descent corporation rather than in individuals who might claim only use. With ambilineal claims on land, the system has produced logarithmic increases in claims over time and a scramble for viable plots within each community. Since particular farmers might hold land in several land divisions, any farming household might claim use rights through a parent's lin-eage or indirectly through children (i.e., a spouse's line). The patchwork quilt of oddly shaped plots obvious to the modern traveler from the air is the landscape lega-cy of generations of individual farm strategies, litigation, and negotiation. Despite the widespread adherence to overarching principles, however, land allocation prac-tice across the highlands have been neither uniform nor static over time. Local prac-tice has responded to specific environmental and demographic conditions. Areas, particularly in Tigray, with low population densities allocated land to all village res-

70. Allan Hoben, *Land Tenure among the Amhara of Ethiopia* (Chicago, 1973), 14-17.

idents regardless of lineage *(chiraf gwoses)* while areas with land scarcity rigidly excluded new claimants.[71]

The evidence drawn from fieldwork and presented in the case studies in part 2 suggests strongly that the most important indicator of a farm's survival has been the nature of its endowment through the founding marriage contract, which establishes capital stock and the initial land holdings, as well as the household's potential land claims through the two spouses' separate descent corporations. Marriage contracts have differed substantially by region, class, and previous marital status of the partners, but the overall guiding principle for highland Christian marriage has been equality of contribution.[72] Land resources could come from parental contributions (gulema) or from claims made through either spouse. For many young households, tenancy—requiring a payment of a quarter or a third of the harvest—has been a common first step in building household resources. New households often expected to build land resources by claims on parental holdings once their parents' estate has been dispersed.[73] Divorce, a frequent cause of household dispersal, has been a common and expected outcome of marriage, the result of early betrothal, weak lineage investment in the union, and lack of any religious proscription in either Islamic or Christian tradition. In northern Shawa in the 1980s, for example, men reported on average, three or four marriages and as many as ten; women frequently reported two or three.[74]

Marriage, divorce, and property on the northern highlands fascinated early Portuguese observers, who provided detailed accounts which indicate strong historical continuity in the social conventions of the ox-plow complex. Alvares in 1520 was shocked by the high incidence of divorce:

> In this country[side] marriages are not fixed, because they separate for any cause. . . . When they make these marriages they enter into contracts, as for

71. For land tenure changes over time, see Giovanni Ellero, "Il Uolcait," *Rassegna di Studi Etiopici* 6-7 (1948), 108-9; Dan F. Bauer, "Land, Leadership, and Legitimacy in Enderta, Tigre," Ph.D. dissertation, University of Rochester, 1972, 218. Local rules of land acquisition in northern Shawa in the 1980s also primarily reflected demographic conditions.

72. Forms of marriage endowment varied in the north. In Tigray dowries were in principle required and among Muslims wives made no contribution at all. In practice many Christian marriages in resource-poor agricultural households made no contribution at all. See Dan F. Bauer, *Household and Society in Ethiopia*, 2nd ed. (East Lansing, 1985), 112-23. Also see McCann, "Social Impact," 262-64.

73. Wolfgang Weissleder, "The Political Ecology of Amhara Domination," Ph.D. dissertation, University of Chicago, 1965, 106, 191, 199-204. See chapter 4.

74. Fieldwork from Tagulet and Ankober lowlands; see also McCann, "The Social Impact of Famine in Ethiopia," in Glantz, ed., *Drought and Hunger*, 262-64. Divorce in Oromo ox-plow regions is much less frequent. See Paul Soleillet, *Voyages en Ethiopie (jan. 1882-Oct. 1884: Nots, lettres et documents* (Rouen, 1886), 262-64 on Oromo marriage.

instance: If you leave me or I leave you, whichever causes the separation shall pay such and such a penalty. And they fix the penalty according to the persons, so much gold or silver, or so many mules, or cloth, or cows, or goats, or so many measures of corn. And if either of them separate, that one immediately seeks a cause of separation for such and such reasons, so that few incur the penalty, and so they separate as they please, both husbands and wives.[75]

Alvares also noted that male farmers maintained "an affection" for their wives less because of strong social sanctions in support of marriage vows, than because their wives helped them to bring up "their [beasts and] sons, and to harrow and weed their tillage, and at night when they come home to their house they find something of a welcome. . . ."[76]

The ox-plow revolution transformed systems of labor by expanding surface area cultivated and by assigning normatively rigid gender roles to cultivation. Recent studies in West Africa, for example, indicate that net savings in plow versus hand hoe cultivation are almost a third higher for men than for women.[77] Historically and across the microagronomic variations, adult and adolescent males have dominated the process of plowing and sowing. Virtually all other processes in cultivation have been subject to local variation, the exigencies of individual households, and change over time: Graham in 1841 observed Shawa and argued that plowing, sowing, and reaping is "the province of the men," while Salt in Tigray concluded that the plow "was guided by men alone; in all other parts of agriculture, the women take an equal if not a greater share. . . . the labour of reaping is thrown entirely on the females."[78]

The fluidity of household estates and the male monopoly of much of the cultivation process (see below) has fostered a marginal status for women in agriculture because divorce, death of a spouse, and household property dissolution inevitably have created within agricultural communities a persistent percentage of fragile, dependent female-headed households. On the one hand, these units have been demonstrably unstable and short-lived because of their inability to control cultivation except through male labor. Without direct control of such labor, women's claims on land, oxen, and capital equipment, though legally recognized, have been weak.

75. Alvares, *Prester John*, vol. 1, 105-7. Abba Gregorious in the seventeenth century told Ludolphus that if divorce was agreeable to both sides property was to be divided equally. J. Ludolph, *Ad suam Historiam Aethiopicam antehac editam Commentarius* (Frankfort, 1691), 440.

76. Alvares, *Prester John*, vol. 1, 107; also see Almeida, *Some Records*, 65-66.

77. Barrett et al., "Traction," 68-69.

78. Graham, "Report," 290; Viscount George Annesley Valentia, *Voyages and Travels to India, Ceylon, and the Red Sea, Abyssinia, and Egypt, in the Years 1802, 1803, 1804, 1805, and 1806*, (London, 1809), vol. 3, 232. In the Ankober lowlands Muslim women did not harvest but Christian women did.

On the other hand, such households—women with young children, divorcees, widows, or several women living together—have been a historically ubiquitous part of rural society. In Shawa of the mid-1980s 15 percent of households in Denki District were female-headed, though fewer in Ada and Gera. Alvares described a large number of such households in the Eritrean town of Barua in 1520.[79]

Women's vulnerability within the ox-plow property system, however, should not overshadow the critical role of their labor in the farm enterprise. Graham, although a cynical mid-Victorian, acknowledged this role in describing the needs of a successful farm: "The possession of a donkey, a pair of bullocks, a slave, and a woman to grind grain being absolutely indispensable." A Tigrayan proverb more succinctly indicates that male perspective: "Negus zayballu aynagad; sebayti zayballu ayharris (He who has no king does not trade; he who has no wife does not plow)."[80]

Women's farm labor in crop cultivation was, however, seasonal; the bulk of their labor concentrated in the processing of crops from the field and livestock (milk) into household food supplies and bringing free goods out of nature (water, wood, and dung) into the domestic domain. Almeida, unlike most travelers, offered a sympathetic view of the labor implied by women's authority in that domestic domain:

> Simple as this food seems it is no small labour to prepare it in Ethiopia, primarily because they have no mills to grind the meal. It is all ground by hand and it is the women's work; men even slaves will not grind at any price. . . . A woman grinds every day enough for 40–50 apas [enjera]. . . . Grinding meal and making apas, grinding more for sava or beer they drink (which uses a lot of meal) and making that, all this is work which calls for many slave women and plenty of firewood and is very great drudgery.[81]

Local exigencies of household labor or short-term need have often blurred gender roles in all parts of agriculture except plowing. In particular microagronomic zones, women have provided the additional labor to intensify cultivation, blurring the lines of gender roles. In Enewari where use of labor-intensive broadbeds to drain waterlogged vertisols allows an August sowing of wheat, it has been women (and children) who have followed the males' plows and built these broadbeds by hand which made this agronomic innovation possible.

The gendered division of labor in agriculture has varied somewhat over time and place. Nowhere in the historical record or contemporary experience of farmers I have interviewed is there evidence of women plowing. When asked my naive question about women plowing, male and female farmers consistently laughed or showed

79. Alvares, *Prester John*: vol. 1, 104-5; McCann, "Social Impact," 263-64, which is based on survey of the Denki area of Ankober district; Weissleder, "Political Ecology," 202-4.

80. Graham, "Report," 260; Conti-Rossini, *Proverbi*, 60.

81. Almeida, *Some Records*, 64.

Figure 2.7. Gendered division of labor in ox-plow agriculture.

76

derision. Women in some cases, however, play a role in managing field preparation by expressing household needs which are counter to their husband's concern for the market or rotation. Kuri Walde, a 45-year-old farmer from Ada argued: "If my husband decides to use the whole land for teff I will insist on having a corner of land for wheat, etc. for children's bread. Such discussions do take place."[82]

Similar to the processes which have affected women, the fluid movement of agricultural property and land between genders, individuals, and households has given shape to distinctive class relations within agricultural communities. At the regional and national level, access to *gult* (income from patrimonial income rights to agricultural land) has been a primary feature of the political struggle between imperial and regional elites. Historically, agricultural land in highland regions newly conquered by the imperial state evolved from temporary imperial land grants to loyal soldiers, eventually devolving to individual smallholder tenure.

On one hand, the lack of enduring family estates, often ambiguous property law, and weak kinship ties have generally promoted generational mobility within local communities and in power politics. On the other hand, the dominant historical metaphor of social, political, and religious culture in Christian theology across the highlands has been vertical patron-client relations, that is, coalitions of the powerful and the weak involving a strong analogy between the Christian ideology of heaven's relations to its earthly clients and the elite's terms for loaning capital resources to local smallholders.[83] Gender relations and the imprint of ox-plow technology have added further divisions in socioeconomic relations in the economics of highland agriculture by producing household labor patterns dominated by male cultivation, female food processing, and weak forms of cooperative labor between household units. Ambilineal inheritance allows women to claim rights to land and property, but powerful social convention prevents their engaging in key agricultural activities such as plowing. This tension has concrete historical dimensions in local agricultural communities and imperial politics as well as in scholarly debate over the nature and import of class in ox-plow social relations.

At the level of the farm, community, and parish, however, the struggle before the 1974 revolution was less one of claiming hereditary privilege than of agricultural

82. Interview with Kuri Walde, Ensalale (Ada), 16 June 1992. Yared Amare has told me of a case of a woman in Wogda (northern Shawa) who insisted on showing that a woman could plow. She cultivated her own plot one year and ensured her fame for generations. I have heard no similar stories.

83. See Bauer, *Household and Society*, 156-59. For ideology of religion, see Taddesse Tamrat, "Feudalism in Heaven and Earth: Ideology and Political Structure in Medieval Ethiopia," in Sven Rubenson, ed., *Proceedings of the Seventh International Conference on Ethiopian Studies* (Addis Ababa, Lund, East Lansing, 1984). An analogous analysis for the household and community levels is made by Weissleder, "Political Ecology," 171 and passim.

households competing for access to the resources directly relevant to agricultural production. Class was less a function of ascribed, hereditary background than success in the manipulation of political and economic networks to control productive farm capital resources (oxen, seed, and local forms of credit) and land. The production equation on ox-plow farms, that is, the balance between factors of land, capital (oxen and seed), and labor—and their aggregate effects on the countryside—depended only partly on political culture. Overall, the importance of oxen to the farming system and the possibility for seasonal transfer of cultivation rights rendered the control over land itself a relatively weak factor in determining who gained access to the full set of agricultural factors of production. The technology of the plow and its exclusive placement within the male domain conditioned the relative importance of land versus other forms of property. Those who did not exercise rights over animal traction through property rights (oxenless sons) or through gender (all women) had little opportunity to exercise the *de jure* rights they might have enjoyed through the land tenure system.

Debt relationships between agricultural households have formed the basis of rural class since they reflect relative control and ownership of productive factors and therefore structure interhousehold relations within local communities and economies. If access to rural capital (oxen and seed) has not been the primary linchpin of production in all localities or farm households, access to oxen labor was at least a *sine qua non*. Moreover, unlike the acquisition of land, which young couples or poorly endowed households obtained through relatively risk-free share-cropping agreements, the acquisition of capital incured debt in cash, labor obligations, or grain. With environmental shocks such as a drought or locust invasions that wipe out a farm's harvest, the only farms that directly incur debt have been those which had borrowed seed or oxen. Agriculture that depends on spring rains, which have failed in many areas one of every three years, has also exacerbated seasonal debt.

Debt has also derived from the unequal distribution of oxen in local farm economies. The ox-plow requires two mature animals to cultivate, and though historical sources are mute on the distribution of oxen per household, contemporary evidence indicates strongly that many households in a given community have fewer than a pair (see chapter 3). Therefore social institutions for the borrowing, sharing, and exchange of oxen appear to have played an important role in household agricultural strategies and in structuring rural debt. The most common form of exchange of draft power have been the practice of "yoking" called *maqanajo* in Shawa, *mallafagn* in Wallo, and *kendi* in Oromo-speaking areas. In yoking exchange, farmers borrow a neighbor's or relative's single ox and, in turn, lend their own for an equal period calculated in plowing days. Oxen exchange thus is a horizontal exchange between low resource households which involves no direct accumulation of debt. Though no empirical research on productivity effects of yoking exists for Ethiopia, the critical importance of timing for planting of annual cereal crops suggests strongly that those farms which exchange oxen are in a structurally weaker position than those that are self-sufficient in draft power. For young households, oxen exchange has been a normative part of the early stage of the development cycle.[84]

The importance of oxen to the farming system and the social practice of seasonal transfer of cultivation rights have rendered the formal control over land a relatively weak factor in determining community stratification. The timing of cultivation and the animals' strength at a given point in the plowing calendar are critical to the quality of seedbed preparation. Borrowers cannot be choosers and such farmers inevitably received oxen after their owners have prepared their own plots. Late planting reduces yields, and empirical research among highland smallholders indicates a strong correlation between oxen ownership and plowing days per household.[85]

Other farm strategies for obtaining oxen have been less benign and represent clear obligations negotiated between households of unequal resources. *Qollo*, an oxen rental agreement which has existed in northern Shawa but not in Harerge, provides for a fixed payment in grain at harvest. *Qollo* rates fluctuate with local market conditions, oxen supply, and expectations in the planting season about the level of the annual harvest. Under these agreements dependence between oxen-rich and oxen-poor households have emerged, especially in years with poor harvests, when fixed payments bite hard into resource-poor households and produce long-standing debt relations within a region or community.

Within each region and within each cultural tradition, institutions have varied, but within these variations unequal relations between agricultural households dominate in the exchange of productive factors. The institution of *magazo*, as practiced in northern Shawa and Wallo, may have been typical. *Magazo* in its ideal form is a seasonal transfer of land use rights on a particular plot from the landholder to a farmer in exchange for a portion of the crop. An elderly or infirm landholder can thus turn over a plot to a younger household in exchange for *erbo*, one fourth of the produce after taxes. *Magazo* thus provides a means of resource exchange between households at either end of the development cycle. Another form allows household with a plot far from its homestead to exchange seasonally for one closer, thus the couplet: "Yaruqun agazto / Yaqarubun tagazto (Let the far one be rented / Rent the one nearby)."

---

84. In his great empirical review of Russian farm economics, Vladamir Lenin, however, refers to yoking as a strategy for "tottering farms." Vladimir I. Lenin, *The Development of Capitalism in Russia* (Moscow, 1977), 80; McCann, *Poverty to Famine*, 80-81.

85. Contrast this perspective with Goody, "Class and Marriage." For effect of late planting, see Gryseels and Anderson, *Research*, 39-41; Noel Cossins, "Day of the Poor Man," unpublished mimeo prepared for the Ethiopian Relief and Rehabilitation Commission, 1975, 44.

In function and in practice, however, *magazo* and other seasonal resource exchanges have provided effective means for capital-rich households to increase seasonal landholdings or to lend oxen to resource-poor farmers who cultivate borrowed land with borrowed oxen. Debt and remuneration vary with the factors exchanged in the transaction. An oxenless household that borrows only oxen pays one-fourth of the harvest to the oxen's owner or agrees to a set number of work days based on the ratio of two days of human labor to one day's use of the oxen. If the lender provides seed as well as oxen, the lender can take back the seed at harvest and then claim *siso,* or one-third of the harvest after taxation. The provision of seed, oxen, and labor by the user yielded only one-fourth of the harvest to the legal landholder. A resource-poor household might often have only its land claims to offer, so might dissolve with only a one-fourth share over time, leaving the land in the hands of the capital-rich tenant, who might then claim the land as *rest.*

The issue of oxen ownership versus rental, sharing, or borrowing appears to be a major determinant of economic class within the ox-plow system. Recent empirical on-farm research suggests there are area and yield effects on farms which own rather than borrow oxen (see Tables 2.1 and 2.2). This same research indicates that two-oxen farms produced 81 percent more than farms which owned no oxen and 18 percent more than those with only one. Moreover, there is also a "cropping" effect: two-oxen farms at Ada produced 63 percent more cereal (as opposed to pulses) than farms which owned no oxen and 19 percent more than one-ox farms.[86] The effect of oxen on rural differentiation has not depended on a farm's access to oxen (farms can borrow, rent, or exchange oxen labor) but on a farm's ownership and ability to maintain its own pair.

The outcomes of seasonal debt are conjunctural and historical, a mix of environmental conditions, farmer skill, and serendipity. Though a sustained drought or epizootic might seem to create a great leveling effect, in fact the structure of seasonal debt places risk on borrowers of capital resources, particularly those with fixed obligations. There is a strong incentive therefore to minimize risk through farm crop choices wherever possible, building an element of rational conservatism in farm strategies. In fact, recent field evidence indicates that economic differentiation increases dramatically during an ecological crisis. Ownership of oxen, rather than seasonal borrowing, appears to have been a crucial factor in the distribution of both risk and debt in communities which experienced drought or related food shortages in the 1970s and 1980s.[87] Unless households owned oxen or had access to them, land had little economic meaning; land claimed but not plowed quickly ended up in the hands of those with the means to cultivate.

86. Bekele Shiferaw, "Crop-Livestock Interactions in the Ethiopian Highlands and Effects on the Sustainability of Mixed Farming: A Case Study from Ada District," M.S. thesis, Agricultural University of Norway, 1991.

87. See chapter 4. Farmers who rented oxen from merchants and had negotiated rental prices in cash or in grain had to renegotiate terms after failed harvests.

TABLE 2.1. Effects of draft power on cereal yield (kg/farm), in Ada, 1978–85.

Oxen owned	1978–80	1983–85
0 or 1	877	636
2+	1597	1390

Source: Guido Gryseels, et al., *Draught Power and Smallholder Grain Production in the Ethiopian Highlands* (Addis Ababa: n.d.).

TABLE 2.2. Effects of oxen ownership on area cultivated (hectares/farm), in Ada, 1978–85.

Oxen owned	1978–80	1983–85
0 or 1	1.09	1.16
2+	1.74	1.67

Source: Guido Gryseels, et al., *Draught Power and Smallholder Grain Production in the Ethiopian Highlands* (Addis Ababa: n.d.).

Feeding the State

The arrival of the ox-plow complex predated the formation of the highland's large overarching states by several millennia; indeed, the labor efficiency and productivity of the system appear to have underwritten the large bureaucratic political and ecclesiastical hierarchies which came to characterize the highlands from Axum through the modernizing twentieth-century imperial state. While the ox-plow complex and the imperial state were not synonymous, this agricultural system seems to have been almost a *sine qua non* of imperial hegemony. Farms transferred tax revenues to the local representatives of the state in the form of tithes *(asrat),* rent, land tax, and an obligation to feed and house guests during special feast days or on military campaign. Beyond direct transfer of goods, the state maintained *hudad,* land for which farmers had to provide corvée labor to plow, weed, and harvest fields directly controlled by local representatives of the state for their own support.[88]

Travelers frequently mentioned the depredations of soldiers and imperial retinues on local farms; they witnessed such extractions first hand since they themselves depended on *dergo,* local contributions of food, during their treks. Nathaniel Pearce, who traveled with foraging troops of Ras Walda Sellase in the 1820s, described treatment of farms in rebel areas:

> We stopped three days on this mountain, where we lived pretty well, there being plenty of corn in a village at the top, belonging to the rebels. . . . Having burnt the town of the above name, we stopped two days, and then marched to the plain of Ardergahso; where the corn was ready to cut, which it took us five days to destroy. We marched thence to the river Munnai, the finest country in that part of Abyssinia for corn and cattle, where we stopped a week to destroy everything.[89]

88. Donald Crummey has estimated such extractions at 30 percent of total farm production. Crummey, "Plow Agriculture," 4; Hoben, Land Tenure, 77.

89. Pearce, Life and Adventures, vol. 1, 68-69.

The long-term effects of such taxation, forced hospitality, and raiding on the local and regional farm economy, however, are not entirely clear, since political stability reduced such direct appropriations. In-kind collections from local farms and the production of state-controlled fields resulted in substantial holdings of grain and livestock within each district, most of which remained in storage locally. Graham in Shawa observed that such royal granaries were "profusely studded over every portion of the kingdom."[90] Although these resources remained available for transfer to the imperial court or to support imperial levies on campaign, most probably were consumed locally to stage local feasting or to engage in seasonal loans to local farmers, activities which cemented social and political relations.[91]

Elite and state income relied upon other prerogatives which did not touch farm production directly. The imperial court and its regional counterparts expected and received revenues in customs fees and market taxes from the movement of middle- and long-distance trade across the zones of its direct authority and tribute from peripheral states which fell under its military domination. These fees, paid in cash and negotiable goods, derived from extractive activities on the highlands' southern and western periphery which provided luxury goods (gold, ivory), human labor (slaves), and exotic goods (civet, musk, myrrh) demanded by Red Sea markets. Those who participated in wars of expansion received such goods as booty, in addition to the rights over annual tribute from those who occupied the land. Farmer-soldiers who accompanied their lords on campaign received tax exemptions in addition to the fruits of a successful campaign.

Ironically, the political economy of these complex patrimonial states provided very little direct state influence on production decisions made in the farm-level agricultural economy. Tax and tribute policies were extractive, not prescriptive in determining how farmers generated payments. Only on their own lands, and not always then, did state representatives exercise control over cropping and agronomy. Abba Bor, the balabat of Gera and a descendant of the old royal family, when asked about who decided what to sow on his lands, argued that the farmer himself decided, since he knew "which plant suits each plot."[92] In other systems closer to urban markets, landowners often did impose specific rules on land use and agronomy (see chapter 6). Where imperial estates were extensive, such as in newly conquered areas, those lands could include plots amenable to irrigation and specialized production. In core areas long under political hegemony of the state,

90. Graham, "Report," 260.

91. This varied by region and by level of political stability. The nineteenth century was particularly unsettled; traveler reports from that period thus may exaggerate effects of extraction. See McCann, *Poverty to Famine,* 57-60. The Lasta region, for example, transferred almost nothing to the imperial court or the court of its governor Ras Kassa at Addis Ababa; taxes were consumed locally to maintain state operations within the district.

92. Interview with Abba Bor Abba Magal, Gera, January 1990.

the royal lands tended to devolve into localized tenure as rest that was controlled by farmers themselves.

The northern state system certainly based itself on surplus from this system, and the southern tier of highland states, independent until the late nineteenth century, did so as well. Kaffa converted to the plow and cereal diet for its royal court in the eighteenth century; the first travelers who described the Gibe states in the mid-nineteenth century reported mixed agriculture, but with clear elite preferences for cereals and pulses over ensete and root crops for both diet and tax revenues. Indeed over time the presence of crops such as cereals, which could be easily collected, divided, and accumulated, may have made initial state formation possible. Teff was the most prominent highland crop; its capacity for long-term storage, value as livestock fodder, and prestige as a food marked it a symbol of elite status. Thus, while states appear not to have forced acceptance of particular forms of agriculture, strong preferences for cereal products as payment provided an indirect pressure for farmers to convert.[93]

At another level, however, the connections between farm and state formed the basis for regional polities, the imperial state, and other smaller kingdoms on the periphery. Cereal-pulse annual agriculture yields products which were moveable from the farm to state storage control, providing state control of a critical local resource. Perennial horticultural crops tend to be perishable and tied to the farm location. If both the accumulation and concentration of food resources were essential to the growth of a bureaucratic state, then state systems would tend to favor the ox-plow complex's annual production of divisible, storable, moveable products.

Conclusion:
Agriculture in the *Longue Durée*

The ox-plow revolution, which spread across the highlands over almost two millennia, brought with it a bundle of proclivities in property relations, gender division of labor, rural debt structure, land use, response to ecological shocks, and management of farm household resources. This revolution changed the way rural populations conceived of nature, of gender roles, and of the natural calendar. It most certainly changed the landscape. The following chapters look for those historical processes on a macro-scale and in specific regional cases.

93. Amon Orent, "From the Hoe to the Plow," 191; Cecchi, *Da Zeila*, vol. 2, 496.

3

Farms in the Agrarian Polity: Historical Trends in Population, Farm Resources, and Specialized Agriculture, 1800–1916

> . . . the poor Abyssinian peasants have not a chance of improving their condition. Every generation adds to the tract of once-cultivated country which is becoming desert; every generation sees villages and churches abandoned, and no others taking their places. If this condition of affairs continues very much longer, the Ethiopian will be, like his elephant, a thing of the past.
> —J. T. Bent, *Sacred City* (1893)

> It seems to us therefore exact to say that Abyssinia for the products of its soil could largely provide for a population at least quadruple the present.
> —Lincoln De Castro,
> *Nella terra dei negus* (1915)

The ox-plow complex in both its collective form and its individual farm management has combined the historical stability of its technology and cropping regime and dynamic forces of population. Its character over the course of the period of this study has resembled what Emmanuel LeRoy Ladurie has described as "a preindustrial society characterized by slow technical change where processes of growth are still dominated by the play between demographic expansion and limited resources."[1] Demographic change has been to a large degree the primary historical

1. Le Roy Ladurie, *The Peasants of Languedoc*, 296.

factor driving the cycle of productivity within local agriculture and accounting for many of the local variations in forms of agronomy, per-capita ratios of key farm resources, and land tenure practice.

The link between population and agricultural practice suggests at least two divergent visions of historical change in preindustrial agriculture. In many ways, the ox-plow complex's historical expansion resembled LeRoy Ladurie's Malthusian "scissors," a model which describes increasing productivity when population expands into relative abundant sources of land, but rapid cyclical decline when population pressure lowers per-capita resources and farm-level productivity stagnates.[2] In this neo-Malthusian vision, technology and agronomy remain historically stagnant; changes in agricultural per-capita output derive from indirect effects of population cycles—and thus indirectly from exogenous factors like disease, fertility, and warfare—rather than structural change in agriculture itself. Thus agriculture is a component of the "Great Agrarian Cycle" but is not its leading edge or a source of economic or social change in and of itself. For France and for LeRoy Ladurie, the break in the agrarian population cycle came with a fundamental transformation in rural mentalité in which literacy and enlightenment ideas of rationalism and empiricism permitted new technology and agronomy.

The alternative model is the classic formula described by the Danish geographer Ester Boserup where population pressure exerts a progressive influence on agriculture. The initial slide in per-capita productivity brought by increasing population in a given area stimulates more intensive land use and an increase in yield per unit of land. In the Boserup formula, population growth therefore serves as the primary engine of genuine agricultural transformation, an intensification of labor and new agronomic practice to raise yield per unit. Boserup's model is diachronic in that it accounts for change over time, but is basically ahistorical in that it does not draw on evidence of the past as its primary rationale.[3]

The history of Ethiopia's highland agriculture allows us to examine the historical basis for these formulae. As one might guess, the interaction of population with agriculture was by no means uniform over time or across the highlands. In some locales, such as lowland marginal zones or along the pastoral southern highlands before adoption of the ox plow, low ratios of population to land and to capital allowed periods of rapid economic expansion, whereas along the midaltitude spine of the long-settled northern highlands, population pressure in the mature stage of the annual cropping regime suppressed both gross product and per-capita production. The evidence of specialized forms of agriculture, especially around political centers such as Axum or Ankober, suggests intensification of labor as a response to concentration of both authority and population.

2. LeRoy Ladurie, *The Peasants of Languedoc*, 51ff.

3. Ester Boserup, *The Conditions of Agricultural Growth* (London, 1965), Introduction.

Figure 3.1. Ox-plow cultivation, c. 1860.

Figure 3.2. Ox-plow cultivation, 1974.

86

The demographic footprint of the plow was distinct and differed substantially from the horticultural and perennial crop or arborial economies visible on the periphery. Everywhere, the plow's dependence on oxen meant an intricate crop-livestock management regime in which new population in the form of land claims by new households threatened the very pasturage that cattle herds and oxen needed to sustain the ox-plow complex. Yet, ox-plow cereal production was relatively inelastic, requiring new land to expand its total output.

This chapter examines the overall evidence of agricultural change in the nineteenth and early twentieth centuries, a period before the existence of large urban markets and urban government and marked by a state with little or no interest in direct intervention in production. This chapter also describes elements of broad, macrolevel change over the 1800–1990 period in specific sectors of the agricultural economy—demography, agronomy, and ratios of farm resources—thus providing benchmarks and a context within which to understand local-level studies in part 2 and chapter 7 which assesses agriculture within the emerging national political economy of the 1916–90 period.

Demography: People on the Land

What was Ethiopia's population in the past and, more important than absolute numbers, how did it change, and what were its local and regional contours? Historical demography has been much debated within African history, partly because the dearth of hard historical data has given an exaggerated power to theoretical constructions as the tools of argument. The dominant "natalist" perspective for Africa, based largely on demographic transition theory, has argued that relatively slow or stagnant population was the result of historically high fertility rates and matching rates of mortality from disease, poor nutrition, and political violence. This theory asserts that Africa's recent rapid population growth in the post-World War II period has resulted from colonial hegemony: improved health care, famine relief, and imposed peace. The natalist approach also holds that at some later date fertility rates will adjust to economic development as they have in the West.[4]

The opposing "anti-natalist" argument derives from the more empirically based historical evidence of preindustrial England, where birth records and death rates indicate that population growth derived from social responses to economic conditions, that is, rural families exercised control over birthrates through social responses such as late marriage. Thus if one assumes a social capacity for fertility control, then

4. For the anti-natalist perspective, see Dennis Cordell, Joel Gregory, and Victor Piché, "African Historical Demography: The Search for a Theoretical Framework," in Dennis Cordell and Joel Gregory, eds., *African Population and Capitalism* (Boulder, 1987), 14-32; John Iliffe, "The Origins of African Population Growth," *Journal of African History* 30 (1989), 165-69.

Africa's historical growth rates have been more than passive responses to colonial rule; they also comprise aggregate local social responses to historical conditions. Where African precolonial data appear, such as John Thornton's eighteenth-century Kongo baptismal records, a slow growth tempered by episodic shocks seems to be the rule.[5] Indeed the earliest descriptions of the highland human landscape from sixteenth-century Portuguese accounts suggest no radical differences from nineteenth-century portraits of the same regions.

For the late-nineteenth and early twentieth centuries, Ethiopia's case nominally stands outside both models' use of the colonial paradigm because of the absence of long-term colonial rule in most of the highlands. At another level, however, the expansion of a Pax Aethiopica, beginning with Shawan expansion in the 1880s and proceeding through the formation of a national political economy in the 1920s, may have brought political shocks to a number of regions analogous to those in colonial societies in the same period, which anti-natalists would argue affected social controls over fertility.[6] Indeed, Ethiopia's annual population growth rate of 2.9 percent measured over the last two decades or so parallels growth rates across the continent, begging questions of comparability.

The historical evidence of demographic variation across the highlands and over the 1800–1990 period is primarily qualitative, though recent population data provide a baseline. Since we do not have a comprehensive data set or historical studies for all regions, we must rely on indicators of demographic change such as correlations between population density and forms of agricultural activity. These include forms of labor, land distribution methods, and household land holding estimates. Such features reveal themselves in oral evidence, in regional survey statistics, and in field observations by travelers and officials.

This evidence supports the argument that, although population has not been distributed evenly, there seems to have been a logarithmic progression of population growth in Ethiopia's northern and central highlands over at least the last millennium. *Prima facie* evidence for this position is the steady expansion of the ox-plow system itself, since that spread involved not only the spread of the technology but also in many cases the migration of its practitioners, particularly in the second half of the period.

5. C. C.Wrigley, "Population in African History," *Journal of African History* 20 (1979), 129-31; for skepticism see Iliffe, "Origins," 167-68. Helge Kjekshus' argument on the environmental effects of imperialism in Tanzania rests upon an argument of a steady precolonial population growth and ecology control devastated by imposed colonial rule. Helge Kjekshus, *Ecology Control and Economic Development in East African History: The Case of Tanganyika, 1850-1950* (Berkeley, 1977), 9-25.

6. Use of a colonial analogy has been a contentious issue in recent Ethiopian historiography. Nonetheless, the extension of central government authority over both northern and southern rural society in the 1920s and 1930s brought substantial changes in taxation and the structure of governance over rural areas. See McCann, *Poverty to Famine*, 127-45.

The nineteenth-century demographic evidence derives almost exclusively from traveler impressions of density and descriptions of forms of cultivation which imply density.[7] For northern Shawa, for example, Douglas Graham described the Ankober highlands as "abundantly inhabited," with "very few forests or wastes," but "unburdened by over-population."[8] Earlier comparative evidence from the sixteenth—and seventeenth-century Portuguese accounts indicates considerable density of settlement in Tigray, around Gonder, and in northern Wallo. Graham's Ankober, in fact, was the recipient of in-migration from Gonder once it had been recaptured from Oromo pastoralists in the late eighteenth century.

Population levels were subject to episodic crises which periodically checked population growth by raising both mortality and morbidity, if only temporarily, through disease or by sustained crises in food supply. Nathaniel Pearce described such an episode in 1811–12, when a smallpox pandemic struck across the northern highlands. The disease, combined with locust swarms that year caused high mortality "so that a great part of the country was left in a state of desolation."[9] Travelers depicted smallpox as the most serious nineteenth-century scourge; by World War I, Dr. P. Merab estimated that 20 percent of all Shawans bore pock marks from the disease.[10] Effects of locusts or endemic diseases such as cholera, malaria, syphilis, and typhus, while devastating, tended to be more localized, with effects impossible to assess.[11]

In the middle of the period of this study, two major environmental shocks struck the highland population as a whole, producing significant mortality and perhaps reducing, if only temporarily, the highland population. The first was *yakefu qan* (Cruel days), the 1889–92 famine. This calamity was the first such disaster to appear in detail in both outside and Ethiopian oral and literary sources. It began in the late fall of 1888 following poor summer rains in the northeast and the great rinderpest panzootic which swept north to south through the mid- to lower-altitude zones, killing most of the region's cattle, including oxen, almost overnight. Alaqa Lamma Haylu, a young man traveling through Gojjam at the time, recalled awakening from an intense fever and finding all the cattle dead.[12] The famine which resulted from the depleted food stocks deepened with the lack of oxen to prepare the next spring and

7. Boserup, *The Conditions.*

8. Graham, "Report," 284.

9. Pearce, *Life and Adventures*, 90-92; Richard Pankhurst, *Economic History*, 216-20.

10. Richard Pankhurst, *Economic History*, 623-25.

11. In 1985 I witnessed a major epidemic of fever in the Denki lowlands. Diagnosed variously as typhus or malaria, the fever struck across an entire altitude zone and resembled the widespread, endemic fevers described for the lowlands by early travellers and highlanders. Mortality was high, since this fever struck at the height of a local famine. Almost 10 percent of Denki's population died. Though tragic, the death figures were in fact much lower than those often implied in descriptions of pandemic diseases.

12. Mengistu Lamma, *Masehafa tezeta zalaqa Lamma Haylu* (Addis Ababa, 1959 E.C.), 100-102.

summer's planting. Local outbreaks of cholera increased mortality when victims gathered around churches, mission stations, and stored government food supplies.

Effects on the countryside shocked travelers, and produced many poignant and sympathetic descriptions, if only impressionistic ones. Pietro Antonelli who had seen the country in 1888, only a year before, wrote, "Previously, the country was inhabited; there were beautiful fields of durra and barley, numerous herds of cattle, sheep, and goats, and the whole area had an atmosphere of abundance and prosperity." Retracing his steps in 1890 he reported: "The countryside was one of continuous desolation . . . absolutely a desert; no more inhabitants, no more cultivation, no more flocks."[13] Impressionistic reportage on the Great Famine masked the important short- and medium-term effects on agriculture and farm economies. As in other famines, the terms of trade of livestock and grain shifted radically when households attempted to exchange remaining goats and sheep (the cattle were already dead) for food. Cattle prices, which traditionally dip early in a food crisis, did so, and then shot up 30-40-fold when demand from farms trying to reinvest in their farm capital pushed up prices and the terms of borrowing.[14] Since it began with sudden cattle death, the Great Famine produced a more radical effect than famines of the 1970s and 1980s, which resulted from protracted drought and allowed farmers to sell oxen and cattle, albeit at depressed prices.

At least one more overarching environmental crisis brought a severe demographic shock to the highlands. The worldwide Spanish flu pandemic, known in Ethiopia as *ya Hadar basheta* (the Disease of October–November) or *ya nafas basheta* (the Disease of the Wind) struck in 1918–19. Though relatively poorly documented in Africa, the Spanish flu killed perhaps 20 million people worldwide and 1.5–2 million in Africa alone. The air-borne virus had debilitating symptoms, causing fever and often death by secondary bacillic infection.[15] The first wave arrived in

13. Camera dei Deputati, *Il Missione Antonelli in Etiopia* (Roma, 1890), 57, quoted in Richard Pankhurst, *Economic History*, 218. Other, less credible reports, largely from Protestant missionaries, reported practices of eating ritually unclean food, infanticide, and cannibalism which have appeared in published historical works. It is very doubtful that such practices occurred, since no such practices have been reported from recent famines of equal or greater severity. For a good summary of local demographic evidence for the Gonder area, see Steven Kaplan, "Kifu-Qen: The Great Famine of 1888-1892 and the Beta Israel (Falasha)," *Paideuma* 36 (1990), 68-77.

14. For cattle prices see Richard Pankhurst, *Economic History*, 218-19; for effects of the grain-livestock price index, see Peter Cutler, "Forecasting Famine: Prices and Peasant Behavior in Northern Ethiopia," *Disasters* (August 1984), 48-56.

15. For the effect of Spanish flu on Africa, see K. David Patterson, "The Demographic Impact of the 1918-19 Pandemic in Sub-Saharan Africa: A Preliminary Assessment," in *African Historical Demography* (Edinburgh, 1981), vol. 2, 402-29; Richard Pankhurst, "YaHedar Besheta," *Journal of Ethiopian Studies* 12 (1975), 103-31; interview with Abba Gabra Masqal Tasfaye, March 1982.

Addis Ababa in the spring 1918, followed by a second, more virulent, wave in the fall. Unlike rinderpest, which moved north to south and had no direct effect on humans, the flu was a disease of the modern world, traveling along lines of modern communication. The infection arrived in Addis Ababa via the newly constructed railway from Djibouti and followed caravan routes north. At least 40,000 died in the capital. Rural mortality and economic effects of mortality and sickness resulted in labor shortages for the harvest and spring plowing. Alberto Pollera, the Italian agent in Adwa, reported that "in Yajju and Lasta there continues a fierce epidemic which seems of a type of typhoid which is hindering the harvest there. Livestock abandoned to pasture without custodians have severely damaged crops."[16]

In demographic terms, the effects of famine and epidemic disease remain ill-described and difficult to assess, since environmental shocks such as disease, drought, or locusts reflected localized patterns of microclimates induced by drastic shifts in elevation, temperature, and vegetation. Not only were shocks localized, but also each had its own character and set of specific effects on the agricultural economy. The presence of such effects, however, does not seem to have derailed the overall evidence of slow growth over time. Doubtless the high mortality reported forced a dip in the population but perhaps with only a momentary hiatus in fertility. As with more recent and better-documented famines, the highest mortality probably took place among the most structurally vulnerable populations—young children and older women—whose deaths would not have affected fertility, though morbidity in child-bearing women might conceivably have slowed fertility in the immediate period of the crisis. The groups in society most vulnerable to starvation, disease, or economic crisis—women beyond child-bearing years and young children—were also those with the least effect on fertility. Female-headed households had few resources in food or goods to liquidate to ride out the crisis.

In local terms and beyond the emotional scars, neither the Great Famine nor the influenza pandemic transformed social institutions or farming practice; rather, recovery from the economic disruption exacerbated debt between households and also reinforced the importance of capital rather than land as the driving force in local farm economics. With the increase of open pasture and new stock drawn from upper altitude zones less affected by the epizootic, cattle numbers recovered. Emperor Menilek himself contributed to reconstructing rural social structure by distributing cattle to his supporters."[17]

At the opening of the twentieth century the population of many regions of the northern highlands appears to have begun a period of recovery after depopulation resulting from the major famine and epizootic of 1889–92 and localized disruptions of Mahdist incursions along the western marches with Sudan. The recovery proba-

16. Pollera to Addis Ababa, 12 December 1919, Archivo Storico delle Ministero Africano Italiano, Roma (Ministero delle Affari Esteri), 54/8.

17. Gabra Sellase, *Tarika Zaman Dagmawi Menilek Negusa Nagast Zaltyopya* (Addis Ababa, 1959 E.C.), 175-79.

bly proceeded slowly, stumbling during regional droughts, locust swarms, and new infestations of rinderpest, such as in 1905–6. Few of the crises subsequent to the Great Famine appear in local memory in such dramatic terms, and none so captures memory across the entire highland zone.

The epidemiological and demographic evidence is insufficient to speculate on the longer-term effects of either the 1918–19 epidemic or the Great Famine. In the longer term a mix of sources provides spotty but generally convincing evidence of a general population increase in the highlands through the 1920s and into the postwar years. The first sources covering the 1905–39 period are Italian censuses from Eritrea (one of our only statistical sources), which indicate that growth of highland populations in that region continued despite faltering aggregate food production. The average annual rate of growth for Eritrea through that period was 2.9 percent, remarkably close to the rate for all Ethiopia and Africa as a whole in the 1980s.[18]

Detailed but localized evidence of population growth more recently comes from comparative aerial photographic data for the period 1955–75 from the Simen region. In Simen, highland population increased by an annual rate of 2.4 percent in the most populous and productive elevations between 1800 and 2500 meters (barely below the natural 2.5 percent increase and the Eritrean figures), but rose 3.58 percent in the lower elevations—strong evidence of local migration. Since 1964 the upper limit of cultivation in the Simen region has climbed 100 meters to just below the frost line, marking a long standing trend of localized movement with population pressure in the more ideal, first-settled altitudes pushing a new generation of households into more agriculturally marginal zones. This evidence from a region which historically and ecologically resembles areas of western Tigray, western Wallo, and parts of Gojjam suggests an overall level of population growth, but also localized migration and the expansion of cultivation in new ecological zones.

This pattern illustrates statistically the much wider trend of immigration to less productive and riskier lowland cultivation evident earlier along the eastern escarpment below the road from Addis Ababa to Asmara. Lowland zones, which had been primarily used for pastoral wet-season grazing, have been cultivated in the past one or two generations by migrant highlanders. The patterns are also similar to those followed in the 1900–20 period in the northwest Setit-Humera region and again for northern Shawa for at least the post-World War II era, if not well before.[19]

18. Istituto Agricolo Coloniale, *L'economia Eritrea*, 43-44, cited in Tekeste Negash, *Italian Colonialism in Eritrea* (Uppsala, 1987), 149. For the decline of food production from 1921 through 1931, see F. Santagata, *La colonia Eritrea nel Mar Rosso davanti all'Abissinia* (Napoli, 1935); and Irma Taddia, *Eritrea-Colonia: Paesaggi, strutture, uomini dei colonialismo, 1890-1952* (Milano, 1986), 209-77. The 2.9 percent growth rate is my calculation based on the Italian censuses cited by Tekeste. Ethiopia's 2.5 percent rate is cited from Ethiopia's Central Statistical Office.

19. B. Messerli and K. Aerni, eds., *Simen Mountains, Ethiopia, vol. 1: Cartography and Its Application for Geographical and Ecological Problems* (Bern, 1978), 34. In the Dabra

Population and Farm Resources

Perhaps the most pertinent proxy evidence of population growth over the course of the twentieth century has been the shrinkage of per capita and per farm resources evident in the post-war period. Although this trend resulted from the slow annual population growth described above, its main effect has been local and uneven because some areas' populations have grown quickly with in-migration or dropped as a result of political or environmental events. In other areas, where conditions sustained high levels of fertility and new household formation, population may have grown within the longer frame of social time. Thresholds of agricultural adjustments resulting from population change have also derived from local variation in topography, soil fertility, and effects of microclimates.

Two measures are the most telling in terms of agricultural productivity and farm management: oxen holdings per household and farm size, both of which by the late nineteenth century had reached critical levels in areas of long settlement (Tigray, Wallo, northern Shawa) and in more recently settled highland zones that have drawn migration (Ada). Across the highlands by the 1980s, average farm size had routinely decreased to less than 2.5 hectares and, more often, ranged between 0.5 and 1.5 hectares per household. Land subdivision into ever smaller plots has resulted from partible inheritance and probably increased as a result of the 1975 land reform proclamation. Fragmentation—the increasing number of plots per farm— has also resulted over time, initially from farmer diversification strategies, but also from competition over the most viable land.[20] This recent evidence of the farm-level effects on population growth offers insights into the longer-term historical problem of localized productivity crises as a natural product of the mature stage of the ox-plow economy and its imperative to expand.

Chapter 2 described the critical role of oxen as capital within the ox-plow system. Ironically, one of the primary results of the maturation of the ox-plow economy and its partible inheritance system is the expansion of cultivated land over natural non-farm lands, that is, open pastureland and forest, resulting over time in a shortage of livestock forage. The chronic shortage of pasturage—and to a lesser degree forest— and consequent scarcity of oxen in the highland plow economy are direct historical products of population growth within the ox-plow complex and thus a fair proxy index of the timing and direction of change. Intensive forage strategies of stubble feeding, lopping browse, and storage of straw have emerged in particular settings over time as an adaptation to reduction and privatization of pasture. The spread of

Berhan area where frost and waterlogging are primary constraints on cultivation rather than drought, a movement up to higher elevations is clear. Gryseels and Anderson, *Research*, 9. For details on the Ankober see chapter 4.

20. For land fragmentation in northern Shawa, see Fassil G. Kiros, "Agricultural Land Fragmentation: A Problem of Land Distribution Observed in Some Ethiopian Peasant Associations," *Ethiopian Journal of Development Research* 4 (1980), 1-12.

the ox-plow complex down the slopes to marginal lands and south into new areas conquered by the imperial state in the nineteenth and early twentieth centuries was thus part of a cycle of maturity.

Like the population growth rate itself, the loss of pasture to population pressure would not have been a steady process, but rather a localized percolation that ebbed and flowed according to ecological and climatic conditions. Areas at the margins of the ox-plow complex initially would have retained pasture, whereas areas of long-term settlement along the midelevation zones (1800–2500) meters would have faced the greatest pressure on open land. Zones along the escarpment east of Wallo, Shawa, and Tigray and in the western zones of Gonder, Tigray, and Gojjam, which historically provided dry-season pasture by the second half of the twentieth century, were under cultivation by highland migrants, areas vulnerable to drought and endemic livestock disease (see chapter 4).

The chronology and spatial distribution of the closing of highland pasture remain murky, since few historical sources note the subtle ebb of open grazing land when demand for new cultivation proceeded. There is some evidence that in the early nineteenth century forage presented few problems to highland farmers. Nathaniel Pearce, who kept his own animals in his Tigray holdings, asserted, "The inhabitants have no notion however in making hay, in any part of the country . . . ," suggesting that available grass sustained local animals. At the same time, he described conflict with a local official who succeeded in claiming prime pastureland from Pearce's holdings. In the same period but farther south, travelers commented on highland *zalan* [pastoralists] in Wallo who found sufficient pasture to avoid agriculture altogether.[21]

By the mid-twentieth century, however, population pressure had severely reduced highland open pasture in areas of longest settlement and areas of intensive in-migration. The effects of this maturity appear in the poor distribution of oxen across highland farms, where well over half of all highland farms have fewer than the required pair of draft animals, increasing rural dependence and potential debt during environmental crisis. In the early 1970s Dan Bauer noted the conundrum that the average farm in Harenya, Tigray, did not have access to sufficient forage for the oxen required to cultivate the average farm holding. Population pressure on land resources in the highland Tagulet District, 100 kilometers northwest of Ankober, after the Italian occupation prompted landlords to protect pastureland and sell grass-cutting rights to their local tenants.[22] As one of the highland areas of longest settlement, Tagulet has experienced steady pop-

21. Pearce, *Life and Adventures*, vol. 1, 204. He noted the *zalan* preference for using straw, which "is but poor food upon a march, unless it is good taff gulliver [*gallaba*, i.e. straw] which much resembles hay."

22. Dan F. Bauer, "For Want of An Ox...: Land, Capital, and Social Stratification in Tigre," in Harold G. Marcus, ed., *Proceedings of the First United States Conference on Ethiopian Studies* (East Lansing, 1975), 242-43. Interviews with farmers at Segatna Jer and Jimana Gan peasant associations (Tagulet, Shawa), June 1985.

ulation growth, though with an apparent acceleration since the postwar period. Pressure from new households' land claims, especially after land reform, to expand the cultivated area has virtually eliminated pasture, and land holdings have dwindled in the postwar period to less than two hectares. In the past two generations even the river flood plains 1000 meters below the plateau have been cultivated. Overall effects on per-capita productivity of such pressure are evident: farmers plow with cows because of the shortage of livestock and pastureland; the age of first marriage has increased dramatically, since there is no land to allocate to new households.

The crisis in livestock management within mature ox-plow local economies exists in long-settled highland areas from northern Shawa through to Eritrea and in the Chercher highlands of Harerge.[23] Table 3.1 indicates the severity of the problem in terms of the distribution of oxen across several highland districts and nationally by the 1980s.

The direct demographic evidence for Ethiopia is far too sketchy to provide definitive support for any single explanation of African demography as a whole.[24] There

TABLE 3.1. Household oxen distribution in the ox-plow highlands, 1980–86.

District	Percentage of Households with:		
	No Ox	One Ox	Two or More Oxen
Denki (Shawa)	40%	40%	20%
Lalo (Shawa)	34%	47%	19%
Segatna Jer (Shawa)	18%	55%	22%
Ada (Shawa)	9%	12%	79%
Dabra Berhan (Shawa)	31%	38%	31%
Wubera (Harerge)	57%	35%	8%
Chercher (Harerge)	32%	37%	31%
National	29%	34%	37%

Sources: Data are drawn from peasant association interviews (1985–86), Gryseels and Anderson, *Research* 15; and the Ministry of Agriculture Survey (1980), shown to me in the Harerge Ministry of Agriculture Office.

23. International Livestock Centre for Africa (ILCA) baseline surveys of the Ada and Dabra Berhan areas report the trend toward reduction of fallow and pasturage. Gryseels and Anderson, *Research*, 10. For a discussion of the needs for intensification, see James McCann, "An Evaluation of Harerge Projects: A Report to Oxfam U.K.," unpublished project report, 1987; and James McCann, "Report to Oxfam America on Evaluation of Ox/Seed Project," unpublished report, 1986.

24. For evaluation of African evidence as a whole, see Iliffe, "Origins," 168-69.

is clear evidence that the significant population growth in twentieth century Ethiopia, especially in the post-World War II era parallels growth patterns elsewhere in Africa. For Ethiopia, however, there can be no argument about decreasing death rates from colonial medical services and famine relief; neither existed before the post-1960 period. Nevertheless, the proxy evidence for a stable population across the highlands in the nineteenth century and accelerated growth in the twentieth century is convincing if not definitive. The evidence here seems to support the anti-natalist argument for changes in fertility, that is, the likeliest cause of increasing population in the twentieth century is not decreases in the death rate, or the advent of better health care or deaths in warfare, but rather increasing fertility when the stable political climate after 1916 gave full expression to the local social institutions of marriage and property transfer, which controlled household formation and thus fertility. The disruptive and often violent expansion in the south and struggles for political hegemony in the north in the early part of the twentieth century subsided under the imperial government's centralization of power in Addis Ababa after 1916. The expansion of cultivation in the years after the Italian occupation probably paralleled the process detailed in the Simen evidence above. Despite episodic crises in climate, disease, and politics, population growth was a slow, cumulative process.

The natalist argument for the effect of peace or more precisely for the role of political change, however, also appears in the Ethiopian case to be a factor in demography, not in its effect on the death rate, but in its effects on social institutions of fertility. The expansion of scale in political authority (see below) which allowed movement between districts and provided highland farmers with state protection to penetrate marginal pastoral lands unleashed rapid expansion in new household formation as well as rural dependency, since new households more often began with fewer resources.[25] Thus state support for the ox-plow complex in its expansion southward and onto new, previously contested lands has permitted new densities to emerge. If populations in the most densely settled areas have slowed in their growth with the rise of age of marriage, the more open economies of towns and marginal lands have continued to grow, playing the role which new agricultural land occupied in previous epochs.

Highland fertility and population growth have not always been agriculturally significant. Even if slow population growth at 2–3 percent could have brought slow cumulative adjustments in the context of agriculture, the more significant and visible short-term shifts in agricultural practice have resulted from population movement rather than growth. In each of the cases presented in part 2, population's effects—either in decline or in rapid growth—resulted from population migration

25. The common argument against a Pax Brittanica as a stimulus for population growth is that precolonial mortality rates from warfare had been exaggerated. Perhaps so, but the argument for the effect of political peace, however imposed, is that it changed fertility practice rather than lowered death rates. See Kjekshus, *Ecology Control*, 18-25.

and localized growth rather than general expansion, that is, migration more than fertility. In the cases of Ankober, Gera, and Ada discussed in part 2, the movement of population into those zones was far more significant than long-term patterns of fertility. Thus, the historical expansion and contraction of imperial political hegemony, the mobility afforded by the annual cropping cycle, and social institutions which favored partible property promoted easy migration to new areas and abandonment of old ones. If the ox-plow complex has been prone to Malthusian scissors, and I would argue that it has been, then the local *movements* of population as much as a long-term growth have been the key forces which have driven the variations of agriculture across the highlands.

Agronomy: Specialized Agriculture

Ox-plow agriculture in its historical development and practice encompassed a broad range of field systems adapted to local, farm-level conditions within micro-agronomic economies. Soil management and water management in the use of irrigation, terracing, broadbed making, and drainage were not anomalies, but part of a range of agronomic strategies historically employed by ox-plow farmers where opportunities existed or conditions demanded. Yet, there is considerable evidence that the techniques such as irrigation and terracing, which nineteenth-century observers reported as practices "in all parts of Abyssinia," have declined over the past century, even in the face of increasing population pressure on resources, which might be expected to stimulate intensification of labor and specialized agriculture.[26]

The most visible example of change in specialized agriculture is what remains of extensive irrigation works at Yeha near Axum. Axumite irrigation existed in close proximity to dryland ox-plow agriculture and reflected both demands from the nearby urban population and South Arabian agronomic traditions. Axumite irrigation did, however, differ from the ubiquitous forms described in the nineteenth-century highlands in that it used permanent features—dams and basins—for water storage in addition to temporary channels and terraces.[27] Nonetheless, Axum's irrigation survived through the sixteenth century. Alvares visited Yeha and reported Yeha's fields "all irrigated with channels of water descending from the highest peaks, artificially made from stone. The crops which they irrigate here are wheat, barley, beans, pulse, peas."[28]

The Axum irrigation system largely disappeared within the period of this study. J. T. Bent in 1895, quoted at the opening of this chapter, exploited the narrative

26. From William Coffin's journal, published in Pearce, *Life and Adventures,* vol. 1, 199-200.

27. Yuri Kobischanov, *Axum* (University Park, 1979), 128-29.

28. Alvares, *Prester John*, vol. 1, 141.

power of ostensible Axumite agricultural decline, using Axumite irrigation and terracing to symbolize his own broader conclusions about the collapse of Ethiopia's civilization:

> A stream runs through this valley, the Mai Veless, and the soil looks extremely fertile; but it is a sad instance of Abyssinian deterioration. Ruined villages are seen in all directions, with the customary church in the middle, almost hidden by its sacred grove, which has turned into jungle. Apparently, at no very distant period, every inch of the valley had been cultivated; now only a few acres at the upper end, where the valley is narrow and irrigation easy is there any cultivation carried on.
>
> All of the surrounding hills have been terraced for cultivation, and present much of the same appearance as the hills in Greece and Asia Minor, which have been neglected for centuries; but no where in Greece or in Asia Minor have I ever seen such an enormous extent of terraced mountains as in this Abyssinian valley. Hundreds of thousands of acres must have been under the most careful cultivation, right up almost to the tops of the mountains, and now nothing is left but the regular lines of the sustaining walls and a few trees dotted about here and there.[29]

Augustus Wylde, who visited the same area three years later was a more sympathetic and insightful observer. He praised the "good water for domestic and irrigation purposes" and attributed evidence of the district's overall decline to recent short-term phenomena: "rinderpest, famine, cholera, and the depredations of the Italian troops."[30] Far from being a sign of systematic civil decline, irrigation around Axum reappeared in the interwar years. In 1989 Axum-born biologist Tewolde Berhan Gebre Egziabher, like Bent, however, saw changes in the area's irrigation as a symbol of long-term environmental degradation, lamenting the decline of the controlled irrigation of his childhood, which he observed had "gone wild in flood and destroyed much of the valley."[31]

The seeming correlation between population increase and decline of specialized agriculture challenges the Boserup equation of agricultural intensification with population pressure.[32] Evidence of historical specialized cultivation also appears in areas of northern Shawa, where extensive lowland irrigation and terracing described in the nineteenth century have disappeared, even while ox-plow agriculture itself has

29. Bent, *Sacred City*, 135.

30. Wylde, *Modern Abyssinia*, 167, 175,

31. Tewolde Berhan Gebre Egziabher, "Land Management and Changes in Land Tenure in Ethiopian History," Asmara University, 1989, 11.

32. For a challenge to Boserup, see Thomas Hakansson, "Social and Political Aspects of Intensive Agriculture in East Africa," *Azania* 24 (1989), 12-20.

spread into the zone. The abandoned terraces, visible at ideal elevations, suggest important changes within the period of study, though their story is absent from oral tradition (see chapter 4).

Irrigation and terracing thus seem to appear within the quiver of ox-plow strategies where opportunity arises and conditions permit, rather than purely as an indication of population pressure and the need to intensify labor to raise production. The evidence of specialized agriculture in the case studies ranges from its use as a part of horticultural, perennial crop systems to its appearance as a part of the ox plow's annual cropping complex. The expansion of the annual cropping system has in some cases led to the decline of specialized forms in lowland Ankober, at Konso, and also at Axum, where permanent structures implied forms of land use and environmental management antithetical to the ox-plow's regime.

The evidence of localized gravity irrigation in Shawa, Wallo, Tigray, Kaffa, and Gojjam in the nineteenth century and into the post-Italian period and its later absence from those locations do not necessarily support an argument for overall decline. Rather, the forms of the small-scale highland irrigation schemes and terracing, which had few permanent structures and relied on annual temporary channel and bund construction, were elusive and based on conjunctural conditions of topography, changing population densities, and local political conditions.[33] Their shifting presence formed part of localized adjustments to conditions of labor, political organization, and population within the annual cropping regime which reflected conjunctural factors rather than progressive, cumulative intensification.

A Barometer of Agricultural Change: The Nineteenth-Century Highland Cropping Map

While seasonal, farm-level decisions are virtually impossible to reconstruct for the past, in the aggregate and over time these decisions appear as changes in the crop repertoire and in the evolution of relative crop mixes. The accumulated effects of these mundane farm-level decisions, however, appear over the *longue durée* as a surrogate measure of trends in production reflecting demography, political economy, and environment. The complex crop repertoire has been one of the ox-plow complex's primary adaptive tools: if the plow itself changed little, the battery of annual cultigens and cultivars provided almost infinite nuances of choice for farmers and reflected both natural conditions (moisture, pests, seasons) and short term, season-by-season farm decisions about labor, potential yield versus risk, household food preferences, storage limitations, and exchange value.

33. See chapter 2; for Konso evidence, see Amborn, "Agricultural Intensification," 71-83. Amborn argues that the permanent terracing structures in that region have declined with the imposition of new labor systems and use of the plow imposed by the Amhara conquest.

Examining the trends in cropping patterns over the past two centuries in Ethiopia may be one of our best methods of understanding change in agrarian economy and society. Without detailed data on labor, fertility, taxation, rainfall, and markets, historians must find other means of assessing the causes and effects of long-term agricultural change. If historians of Ethiopia cannot construct detailed time series data for cropping patterns for either major regions or single localities, historical records (archives, traveler accounts, farm interviews) can provide sufficient evidence to reconstruct general patterns.

Chapter 2 described the highland crop repertoire. This section assesses the nineteenth-century evidence of the distribution of these cultigens and their interaction across the highlands and across time. With the exception of New World crops (potatoes and maize) the mix of cereals and pulses characteristic of the 1800–1900 period was well entrenched according to the early sixteenth-century description of highland agriculture. Alvares mentions wheat, barley, sorghum *(milho zaburro)* chick-peas, lentils, beans, peas, and teff [*tafo*].[34] Three hundred years later European travelers—the first thick description since the Portuguese—described the repertoire in fuller detail. Mid-century travelers Graham, Lefebvre, and Heuglin, among others, explored the richness of cultivars of sorghum (12 varieties), barley (12 varieties), wheat (7 varieties), and teff (3 varieties).[35]

The other major indicator of changes in cropping, the relative balance of crops and their percentage share of total production within the ox-plow complex as a whole, is far less easy to discern. The dominant crops to travelers' eyes and what reached their published narratives depended largely on the regions traversed, their interest in commercial export, and their background in non-European agriculture. Travelers in Tigray, Gonder, and northern Shawa generally lived within the wheat elevation zone, where they tended to draw an exaggerated sense of its importance overall, often reflecting the interests of potential European markets. Observers familiar with India or Sudan readily recognized "juwarry," or *dura* (i.e., sorghum), and its many Ethiopian cultivars. Nathaniel Pearce, among the most acute observers of highland social and political life, habitually described fields of "corn," a nonspecific term which failed to distinguish wheat from barley or even teff; for Italian observers keen to exploit the highlands' potential for wheat as a possible export crop, the vague term *"grano"* more likely meant wheat. The more specific term *"frumento"* came into use only within the twentieth century, when Italian interest in wheat overshadowed reportage of sorghum, millet, or even teff.

34. Alvares, *Prester John*, vol. 1, 190, 251-52. For confusion over millet, maize, and sorghum in Portuguese accounts and their translation, see below and Alvares, *Prester John*, vol. 1, 136 fn. 1. For Lobo, see Lockhart, *Itinerário*, 129.

35. For an excellent survey of crops drawn from travel accounts, see Donald Crummey, "Plow Agriculture," 8-10; Graham, "Report," 269-70 and Pearce, *Life and Adventures*, vol. 2, 344-45.

Given the elevation and topographic range of ox-plow cultivation in the nineteenth century, however, the dominant cereal crop across the highlands was probably barley, with teff second. Sorghum's properties of drought resistance and low labor requirement in weeding and cultivation suit it to the lowlands, especially in western Eritrea, but low nineteenth-century population densities in these zones restricted aggregate totals. Wheat, though cultivated widely, has perhaps the narrowest range of suitable soils and temperatures and lacks the frost tolerance of barley and the resistance to waterlogging of teff. Maize, though introduced by the sixteenth century and known in the nineteenth, has the narrowest range of any crop in its temperature and moisture requirements, and was scarcely reported (see below).

In addition to environmental factors associated with elevation, a key determinant of farmer choice is the relative cost of production of each crop in labor, both human and oxen. Table 3.2 indicates such costs as estimated in a 1968 survey of four midaltitude highland zones which represent conditions similar to those most densely settled in the nineteenth century. The data in Table 3.2 are less important as indicators of monetary value than as a comparative measure of costs in labor, seed, and animal traction relative to yield. The breakdown of activities estimates the distribution of those costs across seasons and gender-based tasks and between environmentally paired crops such as barley and wheat or teff and sorghum, among which farmers made choices for particular farm settings or for individual plots. These data, reflecting primarily the cost of labor and seed, indicate, for example, the relative high cost of seedbed preparation for wheat versus barley. They do not, however, provide a sense of the wider synergies of particular crops within the farming system. Barley, for example, provides subsistence value in both its adaptation to frost

TABLE 3.2. Average cost of production per hectare, 1968.[a]

Crop	Plowing	Sowing	Weeding	Threshing Harvesting/	Total
Teff	$47.09	$17.53	$18.13	$26.37	$119.87
Maize	24.37	10.90	35.28	24.01	102.76
Wheat	38.86	12.63	20.45	25.32	115.32
Barley	27.45	12.05	22.00	16.25	90.27
Chickpeas	33.18	25.48	8.33	23.45	103.95
Millet	55.33	14.17	38.33	33.33	148.56
Sorghum	43.61	11.50	15.65	18.00	102.76
Horsebeans	34.62	13.40	20.72	26.46	104.50

Source: Adapted from Willis G. Eichberger, "A Study of Traditional Farming in Four Areas of the Ethiopian Highlands," unpublished typescript, Institute of Ethiopian Studies, 1968.

[a]Labor costs were calculated at one dollar per day for human and two dollars per day for oxen. The dollar figures are thus useful as relative measures rather than as real production costs.

and its low labor cost. Moreover, barley and wheat compete only at barley's lower altitude range. Pulses have a low labor requirement but poor yields, yet offer nitrogen-fixing qualities, which makes them ideal as a rotational crop.

Teff's qualities as both a subsistence crop and the crop of highest prestige value in the highland cuisine apparently have historically overcome its high cost of production. As a food, teff has represented the height of highland cuisine in both the northern highlands and, by the nineteenth century, in areas like Kaffa, Gera, and Harerge. Teff straw offers the most digestible livestock fodder of any highland cereal, and has also served as the key binding agent in mud-wall housing construction. Teff's small kernels also make it the most storable grain, suffering virtually no weevil damage once threshed and stored. Alvares observed these qualities as well for the highlands' eleusine, which he noted "is highly esteemed because maggots that eat wheat and other vegetables do not eat it, and it keeps a fairly long time."[36]

Evidence from the nineteenth century suggests a remarkable stability in the crop repertoire, and in the conditions under which farmers cultivated across the ox-plow zone. The current bundle of cereal cultigens present in Ethiopia today and in 1900 was also present at the beginning of the nineteenth century.[37] Across the highlands in the nineteenth century, barley and its wide variety of cultivars probably dominated the total food supply, with sorghum and teff in secondary, but important, roles. Barley's historical dominance is reflected in its huge variety of cultivars and "covarieties" (Huffnagel says 170), adapted to virtually every environmental niche within the highland region. Teff showed stability because it mixed strong subsistence characteristics with market appeal as a prestige food crop. Within the data, however, there is some evidence of change. Wheat, especially in Eritrea, probably rose in importance in the twentieth century with urban growth and the strong Italian interest, overcoming its limited ecological range and poor rust resistance. The case studies in chapters 4–7 provide the best evidence of localized adaptations in the cropping system.

Maize in the Nineteenth Century

Maize, rarely mentioned in nineteenth-century accounts, offers an insight into the movement of a crop within the ox-plow regime. In the first half of the period, maize had a foothold in southern systems but only a minor role in total production and food supply. Yet, by the 1990s it has come to occupy the position as the ox-plow system's dominant crop (see chapter 7). Tracing the evidence for its arrival on the scene and early history offers a useful point of contrast between nineteenth- and twentieth-century terms of agricultural production.

36. Alvarez, *Prester John*, vol. 1, 104.

37. Evidence of Italian surveys of 1930s would seem to confirm nineteenth-century evidence of crop hierarchy. See chapter 8.

Maize is a domesticated New World grass which, along with cassava and potato, transformed Old World agriculture because of its range within temperate zones and its high yield to labor. It is the only major New World cereal crop to penetrate the ox-plow repertoire. Portuguese visitors probably introduced it in the sixteenth century, though there is no direct evidence in their accounts regarding when and how. The earliest references to maize in Portuguese accounts were, in fact, nonreferences, that is, mistranslations: *"milho zaburro," "milho"* or *"migli zaburri"* have been translated as "maize," as in the modern Portuguese, but more likely referred to sorghum. Alvares' description of traveling five days through "canas de milho as thick as canes for propping vines" was almost certainly an encounter with sorghum rather than maize or even eleusine. Jerónimo Lobo was more specific, describing a grain, probably sorghum, "coarser than our own fine millet and smaller than Indian corn."[38] Confusion in nineteenth-century accounts, especially from Tigray over references to maize, also stems from European travelers' enjoyment of local inebriates. Travelers confused readers with their references to *mais* or *maize* being served by farmers during their journeys. The allusion, however, was to their transliteration of a nineteenth-century Tigrinya word for honey wine, not the cereal crop *Zea mays*.

The first clear reference to maize as Indian corn in Ethiopian agriculture was Henry Salt's 1805 description of it and its most salient characteristic—its early maturity. He described a valley "well cropped, especially with Indian corn, which is usually more forward in this climate than any other grain." His glossary refers to "E bahr mashella" (the sorghum of the sea) which he carefully glosses as Indian corn. Salt's reference to maize's early maturation, especially over sorghum, suggests in some settings that it had already found a niche in Tigrayan cropping, providing food for the late-summer and preharvest hungry season. Evidence of its presence notwithstanding, it had not effectively penetrated the highlands.

There is little if any evidence for a widely distributed nineteenth-century maize cultivation. Graham mistakenly labeled his list of 12 sorghum varieties under the rubric *Zea mays,* but apparently did not observe maize itself in northern Shawa in his surveys of the 1840s.[39] Combes and Tamisier in 1835–37 noted that "the Galla [Wallo Oromo] harvest wheat, barley, teff, and maize in the valleys," and in 1880 Gustavo Bianchi, reporting on his travel from Tigray to Ankober, noticed that "bahar mascillà, which is our granoturco" is "sometimes found."[40] Antonio Cecchi,

38. Lord Adderly and Beckingham render *canas de milho* as millet. For references to "maize" mistranslations see Alvares, *Prester John*, vol. 1, 69, 88, 104, 136, 190, 391. Lochhart, *Itinerário*, 129.

39. Valentia, *Voyages*, vol. 3, 5; Salt, *A Voyage*, 410; his references to corn are to sorghum or grain in general (274, 278). Graham, "Report," 269-70.

40. E.D. Combes and M. Tamisier, *Voyage en Abyssinie, dans le pays des Gallas, de Choa et d'Ifat*, 4 vols (Paris, 1838), vol. 2, 298. Their reference may in fact have mistaken sorghum for maize. Eastern Wallo Oromo whom I interviewed in 1990 cultivated white sorghum but little maize. Gustavo Bianchi, *Alla terra dei Galla* (Milano, 1882), 36.

our best nineteenth-century observer of southern agriculture described maize in the Gibe region as one of several annual crops, including teff and sorghum. He speculated that merchants had undoubtedly brought it from Abyssinia, that is, the northern highlands. He described, however, the maize as a yellow flint variety only 1.25 meters in height. The presence of only one cultivar probably indicates relatively recent arrival with fairly low yield, that is, "ten rows of small grain."[41] None of the early descriptions of maize in southern cropping systems suggests it had the dominant role it would achieve in the late twentieth century (see chapter 7).

The changing position of maize in the ox-plow repertoire between the nineteenth and twentieth centuries is a visible indicator of a set of more subtle changes described in chapter 7. A major surge of maize cultivation in southern Ethiopia appears to have taken place within the first third of the twentieth century, resulting in maize's relatively strong position in the cropping pattern summarized in Italian surveys in the 1930s, which describe several types, including local hybrids.[42] The rise of maize cultivation in the southwest may well indicate the arrival of new, better-yielding cultivars in the post-1941 period. As chapter 5 argues, it may also reflect the expansion of coffee cultivation and the ideal mix of maize with the coffee labor cycle.

Conclusion

Nineteenth-century evidence of agricultural practice and its demographic context describes a rural economy with few signs of systemic collapse. On the contrary, the ox-plow complex's fullest territorial and hegemonic extension was in its role of consolidating the expansion of the central state in southern, western, and eastern parts of Menilek's empire after 1889. If the ox plow's geographic domain had expanded, its core structure in its technology and its cropping patterns remained remarkably stable from the beginning of the century to its end—patterns which also bore astonishing continuity with our earliest descriptions four centuries earlier.

By the second decade of the twentieth century, however, the elements which would bring fundamental changes in the context, if not the content, of ox-plow farms were already in place. These elements included sustained population growth; political security, which allowed highland farmers to migrate onto marginal lands; a full engagement with horticultural, perennial crop traditions along its southern periphery; and the growth of urban markets as new centers of political power.

41. Cecchi, *Da Zeila*, vol. 2, 382, 278.
42. Conforti, *Impressioni*, 61.

Part II

The Plow and Ethiopian

Historical Landscapes

Part I of this book has presented a description and analysis of the technological, environmental, and social foundations of Ethiopia's highland ox-plow complex. That system in its use of the plow, its historical persistence, and its movement across a vast geographic region has been peculiar, distinctive, and recognizable in its crops, equipment, and methods. Nonetheless, the ox-plow complex's adaptability to new physical environments and social settings has also meant that its specific configuration has varied widely over space and time in a number of microagronomic economies.

Ox-plow farming systems have evolved historically and within diverse environmental niches, which we can at least partly reconstruct and examine. Part II of this book presents three separate regional studies of agriculture in Ethiopia over the 1800-1990 period. Tracing almost two centuries of agricultural change across several historical cases may facilitate description of factors which brought about or inhibited change, that is, the balance among the exigencies of technology, local innovation, diffusion, environmental imperatives, and political climates as determinants in agricultural history.

More often than not, agricultural change is not monocausal but conjunctural, the interaction of all conditions in which social institutions and limits of technology mediate the specific results. The historical cases which follow represent three distinct conjunctures. Although each evolved within the same national political economy and the same technical ox-plow "package," the outcomes have been quite different in terms of the physical effects on the rural landscape and on the fortunes of farmers within it. Thus, chapter 4 describes the Ankober region's agriculture, which formed part of an integrated, prosperous rural economy in the mid-nineteenth century transmogrified into drought-prone sub-subsistence units dependent on international aid by the mid-1980s. Chapter 5 examines the moist southwestern district of Gera, where a diverse cropping system in place by the 1880s shifted first to prosperous forest-based coffee cash cropping in the postwar decades and then slipped to the brink of a subsistence maize monoculture by the mid-1980s. Chapter 6 describes and assesses the transformation of a nineteenth-century pastoral landscape in central Shawa into Addis Ababa's "kitchen": Ada District in the postwar period became and remains Ethiopia's most intensive and specialized example of ox-plow agriculture. Chapter 7 and the Epilogue conclude the book placing ox-plow agriculture within the context of Ethiopia's twentieth-century political economy (1916-90),that is, in the shadow of urban imperatives which have driven state policy toward Ethiopia's agricultural economy and shaped the dissonance in the national political economy which has underlain the economic stagnation and rural unrest of the 1980s and 1990s.

4

From Royal Fields to Marginal Lands: Agriculture in Ankober, Shawa, 1840–1990

Mid-nineteenth-century European travelers to central Ethiopia portrayed the Shawan kingdom they found there as a prosperous, cosmopolitan, culturally diverse, and expanding polity. Their descriptions were not fabrications: Negus Sahla Sellase and his forebears had built a political kingdom upon a strong and growing rural economy which combined a command of Red Sea trade through regional markets such as Aliyu Amba, with a diversified local agricultural economy which fed, clothed, and otherwise sustained the royal body politic. A century and a half later, however, Ankober, which had amply supported the nineteenth-century royal capital, rested on an agricultural economy teetering on the edge of collapse. Once astride the main thoroughfare of commerce to the rich southwest, Ankober by 1985 had become an economic and political backwater tucked away, 40 kilometers off the paved road which leads from the north to Addis Ababa. Though Ankober's rural population remained in place in the wake of the 1984 drought, its rural economy exhibited classic symptoms of highland agricultural crisis: declining agricultural productivity, stagnant rural capital formation, and severe food shortages at the farm and regional levels.

Agriculture in Ankober District took place not only within a specific environmental niche but also within a distinct and evolving political economy. This chapter will describe and examine Ankober over time as a specific historical and geographic setting for ox-plow agriculture in the 1840–1990 period.

Figure 4.1. Ankober c. 1840.

The Setting

I have defined Ankober District in this study as those lands within a day's mule ride of the former royal capital. This area was not a homogeneous agroecological niche, but an environmentally and ethnically diverse unit which encompassed several complementary agricultural systems that have interacted historically. It includes the setting of the town of Ankober itself, a royal capital perched at 2400 meters atop a volcanic cone above the steep escarpment of the eastern highlands, and an expanse of lowlands which stretch below to the east and descend to the market town of Aliyu Amba at 1800 meters and into the Denki River Valley at 1500 meters elevation. Behind Ankober town to the west is a high plateau (2300–2500 meters) which extends southwest to Dabra Berhan and Angolela, auxiliary royal towns, and their hinterlands at the southern edge of Shawa's nineteenth-century suzerainty. The town of Ankober, founded as a frontier outpost in the late-eighteenth century, served as a base from which Negus Sahla Sellase (r. 1813–47) extended his control over the eastern lowlands inhabited by Argobba Muslims and south over Oromo pastoralists, who had conquered the Amhara highlands as pastoral raiders three or four generations earlier, but had been long settled as farmers on the highlands by the opening of the nineteenth century.

The physical setting of agriculture in this district mirrored relations in the 1840 political economy. To the west of the town of Ankober, the Shawan highlands composed the heartland of Amhara settlement and ox-plow agriculture. The key element

Figure 4.2. The royal table: interior of an elite house at Ankober, c. 1840.

of Ankober's economic success, however, lay to the east below the escarpment. By standing in front of his palace atop Ankober's rocky cone and facing east, Sahla Sellase commanded a spectacular vista of his dominions on the rolling hills 1000 meters below, which included key trade routes to the Red Sea as well as smallholder agricultural lands and his own royal farms, which produced teff, wheat, and fruit for royal banquets as well as the cotton cloth which clothed his entourage.

Reversing the historical lens, however, we see that the view from the lowlands below provides a clearer picture of the region's political and agricultural economy. The farmers and weavers of the Denki Valley who, lived a hard six-hour walk below the royal compound, could see the royal palace's outline above them on the horizon even on a cloudy day. While they plowed their scattered sorghum fields, harvested cotton, or tended their irrigated banana and citron gardens, the farmers could easily see the bluish green of the old junipers in Ankober's compounds and must have felt themselves constantly under the royal gaze. The omnipresence of the royal compound above thus served as a daily physical reminder of the state's ability to extract both taxes and the fruits of the fields.

The chronology of the shift of the political center of Shawa and the national political economy away from this district to Entoto and then Finfine (later renamed Addis Ababa) is well documented in the historical literature. Before he claimed the imperial throne in 1889 as Emperor Menilek II, Negus Menilek, Sahla Sellase's grandson, used northern Shawa as a base for an aggressive expansion of Shawan hegemony to the southwest and east. He thus firmly established Shawan dominance over trade

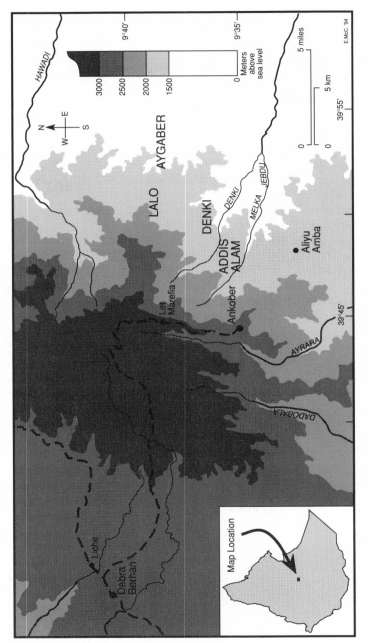

Map 4.1. Ankober

112

from the slave—and ivory-rich southwest and the easterly trade route to Harar. This expansion of scale drove Menilek's decision to relocate his capital more directly astride the new trade networks and Finfine's warm mineral baths. In 1886 he founded a new capital at Addis Ababa, a site which afforded direct oversight of the new political economy based on the extraction of wealth from a vast southern hinterland.

By the turn of the century, after bouts with cholera and attempts at rebuilding by Menilek, Ankober and its environs reverted to what the Frenchman J. G. Vanderheym described in 1894 as a dead town. Ankober's population, estimated by Cecchi in 1878 as 6000, declined dramatically as a result not of disease, but of a fall from political grace. In 1900, Augustus Wylde described Ankober as "the old capital of Shoa, now a place of only second rate importance." In his travel through northern Shawa, Wylde passed within a day's ride of the old capital but did not deign to visit, hurrying on to Addis Ababa instead.[1] Ankober in its twentieth-century role has never merited an asphalted road, a high school, piped water, or a place on the national electrification grid, modern Ethiopia's markers of political stature and economic value.

If the political consequences of this shift of center are well documented, the corollary change which affected Ankober's agricultural base and therefore the experience of the population itself has remained unexamined. By the 1960s the former capital had become a ruin superseded by nearby Gorebela, its former market and now a backwater *warada* (district) town; Aliyu Amba, Shawa's major mid-nineteenth-century entrepôt had declined to a minor food market and part-time center of cotton weaving and contraband trade. Ankober District was connected to the national economy by a poor gravel road crossing a largely eroded and treeless highland plain which linked Gorebela to Dabra Berhan. The diverse nineteenth-century population had changed as well: the Oromo had assimilated or been pushed out and the distinctive Argobba culture had largely disappeared, with its unique language within a decade of extinction. By 1985 three-quarters of Ankober District had fallen victim to famine, lacking both the agricultural productivity and political clout to feed its population.[2]

1. J.G. Vanderheym, *Une expédition avec le négous Ménélik* (Paris, 1896), 59 cited in R. Pankhurst *History of Ethiopian Towns* (Stuttgart, 1985) vol 2, 167. A. Wylde, *Modern Abyssinia,* (London, 1901), 403. Prior to the shift to Entoto Menilek had also occupied Enewari, a site on a rich vertisol plain near his birthplace northwest of Dabra Berhan.

2. For twentieth century decline, see Rosita Forbes, *From Red Sea to Blue Nile: Abyssinian Adventure* (New York, 1925), 131-32. She noted that "No trade comes to Ankober. . . . The people of Ankober are an isolated, self-supporting community dependant on the wood from their hills, the grain and the herds from their slopes, and produce from their gardens." Also see Richard Pankhurst, *History of Ethiopian Towns in Ethiopia from the Mid-Nineteenth Century to 1935* (Stuttgart, 1985), vol. 2, 315-16. For an estimate of drought impact in the district see Guido Gryseels and Samuel Jutzi, *Regenerating farming systems after drought: ILCA's ox/seed project, 1985 results* (Addis Ababa, 1986), 3.

Historical Sources on Agriculture:
Ankober and the European Witness

The mid-nineteenth-century evidence describing Ankober's physical and cultural landscape is remarkably rich, deriving from a resurgence of interest in Africa on the part of both commercial interests in the Red Sea region and nascent European geographic societies, which encouraged the "scientific exploration" of the non-Western world. In the 1840s in particular, the political chaos of northern Ethiopia prompted many to choose the relatively stable Shawan kingdom as a point of entry to the highlands. Travelers to Ankober in the 1840s were an eclectic and eccentric collection of observers who brought personal ambition along with considerable observational and narrative skills to bear on their surroundings. The Frenchman Charles Rochet d'Hericourt visited Ankober twice; the missionary Johan Krapf stayed in residence two and a half years; the Cornwallis Harris expedition of over 30 members remained almost two years in Shawa and included Douglas Graham, a specialist in tropical agriculture and Johan Bernatz, a talented illustrator. Still others passed through Ankober before stepping off on wider journeys to the south and west.[3] Their published narratives provide a lively, "thick" description of the physical environment and accounts of Ankober in the 1840s as a cosmopolitan social and economic center sustained by a rich agricultural base.

Shawa's royal towns in the mid-nineteenth century were cosmopolitan in an Ethiopian and an international sense. Royal patronage brought Muslim merchants from the southwest and Orthodox theologians from around the empire, contributing to a lively economic and intellectual life and demand for luxury goods and a rich diet for entertaining visitors, who included Europeans and members of the Mediterranean mercantile world. Sahla Sellase's two-story palace at Angolela had been built by an Armenian visitor; a Greek mason had constructed a water mill near Ankober. The spurt of published accounts of Ankober's life and economy, however, was short-lived. Foreign travel to Ethiopia after 1855 concentrated on the court of the charismatic Tewodros to the north, leaving few notions of economic trends in Shawa. While the oral tradition of the political life of Shawa and of its personalities is strong, Amharic documentary sources provide no basis for reconstructing the environment or agriculture.[4]

3. For European residents and travellers, see Svein Ege, "Chiefs and Peasants: The sociopolitical structure of the Kingdom of Shoa about 1840." unpublished Hovedoppgave dissertation, University of Bergen, 1978," 14 for list and schedules of 1830s and 40s travellers.

4. Isenberg and Krapf, *Journals,* 56-7; Buxton, "The Shoan Plateau," 163. Amharic sources such as chronicles and local oral tradition allow a reconstruction of events in the political sphere with the rise of Hayla Malekot as successor to Sahla Sellase and the imprisonment of his son Menilek at Tewodros' court. On sources see R. Kofi Darkwah, *Shewa, Menilek and the Ethiopian Empire 1813-1889* (London, 1975), xi-xviii.

A new set of precise narratives on Ankober's rural economy appeared again in January 1877, when Shawa's new ruler, Negus Menilek granted 90 hectares of neglected royal pasture to Società Geografica Italiana at Let Marefia (lit., "a place of rest"). While Menilek told the Italians he intended the concession only as a site *"al vostre repose,"* the marchese Orazio Antinori seized the opportunity to turn the site into a permanent research station and jumping off point for Italian expeditions to the south. Located on a "difficult and tiring" two hours' walk northeast of Ankober at the foot of the district's only primary forest, Let Marefia by 1880 had become not only a rest spot but also the beginning of an agricultural research station. Between 1877 and 1895 (i.e., just prior to the Battle of Adwa), Let Marefia and the Società Geografica provided detailed records of political events, the physical environment, and agricultural conditions at Ankober. The station experimented with new techniques and crops and boasted major increases in yield by its tenant farmers. The descriptions of these European residents (at one point in 1879 some 15 Europeans were in residence in Ankober) compose an important documentary record.[5]

The departure of the royal capital in 1886 (the time of Addis Ababa's foundation) and Let Marefia's abandonment in 1895 (prior to the Battle of Adwa) cast a shadow over Ankober for the late nineteenth century and most of the twentieth century; only a few passing travelers mention it and Ethiopian documents had long shifted their attentions to Menilek's imperial court at Addis Ababa, four days' ride to the south. The modernizing state of the twentieth century created a new bureaucracy, and it focused its attention on the national economy's mercantile base rather than on agriculture.

Ankober's cloud of historical anonymity lifted again in the early 1980s. The foundation of a new agricultural research station on the highlands at nearby Dabra Berhan in 1980, two pieces of my own fieldwork, and a drought relief project in 1985–87 have provided a series of snapshots for the period after 1960. My 1985–86 fieldwork in the Ankober lowlands, especially the Denki River valley, provided data on household development-cycle histories, cropping patterns, migration/settlement trends, land use, and marriage-property practice reaching one or two generations back into the twentieth century.[6] These data and the opportunity to work with farmers in the physical setting under the shadow of the old royal capital have offered an important complement to nineteenth-century narratives.

For those Europeans who arrived at Ankober via the direct route from the Red Sea across the Danakil depression, and the long hot journey through the Awash Valley, the first view drew an alpine metaphor in terms of their limited geographic vocabulary. A gaze upward to Ankober from the desert floor afforded,

5. Traversi, *Let Marefià,* 54, 59; Gustavo Bianchi *Alla Terra dei Galla* (Milan, 1884), 188.

6. For project details, see Gryseels and Jutzi, *Regenerating farming systems,* 2-12 and James McCann, "Preliminary Report on Oxfam/ILCA Oxen Seed Project," unpublished report to Oxfam America, July 1985.

. . . a magnificent view of the Abyssinian Alps . . . . Hill rose above hill, clothed in the most luxuriant and vigorous vegetation; mountain towered above mountain; and the hill-clad peaks of the most remote range stretched far into the cold blue sky. Villages' dark groves of evergreens and rich fields of every hue, chequered the broad valley; and the setting sun shot a last stream of golden light over the mingled beauties of wild woodland scenery and the labours of the Christian husbandman.[7]

Rochet d'Hericourt called the view of the highlands "comparable to the most beautiful sights of Switzerland."[8] Dr. Charles Johnston, who arrived in 1842, however, decried his predecessors' analogy of the alps as "ridiculous to name so a succession of low, denuded hills," but nonetheless admitted his relief at reaching a semblance of a physical and cultural world similar to his own: "I looked upon the lovely scene so long, and felt so strongly my return to civilized life, that, like a worthy friend of mine relating to me his feelings on reaching the self-same spot, I could have found relief in a good flow of tears, so sincere was my joy."[9]

Johnston's subjective valuation of the royal capital's cool setting would have contrasted significantly with the interpretation by lowland Argobba farmers and weavers standing on the same spot. For them, the "dark groves of evergreens" were the home of tax collectors, a punitive soldiery, and demands for farm labor. Changing relations between highland and lowland agricultural economies would, in fact, be a major theme of Ankober's economic base.

Arenas of Change and Continuity

Human Settlement and Demography

Over the past century and a half Ankober's movement from royal granary to famine-prone periphery has involved marked shifts in the relationship between the physical environment and human activity, but also surprising elements of continuity. Embedded among the emotional narrative accounts of first glimpses of Ankober are indications of demography and land management. The physical and human landscape which greeted the English traveler Douglas Graham in 1840 would not be unfamiliar to a first-time visitor in 1989:

. . . passing in an instant from the burning plains of the Adaiel [Adal] to a rich landscape in which flocks and towns and villages abound, the strange sight is

7. Harris, *Highlands,* vol. 1, 314-15.

8. Rochet d'Hericourt, *Voyage,* 115, 133.

9. Charles Johnston, *Travels in Southern Abyssinia, through the Country of the Adal to the Kingdom of Shoa during the Years 1842-43,* 2 vols. (London, 1844), vol. 1, 469, 471.

afforded of regularly marked fields mounting in terraces from the very base of the Abyssinia mountains throughout a steep ascent of five thousand feet which leads the traveller to an unlimited tableland where the eye is perfectly satiated with the endless succession of waving crops and rich green meadows.[10]

The presence of terracing as a dominant element of Ankober agriculture in 1840 indicates a highland population density sufficient to generate the need for new arable land and sufficient labor to build and maintain terraces. At the same time, as Graham noted, cultivation had not yet impinged on pasture ("rich green meadows") which sustained livestock. There are, however, no demographic statistics for this critical feature of agrarian history, and the nature and pace of population growth must be surmised from comparative historical evidence of density. In the early nineteenth century the population of Shawa as a whole appears to have been in balance among cultivated land, pasture, and infertile or inaccessible land. In 1840 Graham observed:

> From a careful observation during many journeys in every direction, I have calculated that one-fifth of the whole surface of Shoa may be fairly considered to be under cultivation, whilst two-fifths are preserved as good meadow land and the remaining two-fifths may be stated to be very indifferent soil, forest, or impractical rock. . . private property in land is everywhere sanctioned, allowed, and established [i.e., smallholder farms]; there are few forests or wastes, excepting the impractical for pasture or cultivation.[11]

Graham's estimate, although not empirically based, nonetheless leaves a clear impression of only modest population pressure on the land, but it does not suggest the low density of population evident in other historical highland settings.[12]

Like most demographic change which has affected agriculture in highland Ethiopian history, population expansion in the Ankober region was much more a result of movement than fertility. In 1840 Sahla Sellase was the leader of an active expansion of domination over existing populations but he also attracted migrants from other highland zones, especially the Gonder region, who brought with them ideas about land tenure, agriculture, and state-peasant relations. By opening new lands, Sahle Sellase held the power to grant land use rights (*gasha maret*, lit., "land of the shield") to local residents and his soldiers and to exercise substantial control over distribution of income rights to local elite *(malkanya)*.

10. Graham, "Report," 256.

11. Graham, "Report," 259; 254.

12. An excellent account of the effects of low population density on land tenure for Walqayt is Giovanni Ellero, Il Uolcait," *Rassegna di Studi Etiopici*, 6-7 (1948), 108-09; also see Dan Bauer, "Land, Leadership and Legitimacy among the Inderta Tigray of Ethiopia," Ph.D. dissertation, University of Rochester, 1973, 218-20.

Evidence of demographic growth based on in-migration reaches back to the mid-nineteenth century, when land tenure within Ankober District fell within the broad category of areas conquered (or reclaimed) by the expanding Amhara polity, which gave the king and his representative *(meslene)* the right to allocate agricultural land to new migrants or cooperating local tenants. Graham described mid-nineteenth-century land tenure arrangements, which foreshadowed other areas that served to feed Addis Ababa's population in the twentieth century:

> The ground belongs partly to the king, partly to the temporary resident governors in the district *[meslene]* and partly to the inhabitants themselves. Where there is no previous right existing, a field can be purchased by a private individual on payment to the government of a regulated present of honey, cloth, or pieces of salt and subject to the annual taxation of produce; but all the more favored parts of the country already appertain to his majesty, whether in pasture or arable land, and the royal magazines for grain and farm produce are profusely studded over every portion of the kingdom.[13]

As a royal town, Ankober had special access to nearby resources, including food, produced on surrounding land called *madebet*. In essence, *madebet* (lit., "kitchen") represented income rights in labor and food, which accrued not to loyal supporters but directly to the imperial court. Royal prerogatives over income, particularly during times of military expansion, superseded existing elite income rights. Graham and Harris in 1840 described this form of tenure when they noted that Sahle Sellase held land as royal farms managed by his representatives, protected royal forests, and strategically located royal pasture to supply the prodigious meat consumption of his palace. Royal granaries stored the accumulated cereals from the royal *hudad* (farm) and from local tax collections.

At key points around the district, the king maintained royal farms and granaries from which he provisioned his army and guests. Local farmers owed labor and use of their oxen in the cultivation, harvest, and threshing of the royal crops.[14] The departure of the imperial court for Addis Ababa meant a devolution of king's land into individual holdings.[15] Over time, however, highland Christian cultivators, many of whose ancestors had migrated to Shawa from the Gonder region, asserted their residence as claims of hereditary land *(rest)* which resembled the rights *(atsme rest)* enjoyed by core cultural areas farther west, such as Tagulet, Manz, and Gonder. The imperial frontier expansion gradually reduced the flow of emigrants from the district and allowed a gradual population increase under the political stability of the

13. Graham, "Report," 260-61.

14. Graham, "Report," 260-61. Harris, *Highlands,* vol. 3, 221-23. Borelli reports the same labor obligations owed to Menilek almost fifty years later, Borelli, *Éthiopie Méridionale,* 94.

15. The most obvious case was the royal pasture at Let Marefia which by 1877 had been occupied and cultivated by local farmers. Traversi, *Let Marefià,* 44-46.

post-1941 period. By the 1960s highland areas of Ankober district exhibited land-tenure customary law which strongly resembled core *atsme rest* (hereditary land claim) areas. Lowland areas, however, retained the characteristics of *gasha maret;* that is, the distribution was controlled by local landlords.[16]

We know more about Ankober's royal kitchen *(madebet)* land from the early years of Menilek's reign. Provisioning of the royal court and town of Ankober was the specific responsibility of a certain Walda Giyorgis who, according to Guglielmo Massaja, in 1868 "had custody of the royal granaries and the stores of comestibles and provided all the necessities of the king's table, of the court, and of those maintained by the government." The stores and provisions were drawn from the entire range of ecological zones to provide the variety of food appropriate to a royal table.[17]

Massaja's account also informs us about the temporary nature of *madebet* tenure and the effects of the capital's movement south. When Menilek decided to move his capital from Ankober to the more defensible site at Enewari, *madebet* lands and rights around the Ankober area reverted to the local aristocracy.[18] Over the next generation or so, the absence of nearby royal authority permitted those lands that were occupied by soldier-settlers as *gasha maret* to evolve under local customary practice into land rights that they conceived of as ambilineal property rights *(atsme rest)*, a concept imported from Amhara heartlands at Manz.

Natural Resource Use: Forest and Pasture

Historical evidence of resource use and settlement in Ankober also challenges a once widespread assumption about land use in the past. Forests and wood fuel have recently been used as a measure of historical population pressure and land degradation; many influential development studies allege a dramatic reduction in northern Ethiopia's forest cover over the course of the twentieth century. Indeed historians and geographers have often argued, or assumed, that the movement of the royal capital from Ankober was driven by depletion of its fuel supplies.[19] The treeless plain over

16. For distribution of *gasha maret*, see Ege, "Chiefs and Peasants," 60-63. Wolfgang Weissleder who did field work in Ankober in the early 1960s notes the presence of rest in the highlands but tenancy in the lowlands. Weissleder, "Political Ecology," 172-74. In interviews, elder Argobba often referred to their pre-1975 land as rest, though the prevalence of tenancy in the area suggests a loose use of that term. For Gonder influence, migrations, and role of gasha maret, see Volker Stitz, "The Amhara Resettlement of Northern Shoa During the 18th and 19th Centuries," *Rural Africana*, 11 (1970)," 78-79, who cites Asme Giyorgis' history "Ya Galla Tarik."

17. Isenberg and Krapf, *Journals*, 276-77; W. Cornwallis Harris, *The Ethiopian Highlands*, (London, 1844), vol. 3, 222-23; Graham, "Report," 260-61; Guglielmo Massaja, *I miei trentecinque anni di misione nell'alta Etiopia*, vol. 15, 204-5.

18. Massaja, *I miei*, vol. 10, 9.

Figure 4.3. Nineteenth-century deforestation, view between Entoto and Ankober, 1888.

which the modern road from Dabra Berhan to Ankober passes has reinforced the conclusion that Ankober's settlement denuded the hillsides over the course of less than a century of intensive occupation. The nineteenth-century evidence, however, indicates that the condition is not new; Ankober and much of northeastern Shawa has been devoid of trees and wood fuel for at least 150 years and perhaps longer.

If the authors of the early narratives of Ankober were enraptured by its idyllic beauty above the lowland wastes, they were nevertheless shocked and often appalled at the lack of forest and wood fuel on the highlands behind the royal capital. In 1838 Rochet d'Hericourt noted that forests in Shawa were "quite rare" and Harris referred to areas west of Ankober as 'long naked sweeping plains. . . destitute of wood." Traveling for several days in Oromo country on Shawa's southern frontier he recalled that "not a single bush or tree was visible during the long ride." Johnston, writing in the 1840s described the area as a "bleak, moor-like scene."[20]

At the turn of the century, Augustus Wylde looked south from the edge of the Adabai River gorge and reported no trees at all on the vertisol plain a day's march

19. See for example W.E.M. Logan, *An Introduction to the Forests of Central and Southern Ethiopia* (Oxford, 1964). For a further refutation of the standard arguments about declining forest cover in Ethiopia, see chapter one; also Christopher Clapham, *Transformation and Continuity in Revolutionary Ethiopia* (Cambridge, 1988), xi-xii. For moving capitals argument, see Ronald Horvath, "The wandering capitals of Ethiopia," *Journal of African History,* 10 (1969), 205-19, who cites Portuguese accounts as major sources for this argument.

20. C. Rochet d'Hericourt, *Voyage sur la côte orientale de la mer Rouge dan le pays d'Adel et le royaume de Choa.* Paris, 1841, 219; Harris, *Highlands,* vol. 2 46-47, 150-51;

Figure 4.4. Faggot sellers, northern Shawa, 1880s.

west of Ankober District. Jules Borelli's late-nineteenth-century engraving (see figure 4.3) based on a photograph shows the area between Ankober and Entoto (four days' ride southwest of Ankober) as a barren plain. The blue-green Entoto forest, familiar to visitors to Addis Ababa in the 1970s, consisted of transplanted twentieth-century eucalyptus. Like their predecessors in the 1840s, travelers of the late-nineteenth century also noted the lack of trees on the high plateau as well as the dearth of forage for their mules. In 1925 Rosita Forbes confirmed these observations, describing an absolute lack of wood fuel on her route between Ankober and Addis Ababa. She noted, as had Wylde, that cattle dung was the dominant household fuel. Eighty years earlier Graham had already complained about the stench around Shawan houses deriving from the farmers' habit of stacking dried dung fuel close by their hearths.[21]

The town of Ankober and the royal compound managed their own fuel supply and domestic forest. Ankober's domestic and church compounds protected juniper (*tid*) as shade trees and non-fruitbearing banana plants, whose leaves served as

Charles Johnston, *Travels in Southern Abyssinia through the Country of the Adol to the Kindom of Shoa during the years 1842-43* (London, 1844), vol. 2, 63-64.

21. Graham, "Report," 289; Forbes *Red Sea,* 138. The lowland environment around and below Aliu Amba has never supported extensive forest cover, except occasional acacia, *Euphorbia Arborea,* and tamarind. Borelli, *Éthiopie méridionale: journal de mon voyage aux pays Amhara, Oromo et Sidama septembre 1885 á novembre 1888* (Paris, 1890), 98-101; Bianchi, 158. Bianchi had to request forage for his pack animals directly from Menilek.

Figure 4.5. Rural Shawan house, 1888 (note dung stored at left).

wrapping for bread.[22] The king also maintained a group of 300 slaves who were specialized as cutters of wood and who supplied the royal palaces at Ankober, Dabra Berhan, and Angolela with juniper and wild olive logs cut from royal forests at the heads of the steep river valleys below the escarpment. The king allowed each slave three days to gather and deliver each load, an indication of the remoteness of the royal forests even at the apogee of Ankober's power. When Tewodros' troops occupied Ankober in 1859–60, such environmental management mechanisms had apparently broken down, since his occupying troops cut down about half of the royal town's trees.[23]

Ankober's wood resources were never its own *Juniperus* shade trees, but forested niches in surrounding zones. The most extensive of the escarpment forests was the Feqre Gemb, a primary forest located north of the town of Ankober, adjacent to the Let Marefia royal pasture, and still visible to the left below the Dabra Berhan-Ankober road. In 1879 Antinori reported over 20 species of trees, 22 species of shrub, and abundant game. The Feqre Gemb is a superb example of primary forest,

22. John Martin Bernatz, *Scenes in Ethiopia,* 2 vols. (Munich-London, 1852), vol. 2, Plate VI (see illustration) indicated that Ankober households planted banana trees for decoration and to use their leaves for baking and storing bread. Banana trees only bore fruit in the lower altitudes like Denki.

23. Massaja, *I miei,* XI, 39. Most of the town was destroyed, with only a single church left standing.

Figure 4.6. Non-farm income: male Argobba weaver, c. 1840 (note banana trees).

but its setting suggests it was not a remnant of any forest cover which existed within the last two centuries. That it has survived is attributable to its inaccessibility in a steep valley and, at least in the nineteenth century, its protection by royal decree and the presence of a nearby monastery.[24]

Seasonal forage and pastureland were as critical to the domestic farm economy as land in general, since they sustained the livestock—especially oxen—which formed the economic core of plow cultivation. Crop-livestock interaction, in fact, has come to serve agricultural economists as an index of intensification in smallholder agriculture. Pressure on cultivable land from population increase in turn creates declining pasture and the need to develop on-farm forage resources to sustain livestock through seasonal forage shortages and prolonged drought. Indicators of pressure on forage resources include increasing practice of stubble grazing and stall feeding of crop residues as well as privatization of pastureland and local markets for crop residues.[25] By 1840 in Ankober, the concentration of land use appears to have created the need for private pasture holdings as part of the *madebet* system. Shawa's political economy did not provide for a commercial market for forage, nor for grain, but did have means to distribute scarce goods. Graham noted the presence of royal pasture as a part of the royal land prerogatives exercised by Sahle Sellase. Krapf lists 17 sites of royal pasture in the region where livestock accumulated from tribute and southern raids was under the care of *"abellam,"* professional herdsmen.[26]

Like *madebet* (kitchen) lands, rights in forage land devolved to public ownership with the movement of the royal capital and pressure on natural resources. Evidence from the last quarter of the nineteenth century supports the presence of a decline of pressure on land resources well before the 1889 rinderpest epizootic and famine. The 1877 Let Marefia land grant from Menilek to the Società Geografica

24. Orazio Antinori, "Lettera del M.O. Antinori a S.E. il comm. Correnti Presidente dell Società," *Bollettino dell Società Geografica Italiana,* 16 (1879), 403-4.

25. For crop-livestock integration, see J. McIntire, D. Bourzat, and P. Pingali, *Crop Livestock Interactions in Sub-Saharan Africa.* Washington D.C., 1990.

26. Isenberg and Krapf, *Journals,* 256.

Figure 4.7. Elite Shawan women, c. 1840 (Amhara in foreground, Argobba in background).

Italiana consisted of 90 hectares of prime, well-watered bottom lands only two hours' walk northeast of the royal capital. Prior to its occupation by the Italians the plot had been a royal pasture and irrigation scheme which Menilek had abandoned after his 1878 departure to Enewari and then Entoto. After the land's de facto reversion to the public domain, Leopoldo Traversi observed that it had been cultivated illicitly by some of the king's local servants. Though cropped, it had yielded only about four or five *dawla* of grain prior to its establishment as a research site, indicating quite unintensive use.[27]

Reasons for the lack of intensity in natural resource use in the last quarter of the nineteenth century are not immediately clear in purely demographic terms. Much of Ankober's highland population derived from a major in-migration from Ethiopia's central highlands, especially Gonder, rather than from changes in local fertility. By the 1840s much of this migration was already in place. In contrast, by the 1870s, Sahla Sellase's and Menilek's expansion policies had already provided a new settlement frontier on attractive land farther south, which drained populations from areas of older settlement.[28]

27. Traversi *Let Marefià,* 71-72. In its first season of intensive cultivation under the direction of the Italians but with local labor and implements, the area yielded 200 dawla (1-1.5 metric tons).

28. Area included northern Shawa region, i.e. area under Menilek's direct control. Italian observations, though certainly guesswork, held that less than 10% of Shawa was under cul-

Agriculture:
Tending Royal Fields

Cropping

Crops can be a bellwether of agricultural change since farmers' crop choices reflect cumulative responses by individual farmers to natural conditions (moisture, pests, seasons) and short-term, season-by-season farm decisions about labor, potential yield versus risk, and markets. Choices of major crops drawn from evidence of Ankober's history indicate the importance of long-term trends of economic, demographic, and environmental change which might otherwise be invisible in the historical record of events.

Nineteenth-century narratives provide a fairly detailed synchronic description of the 1840s and 1880s, benchmarks from which longer-term patterns emerge. The first comprehensive account of crops in Ankober District belongs to Douglas Graham, who identified 25 different cultigens from Shawa, including 12 cultivars of sorghum and 3 each of wheat and barley. The Swiss landscape artist Johan Bernatz claimed an even greater variety: "twenty-eight sorts of morschela [sorghum], twenty-four of wheat, sixteen of barley, two of spelt, four of teff (Poa Abyssinica), two of oats, two of rye, and an abundance of peas, beans, and pepper."[29]

Graham's elaborate descriptions of cotton and sorghum cultivation within his 1844 report indicate that he was familiar with lowland systems as well. The use of 12 cultivars of sorghum strongly indicates a sophisticated cropping system in lowland areas, and the use of two sorghum cultivars exclusively for beer making indicates the involvement of non-Muslim farmers in areas in the 1500–1800 meter elevation range. Graham was particularly impressed with lowland cotton, which he found "more healthy and luxuriant than any of the species I [he] ever saw in India." Lowland cotton was a perennial crop yielding two harvests over a period of three to seven years.[30]

A larger body of evidence on Ankober district cropping derives from the Società Geografica Italiana records from the period 1877–95 when Let Marefia served as a fledgling agricultural research station. In a July 1881 letter to his family, Pietro Antonelli noted the principal crops and average prices at Ankober District's three major markets (see table 4.1), and in 1879, station founder Marchese Orazio Anti-

tivation, half of Graham's 1840 estimate. See Pietro Antonelli, *Rapporti sullo Scioa del Conte Pietro Antonelli al R. Ministero degli Affari Esteri (dal 22 Maggio 1883 al 19 giungno 1888)*. (Rome, 1889); and Graham, "Report," 259.

29. Graham, "Report," 263. Though he claims to describe all of Shoa, Graham drew on Ankober for the majority of his experience. Bernatz, *Scenes,* Plate 14. See also Harris, *Highlands,* vol. 2, 395-409.

30. In the final two years, local cultivators pruned back the branches and intercropped it with wheat or other cereal. Graham, "Report," 272-73.

nori estimated the relative yield per unit of seed of major crops at Let Marefia (see table 4.2). Together, tables 4.1 and 4.2 indicate types of grain and pulse as well as the relative value of each in the district. Antinori also collected and published a comprehensive description of the Shawan diet, which demonstrates a remarkable stability of food types and commensural custom.[31]

Though observers' gazes invariably favored the "salubrious" highlands, the lowland river valleys below Ankober were occupied and cultivated, though less intensively than the town of Ankober immediate environs, as Graham noted during his first day's descent from Ankober east to the Awash:

> The cliffs were flat and naked and the prevailing scrub was a small species of acacia. . . but wherever the plow could be held, there the hand of industry had been busy, and little waste or uncultivated land was to be seen. The valleys were thickly populated, villages peeped out in every direction, and invariably, as in all other parts of the country the land belonged to the king.[32]

Plow agriculture, however, reached its limit below 1500 meters elevation. On the second day's march out of Ankober, Graham described "a wild country, thinly populated." In the 1920s Rosita Forbes, traveling in the opposite direction, noted the same settlement pattern, though by 1925 the buildings at Aliyu Amba's market sported a few tin roofs.[33]

While these lowland zones below 1800 meters held the trade routes critical to Shawa's prosperity, highlanders historically feared these zones and preferred to keep them in the hands of local peoples, such as the Argobba, rather than extend highland cultivation there. The king himself reflected the dominant highland attitude when he argued to Harris that: "the water of the Kwálla [*qolla*] is putrid, and the air hot and unwholesome. Noxious vapors arise during the night and the people die from fever. We fear their [the lowlanders'] sultry climate and their dense forests,

---

31. To this list should be added other non-cereal crops reported by Antinori, Cecchi and Chiarini, and Traversi. These include lowland fruits such as banana, ensete, citron, limes, and grapes as well as oil seeds flax (talba), castor bean (gulo), niger seed (nug), and safflower (suf). See A. Cecchi and G. Chiarini, "Relazione alla Presidenza della Società geografica," *Bollettino delle Società Geografica Italiana*, 16 (1879), 396, 400; Orazio Antinori, "Sul vitto e sul modo di aggiustare i cibi presso il popolo di Scioa," *Bollettino delle Società Geografica Italiana*, 16 (1879), 388-403.

32. Graham "Report," 54-56.

33. Forbes, *Red Sea,* 129; Graham, "Report," 56; Cecchi and Chiarini in 1879 described the lowlands as poor, dry, barren, and arid with no trees save the Euphorbia Arborea and Digitata and cultivating only sorghum and cotton. Cecchi and Chiarini, "Relazione,"417.

34. Harris, *Highlands,* vol. 2, 226-27.

TABLE 4.1. Principal grains from markets at Liche, Bolloworkie and Aliyu Amba[1]

Amharic Crop Name[a]	English Crop Name	Price[b]	Amount[c]
Sindie	wheat	1 MT	10–12 kunna
Gheps	barley	1 MT	18–22 kunna
Tief	teff	1 MT	18–22 kunna
Mascillà	white sorghum	1 MT	18–22 kunna
Zengadà	red sorghum.	1 MT	20–24 kunna
Atàr	field peas	1 MT	15–20 kunna
Bakelà	horsebeans	1 MT	18–25 kunna
Messer	lentils	1 MT	12–15 kunna
Talbà	flax	1 MT	10–12 kunna

Source: Adapted from Pietro Antonelli, "Lettera del Conte Pietro Antonelli alla sua familglia," Bollettino dell Società Geografica Italiana, 19 (1882), 80.

[a]Amharic spellings are Antonelli's.
[b]MT= Maria Teresa talers.
[c]Kunna = 5.31 litres.

TABLE 4.2. Antinori's estimates of yield per unit of seed in Ankober District, c. 1880.

Type of seed	Yield per unit
Teff	30–32
Wheat	30–32
Horse Beans	8–10
Lentils	14
Barley	20
Field Peas	5
Sorghum (white and red)	40–50

Source: Traversi, Let Marefià, 63.

Note: The units of yield were not specified and should be considered in relative terms.

and their mode of warfare."[34] Travelers often adopted this prejudice. In 1838 the Frenchmen Combes and Tamisier commented that the lowland zones were "little inhabited, [and] ravaged by terrible diseases."[35]

The river valleys below Ankober and Aliyu Amba, though not densely populated, in fact supported a sophisticated and specialized economy substantially different from highland ox-plow traditions. Though Shawan political expansion in the late-eighteenth and early nineteenth centuries controlled lowland markets and created local land grants to soldiers *(gasha maret)* administered through state-appointed elite *(malkanya)*, it did little initially to change economic and demographic patterns. Rural Argobba practiced an apparently lively mixed economy of irrigated perennial crops, ox-plow rain-fed agriculture, and cotton weaving (men were weavers and women spinners) producing the renowned *shamma* cloth of the Aliyu Amba market. Denki fields produced a distinctive variety of triennial cotton and long-maturing white sorghum. In 1841 Graham described there an effective semi-arid agricultural system in which valleys were "thickly peopled" and "villages peeped out in every direction," and Combes and Tamisier had observed that "at one half-day southeast of Ankober, the environs of the village of Denki, arising by a river of the same name, are covered with fruit trees. . . " furnishing the highlanders with "a great quantity of oranges, citrons, bananas, and sugar cane." Massaja in 1868 arrived at the customs post at Denki, where he received bananas, sugar cane, and limes from Sahla Sellase's resident governor. Ten years later Antonio Cecchi and Giovanni Chiarini remarked on the lowland bananas, limes, citrons, and sugar cane produced in irrigated and fertilized gardens along lowland water courses.[36]

Significant for its absence among the crops described by Graham and Bernatz is maize, which by the early nineteenth century was well known in Tigray and Wallo, and already dominant in much of Harerge. Of the early observers only Harris mentions maize *(Zea mays)*, "principally eaten when fresh and milky," that is, as a garden crop eaten in the preharvest hungry season.[37] The evidence of surveys of Shawa by Italian observers suggests that maize production was no more widespread by the end of the century. Antinori in 1879 noted *mar mascillà,* (lit.,"honey sorghum"), probably a reference to a local maize cultivar, as a minor lowland crop

35. E. Combes and M. Tamasier, *Voyage en Abyssinie, dan le pays des Galla, de Choa et d'Ifat.* 4 vols. (Paris, 1838), vol. 3, 20.

36. Graham, "Report," 54-56; Combes and Tamasier, *Voyage,* vol. 3, 20. Cecchi and Chiarini, "Relazione," 415-17. Also see Orazio Antinori "Relazione del M.O. Antinori," *Bollettino delle Società Geografica Italiana,* 16 (1879), 396; Massaja, *I miei,* vol. 8, 203, vol. 10, 113-115.

37. Graham, "Report," 275-76. Harris who claimed two varieties of maize at the tail end of his list of crops elsewhere mistakenly glosses Mashila (i.e. sorghum) as maize. Harris, *Highlands,* vol. 2, 313, 396.

harvested twice a year, once in May and again in November, but he did not note is as a part of the local cuisine. A decade and a half later Traversi mentioned crops which did well in the middle altitude *(wayna daga)* environment, including numerous varieties of barley and wheat, red and white teff, and "poi fave, piselli, peperoni, lino, granoturco [maize] ecc." Maize's minor role may have resulted from the poor quality of the local variety which Traversi called weak and emaciated. He noted, however, that he had introduced improved maize seeds ("seme di granoturco nostrano") with "perfectly straight ears." The only other regional reference in this period is that of Augustus Wylde, who noticed maize growing above the level of *dura* cultivation on the slopes of the Adabai Valley, two days' journey from Ankober.[38]

This evidence suggests that maize was known, but that, under the prevailing conditions of the mid-nineteenth century, farmers found it a poor choice as a staple crop. Thus, the later widespread adaptation of maize as a major subsistence crop suggests a substantial change in agricultural conditions, judgments about crop choice made by farmers themselves, and postwar availability of improved maize cultivars.

Agronomy

The Ankober evidence suggests not only significant continuity in the agronomy of local agriculture, but also some important changes in the management of soil and cultivation indicating broader patterns of economic change. At its most basic level, Ankober's crop and soil management reflected the farmers' attempts to merge their own labors in preparing the earth with the inexorable and often inscrutable movements of the natural world. Within the farmers' repertoire of soil preparation techniques, use of crop rotation, and judgements about key "signs" from nature was the accumulated experience with particular tools, soil types, and climatic cycles. Farmers in 1841, as in 1990, normatively expected the *meher* rains to begin on St. Mikael's Day (Sene Mikael, i.e., 19 June) but waited for clear signs of soil moisture before sowing. Despite the overall persistence of agronomic patterns in the ox-plow complex and in Ankober in particular, broader changes took place in response to changes in the political economy and to the movement of population to new natural settings.

38. Traversi, *Let Marefià,* 75. These crops are horse beans, field peas, peppers, flax; granoturco can be translated as either wheat or maize, though the more commonly the latter. Given its place in the list it would appear to have been a quite marginal crop with few cultivars available. Antinoris' reference to "mar mascilla (Zea Mais)" was later amended by Cecchi to be "bahar macillà" a clear reference to maize. It is the only reference to maize in any of his agricultural notebooks. Antinori, "Relazione," 394-96; A. Cecchi, *Da Zeila alle frontiere del Caffa* (Rome, 1886), vol. 1, 452; Wylde, *Modern Abyssinia,* 399.

The technology of soil preparation has probably been the most agronomically conservative part of the smallholder production systems in Ankober. The single-tine scratch plow based on oxen power and supplemented by the hand hoe, sickle, and chopping tools has remained the technological basis for all soil preparation throughout the period. The iron plow tip, the only piece of equipment not always produced locally, was already present in 1840 and has not changed in form over the 1840–1990 period. The "quaint and primitive" plow only "slightly armed with a tiny bit of iron" was exactly the same tool in materials and engineering in 1840 as that of 1990. In the early nineteenth century a good plowshare cost 15–20 pieces of salt and had to be resmithed once or twice a year. Plowshares made from iron smelted at the king's ironworks were replaced in the twentieth century by imported steel from Eritrea or salvaged from automobile leaf springs.[39]

If the technology of the plow was historically stable, the availability of rural capital in the form of plow oxen was subject to uneven distribution by social class and local environmental variation. On the one hand, nineteenth-century accounts of vast royal pasture on the highlands suggest an abundance of forage for oxen. On the other hand, the need to control the commons—in this case the royal court's need to monopolize prime all-season pasture close to the capital—denotes a scarce good. For others, oxen and other livestock had to be moved seasonally to available pastureland. The lowlands' open pastureland in the dry season, for example, offered primarily desiccated grasses, which had little value for livestock. Overall, observers of the mid-nineteenth century did not describe pasture as a major constraint to Ankober's agriculture. Harris' description of a royal farm, for example, indicates that local farmers used their own pairs of oxen as a part of their corvée obligations.[40]

In recent generations the loss of pastureland to cultivation on the highlands, epizootic disease such as the 1889 rinderpest, and endemic disease and intestinal parasite problems in the lowlands radically reduced the ratio of livestock to population. By the 1980s, levels of oxen ownership among smallholder farms meant that over 75 percent of rural households in the Ankober area—as across most of Ethiopia's highlands—were not self-sufficient in traction animals (see table 4.3). Over time, the oxen rental and exchange conventions which once had been a strategy characteristic of the early household development cycle grew to be chronic conditions symptomatic of reduced per-capita resources. Reduction of total oxen labor meant late seedbed preparation, and reduction of labor for weeding one's own plots, and thus it determined crop decisions.

39. Pearce, *Life and Adventures,* Vol. 2, 202. Graham describes the plow as "extremely rude" while Michael Goe, a livestock scientist writing in 1991 begins from the premise that it is ingeniously simple and versatile. Graham, "Report," 265;. Goe, "The Ethiopian Maresha," 71-112.

40. Harris, *Highlands,* vol. 3, 221-23.

TABLE 4.3. Percentage of oxen ownership on northern Shawa farms, 1985.

Number of Oxen	Denki	Lalo	Segatna Jer	Dabra Berhan
0	40	34	18	31
1	40	47	55	38
2 or more	20	19	22	31

Source: Gryseels, and Anderson, *Research;* McCann, Ox/Seed (1985). These data are drawn from post-drought years, i.e. at the low end of oxen ownership cycles.

The evidence of farmers' techniques of maintaining soil fertility in both short-fallow and long-fallow field systems is clear. In the middle altitude zones immediately around Ankober, short-fallow systems, which included rotations of cereal crops, were "scarcely ever departed from; founded on the principle of preserving the soil from becoming utterly impoverished." In the valleys, teff, sorghum, and wheat appeared in alternate seasons; in the highlands wheat and barley rotated. Graham's description of Shawan crop rotation indicates variations in crop regimes at different elevations:

In the valleys, teff, jewarree [sorghum], cotton, oil and wheat follow in succession. On the high country, barley and wheat in alternate seasons, and in the cold moors of the tableland, the ground is left fallow for a year to recover itself, before a fresh crop can be taken from the exhausted material. Every quality of soil, however, is not adapted to the growth of wheat, nor would the crop arrive at maturity in every situation owing to the bleakness of the elevation and the temperature blasts and vapours which cover the crest of these high mountains; and in these districts, peas, beans, and barley form the successive crops.

Cereal-pulse rotation was apparently practiced only on the most easily leached, high plateau soils, which also required long fallow periods.[41]

The use of manure to preserve soil fertility has often been assumed to have declined in recent years because of its increasing use as fuel. In fact, in the entire 1840–1990 period the historical evidence indicates that manuring has been limited to a few selected fields and never applied to fields in a systematic fashion. Graham, for example, noted in 1841 that difficulties of transport (e.g. no wheeled carriages), meant that manure was never used in fields, save that from animals placed on har-

41. Graham, "Report," 262-63.

Figure 4.8. Gaye (soil burning), northern Shawa, 1982.

vested fields (for stubble feeding). Antinori in 1879 described the use of dung on nonfallow land, but only on some fields and in some parts of Shawa. Indeed, far from any dramatic changes in the use of dung and wood for fuel, there has been little change visible in the use of dung for fertilizer over the past two centuries. Both Wylde and Forbes, traveling in northern Shawa in the late-nineteenth and early twentieth centuries, noted the use of dung exclusively for fuel.[42] The carefully collected and dried piles of dung around northern Shawa homesteads were as common a sight to nineteenth-century travelers on muleback as they have been to vehicles traveling on the main tarmac road in recent decades.

Other forms of fertility management also show considerable continuity. Throughout the 1840–1990 period farmers on Shawa's high plateau engaged in a peculiar form of soil enrichment called *gaye,* or soil burning. This labor-intensive practice raises the fertility of fallow land left as pasture for as long as 15 years. Shortly after the rains in September, farmers break the *edari* (fallow) with the plow set at a shallow angle. After the soil dries family labor breaks the clods and builds

42. Graham, "Report," 263, Forbes, *Red Sea,* 138; Cecchi, *Da Zeila,* vol. 1, 447-48; Lefébvre, vol. 3, 253 cited in Ege, "Chiefs and Peasants," 52. Helen Pankhurst, "The Value of Dung," in Ethiopia: *Problems of Sustainable Development: A Conference Report* (Trondheim, 1989), 75-88.

soil, straw, stubble, and dung into small mounds. In the late spring (i.e., just before the rains and the major sowing period), farmers then ignite the mounds by placing burning dung at the center and scatter the ashes over the field. The soil mounds, up to 1200 of them per hectare, resemble small smoking volcanoes, which may smolder for as much as two weeks. Graham observed *gaye* in 1841; Cecchi's 1886 account described *gaye* as being used in "the majority of cases" in cultivating fallow; Buxton in the late 1940s saw soil burning as standard dry-season practice for preparing fallowed grassland.[43]

Most observers were impressed by the practice as a means of achieving increased fertility. Wylde in 1900 noted that the plots where *gaye* was practiced "can always be seen by a richer growth of the crop." Recent soil science research suggests, however, that the gains are short term only. *Gaye* is indeed effective in raising fertility in the short term (two to three years) by increasing available phosphorus and potassium while raising soil pH. In the long term, however, *gaye* changes the soil mineralogy and fuses clay into sand-sized particles. The result is a steep decline in total productivity over a decade and a need for even longer fallow periods. Thus, as land use has intensified through the twentieth century and fallow has decreased to well below 10 years, *gaye* has reduced overall crop productivity. Its persistence as a strategy over time suggests the extent to which short-term versus long-term land use strategies have dominated ox-plow agronomy: farmers in northern Shawa have been willing to gamble on high yields in one or two seasons at the risk of long-term productivity decline.[44]

Evidence of large-scale agronomic specialization, terracing, and irrigation appears frequently in the historical record. Terraces, for example, have been widely reported in all the environmental niches in the Ankober area from the 1840s to the present. Harris described the entire range of Ankober hills as being "broken by banks supporting the soil . . . a succession of highly cultivated terraces," and Graham reported that the soil on the mountains needed "artificial support" to keep from being washed away. In the southwest of the district around Angolela, farmers in the valleys below the high plateau built terraced embankments "not to check erosion but to ease the work" on steep hillsides dangerous to oxen. In the late 1940s David Buxton described terraces and their effects on the landscape. He pointed out that

43. Graham, "Report," 262; Cecchi, *Da Zeila*, vol. 2, 448; Buxton, Shoan Plateau, 166.

44. Wylde, *Modern Abyssinia*, 257. Research on soil burning in the past decade indicates that fertility in the first three years after *gaye* rises, but declines sharply in subsequent years. Over ten years the average productivity fields treated by *gaye* actually declines. See Roy L. Donahue, *Ethiopia Taxonomy, Cartography, and Ecology of Soils* (East Lansing: Michigan State University, 1972), 34-35; Goe, "Maresha," 144-45; Roorda, "Soil Burning," 124. Gryseels and Anderson, *Research*, 12. For decrease in fallowing in northern Shoa, see Gryseels and Anderson, *Research*, 10.

although northern Shawa's terraces were "nothing like the impressive and elaborate terracing practiced in some eastern rice-growing countries. . . many slopes are nevertheless unobtrusively terraced, both in the gorges and on the plateau."[45]

Whereas in many areas of Ethiopia terracing has served in water conservation, in Ankober its purpose has equally been to prevent erosion and expand cultivable land in areas of high fertility close to the town of Ankober. Terraces have the advantage of catching alluvial soils above the flood-prone bottom lands in Ankober's valleys. Ankober's terraces have not been large-scale integrated structures, but farmer-built stone walls so modest that "it is often impossible to tell at first sight whether a given feature is natural or wholly due to terracing."[46] In 1985 in the still sparsely settled Denki Valley, abandoned terraces were visible along a low ridge overlooking the valley itself. The former *malkanya* in Denki recounted to me that the abandoned terraces along the Denki Valley were constructed by individual household labor in the late-nineteenth century but abandoned by the 1920s. Farmers around the Denki river valley today continue to build low terraces even within single farm plots to reduce runoff and to level the slope near homesteads or where land demand is high (as with the land near the royal capital in the nineteenth century).[47]

Irrigation, another labor-intensive practice often associated with intensification, has been widespread in Shawa, particularly in Ankober District. Bianchi, who reached Shawa by traveling through the north, argued that the simple irrigation systems common in southern Wallo and northern Shawa were not in evidence farther north in Tigray and northern Wallo. Graham's 1844 description, however, portrays gravity-fed irrigation as a ubiquitous feature of the Ankober landscape:

> Artificial irrigation is resorted to in every situation when a supply of water can be obtained without much trouble, and crops of onions, chillies, and gourds are grown in patches by the river sides, where water can be easily diverted from its bed. The king's farms are in general, for their choice localities, well watered and clearly cultivated . . . The valleys of the low country are completely intersected with tiny canals . . .

45. Woobeshet Shibeshi, "A Regional Study of Angolela," 4th Year Essay, Haile Sellassie I University, 1970, 7; Harris, *Highlands*, vol. 1, 354; Graham, "Report," 256. Buxton, "Shoan Plateau," 167.

46. Buxton, "Shoan Plateau," 167-68.

47. Graham, "Report," 256. The Swiss geographer Hans Hurni who has worked extensively on conservation in the Simen mountains has told me that most terracing in Ethiopia historically has been for water not soil conservation (Personal communication). Sheh Abo Falaqa interview 6 June 1986. Terraces in the past few years have been constructed through food for work programs organized by non-governmental organizations or the Ministry of Agriculture exclusively to reduce erosion. Long-term effects of terracing on fertility remains a question since it reduces total land availability as much as 25% in the short run. Field interviews and participant observation Denki, 1986.

Figure 4.9. Ankober terraces, c. 1888 (from a photograph).

In addition to the gravity-fed irrigation of lowland irrigated banana gardens, key river systems have also provided highland irrigation. Traversi described an irrigation channel at Let Marefia which ran parallel to the left bank of the Aygeber River and which may have provided dry-season forage for royal herds. Johan Krapf also noted that irrigation was "not uncommon" in the country and that Logheita north of Dabra Berhan was *"bala-meseno";* that is, it was irrigated land, and owned by an important monastery. A century later Buxton described an elaborate system of aqueducts and canals drawing water from a dam on the Baressa River south of Dabra Berhan.[48]

Shawa, however, was no hydraulic society. Its irrigation was gravity-based, small-scale, and highly localized, that is, practiced only where topography allowed the diversion of river water to lower ground. There were few permanent structures and no water-lifting devices: the canals of stones and heavy sod had to be rebuilt every year after the rains; women's and family labor appears to have been particularly important in such specialized field systems. Moreover, irrigation as an agronomic strategy serviced two widely divergent cropping philosophies. In the nineteenth and early twentieth centuries, highland irrigation systems found at Angolela and in the Baressa Valley allowed highland farmers dedicated to an annual cropping regime to cultivate an additional cereal crop, usually barley, during the dry season for a June harvest. Irrigation in the gorges below the plateau and on lands below the escarpment to the east, however, concentrated on horticultural crops and

48. Graham, "Report," 263-64; Isenberg and Krapf, *Journals,* 282; Wylde, *Modern Abyssinia,* 261-62.

perennial tree crops—bananas, citrus, coffee, tobacco, and chat—destined for the market and, in the nineteenth century, for the royal capital.[49]

Ankober's "specialized" forms of agriculture did not, apparently, derive directly from Boserupian population pressure. Rather, such forms seem to be an indicator of a wide range of stimuli, such as localized demand, defense, and labor management. In Ankober's case, the nineteenth-century evidence cited earlier suggests that pressure on agricultural land and pasture in Shawa as a whole was minimal. *Madebet* lands and needs of the royal towns, however, implied concentrated demand and close management. The presence of the royal court provided a year-round demand for normally seasonal agricultural products and livestock beyond the usual rural diet in both amount and type. The sumptuary needs of royal banquets and hospitality required huge amounts of livestock and cereals, especially teff. Highland plateau agriculture could not provide the chilies, fruit, and even teff required at the royal table and at aristocratic households in the royal towns; these were the products of the middle- and low-altitude zones below the escarpment, where royal farms and tenants were settled and where irrigation defied seasonal constraints.

The royal capital's political presence also played a role. The extensive royal prerogatives over settlers who became tenants and over royal farms and the actual physical presence of the royal course lent a control over rural labor that was peculiar to highland agriculture and perhaps unique in Ethiopia's modern agricultural history. The high labor requirements of seasonal maintenance of irrigation canals were provided by royal claims on corvée, and the political stability offered by the king's presence promoted the local cooperation needed to maintain channels and apportion dry-season water supplies to specific fields. These specialized prerogatives allowed the application of labor to concentrate on specialized agriculture, which transformed the rural landscape in the case of terraces and overcame the vagaries of climate in the case of irrigation.

Ankober's peculiar topography and crop sensitivity also account for the presence of specialized forms of agriculture historically. Highland plateau zones above 2400 meters are both frost and flood prone during main cultivation season; rainfall supports a dry season crop of barley or a pulse in only one of every three years.[50] Irrigation in the Baressa Valley therefore has concentrated on producing a consistent dry-season barley crop rather than horticultural cultivation. It was this system which highland migrants imported to the Denki Valley in the post World War II period. Those same climatic restrictions on high plateau cultivation also stimulated the building of terraces to create alluvial plots on hillsides, which could then produce mid-altitude crops like teff, wheat, and pulses. In lowland valleys, terraces provided a method of maximizing fertile outcroppings, which could be cultivated by highland tenants.

49. For gorge cultivation, see Buxton, "Shoan Plateau," 166; for lowlands, see Cecchi and Chiarini, "Relazione," 417.

Remnants of small-scale irrigation visible today indicate that that system was far more extensive a century ago and cultivated a variety of perennial cash crops such as *chat,* coffee, and fruits. The unpredictability of the rains meant that such rain-fed agriculture as existed focused on drought-tolerant cultigens like sorghum and cotton.[51]

Cultivating Marginal Lands:
Ankober in the Late Twentieth Century

The most dramatic evidence of expansion of cultivation and pressure on land comes from the post-Italian occupation era. For this period the evidence from life histories, household development cycles, and new land use patterns depicts a major population movement onto marginal land below 1800 meters. The recent demographic history of the Denki River Valley illustrates not only trends within lowland areas of Ankober District, but also the effects of a population expansion on the highlands. Ecologically, Denki is typical of the lowland zones which lie between 1500 and 1900 meters elevation below the escarpment of northern Shawa and which occupy the interstitial zone between the highlands and the Awash Valley far to the south and east. Though rainfall with this district may average 500 millimeters per annum, recent experience has shown seasonal distribution and interannual variability rather than average annual levels to be the critical constraint on local agriculture. Such instability of moisture availability and the presence of rich alluvial soils in the bottom lands along the Denki River and its tributaries contributed to the historical concentration of cultivation within irrigable river valleys rather than on rain-fed plots.

Within the Argobba population, historical evidence of population change is circumstantial but intriguing. In areas above the alluvial valleys, abandoned rain fed terraces suggest a denser settlement and perhaps a modestly wetter climate at some point in the recent past. One local elder recalled these terraces as being built by tenant farmers late in the reign of Menilek and abandoned sometime after the Battle of Sagale (1916), perhaps as a result of the *Ya Hadar basheta* (the 1918–19 influenza pandemic), which had a severe effect in rural Ethiopia. A wider depopulation among the Denki Argobba may also have resulted from a major out-migration of Argobba-speakers to the market towns of the Awash Val-

50. Gryseels and Anderson, *Research,* 9.

51. On cotton, see Graham, "Report," 272-3. My 1985-86 field work consisted of over one hundred interviews involving fifty households drawn from five peasant associations within a day's walk of Aliyu Amba. My interviews focused on the life histories of men and women, particularly the key points at which households formed, dispersed, capitalized, and disinvested.

ley after the building of the Djibouti railroad and the economic decline of the Aliyu Amba route.[52]

The population of the highland agricultural economy, by contrast, was expanding against fixed land resources: a spontaneous post-World War II in-migration of highland Christian agriculturalists transformed the Denki Valley and adjacent areas into a mixed agricultural, though increasingly less economically diversified, zone which in terms of technology and farming systems more closely resembled classic forms of highland agriculture than lowland forms described in the nineteenth-century sources. This movement was a direct product of the operation of indigenous social institutions—farming systems, property rights, and social customs of marriage—which channeled local population growth into a broad intraregional migration. An important contributor to the major in-migration to Argobba land was the relatively stable period of rains in the decades prior to the 1974 revolution, which encouraged not only the in-migration of new agricultural households from highland districts but also the conversion of many Argobba cotton weavers to first-time farmers.[53]

Highland demographic growth had a specific set of effects on land tenure and per capita farm resources in land and livestock. The descent-based inheritance *(rest)* which had evolved after 1878 in formerly *gasha maret* highland zones, normatively prescribed equal shares of resources (land, livestock, equipment) for each offspring and produced subdivision and fragmentation of the highland land in Ankober highland district lands above 1800 meters elevation. When average household land holdings reached a critical stage (perhaps, on average, three-quarters hectare) important shifts in normative inheritance practice resulted. Land divisions on the death of the household head increasingly favored older sons at the expense of daughters and junior sons. In other cases, relatives within the descent corporation bought off junior sons' potential claims with cash payments and consolidated their holdings into more agriculturally viable units. Within marriage, parents adjusted their obligations to endow the marriages of their offspring. Those progeny who inherited highland land tended also to receive the lion's share of endowments of oxen, seed, and equipment while the disinherited, especially daughters, entered marriage *bado* (empty). The result of these social processes was a new, disinherited generation who reached maturity in the two generations prior to the revolution and resorted to new forms of marriage and household formation strategies.[54]

52. Interview with Sheh Abo Falaqa, Denki, 6 June 1986; Jamilal Shehnuri, Denki, 10 June 1985.

53. Interview with Abdullah Sawmaweded, 13 June 1985 (Abdullah was chairman of Denki peasant association) and Ahmad Badru, 13 June 1985. My data on patterns of migration derive from interviews with Christian migrants and their second-generation off-spring.

54. These patterns of inheritance emerged clearly from interviews with Christian migrants; for description of highland household formation in the early 1960s, see Weissleder, "Political ecology," 122-23, 172-75, 190-95.

The displaced class of poor Christian farmers in the Ankober highlands, however, found a natural outlet within a few hours' walk of their birthplaces in the Muslim Argobba areas such as Denki, Addis Alam, and Tach Lalo, where rain-fed land was available through tenancy. Though ecologically and culturally distinct, these lowland zones had long been part of the Ankober political economy. The district governor at Ankober appointed local elite *(malkanya)* to collect taxes and cooperate with Argobba landlords, who lent large bottom land holdings out to tenants. Though Sheh Abo, a former Denki *malkanya*, recalled a few Christian ("Amara") tenants in the area during Menilek's reign, the complete absence of churches below the 2000 meter elevation range testifies to the historical separation of Amhara and Argobba populations.[55]

The arrival of highland migrants into Argobba land in the two generations before the revolution therefore represented a historically new phenomenon. Their arrival, however, was neither conquest nor an imposition of highland will. New migrants found tenancy arrangements in Denki and nearby districts similar to those traditionally available in the highlands. Muslim landlords offered land on the basis of *magazo,* a seasonal tenancy agreement where newcomers cultivate in exchange for a portion of the harvest (usually a fourth). Rental of oxen *(qollo)* or oxen-labor exchanges *(balegn)* and seed borrowed from landlords, neighbors, or Aliyu Amba merchants provided the basis for their initial investment in agriculture. Under the fairly stable climatological conditions of the three decades of the postwar period, these highland immigrants fared well, adapting their highland farming systems to local conditions, and encouraging successive generations of new migrants. By the time land reform was proclaimed in 1975, almost 40 percent of those who applied for the land division within the new peasant association of Denki were Christian highland migrants, whose ideas about agriculture derived from a farming system based on annual cropping and another ecological setting.[56]

This new agricultural community in Denki was, however, built on a weak foundation. First, long-term patterns of climate and endemic disease that had been reported in the nineteenth-century accounts and had historically limited the expansion of rain-fed agriculture in the Ankober lowlands were still in play; the conditions of reliable rainfall of the 1950s and 1960s, which had supported migrants on rain-fed lands, proved to be more an aberration than a climatological trend. Highland farming systems based upon drought-sensitive crops like wheat, teff, and maize were poorly adapted to unreliable rains. Moreover, few of the new migrants had access to the irrigated plots devoted primarily to perennial Argobba crops.

55. Ankober *warada* had 79 churches in 1960 but none below about 2000 meters. Interview with Sheh Abo Falaqa, 6 June 1986, Denki. The nearest churches to Denki are Lalo Giyorgis two-hour's walk and Garo Mikael, two-three hours' walk up the escarpment.

56. Forty percent is an estimate of current population based on household lists provided by the Peasant Association chairman.

Second, new households from the highlands tended to be young, cultural outsiders and poorly endowed with the means of production, relying heavily on tenancy and borrowed capital. Their ability to manage the risk of harvest shortfalls was limited by their lack of local social networks, the fixed rental fee required to hire an ox, and to forms of drought-year income, such as weaving and irrigated gardens that were available to their better-established Argobba neighbors. Thus the Denki Valley, a synthesis of the highland and lowland agrarian economies of Ankober District, constituted a new and particularly vulnerable population. Not surprisingly, the 1984 drought hit this population especially hard and brought social responses strongly determined by patterns of agriculture and settlement.

Contemporary Ankober cropping as a whole exhibits the symptoms of a subsistence-oriented risk-aversion rural economy where both cultivated land and pasture resources have diminished over time with little evidence of increasing per-capita productivity. Changes have been relatively slow, reflecting Ankober's distance from major roads, the failure of major grain markets to penetrate the district, and its position on the periphery of the national political economy. There is no evidence of commercial specialization by area, though there is a substantial difference in cropping and rotation patterns between highland and lowland zones. Farmers in highland areas above 2000 meters have continued the nineteenth-century pattern of planting a wide variety of cereals and pulses, which include barley, wheat, horsebeans, and chick-peas cultivated in tight rotations. Highland farmers who maintain plots at altitudes below 2000 meters also rotate teff and, less frequently, *zangada* (early-maturing red sorghum) with pulses. In many highland areas of northern Shawa oats have begun to replace barley because of their better agronomic characteristics on marginal land where animal traction is in short supply: oats tolerate waterlogging, require less land preparation than barley, and their straw provides better livestock fodder.[57]

In lower elevations of the highlands, between 1800 and 2000 meters, the cereal-pulse rotational system operates, though teff is the cereal crop of choice, which farmers alternate with *zangada,* white sorghum, or wheat, depending on plot elevation. The most significant addition to this cropping system in the past generation has been maize, a crop whose high yield per unit of land and low labor requirements have made it attractive in areas poor in cultivable land or oxen. More important, maize is the only cereal crop which farmers can consume at the green stage: its expansion in the cropping regime is therefore a telltale sign of low food reserves from previous harvests. A further trend across all highland areas is toward pulses (horsebeans and

57. Harris notes that oats were disdained as human food except that "In times of scarcity the poor are compelled to resort to it." Harris, *Highlands,* vol. 2, 397. Evidence on highland cropping patterns derive from my 1985 and 1986 interviews with farmers in two peasant associations of Tegulet district, as well as Ankober farms above 1800 meters where I gathered crop data on each farm plot for crops and yields 3-5 years previously. Information on oats derives from Dr. Samuel Jutzi, agronomist with the Highlands Programme of ILCA.

chick-peas) and away from all cereals because of the low animal traction requirements of the former, that is, one plow pass rather than three to five plowings.[58]

In lowland farming systems a much different regime has emerged, one in which
historical change is more evident. Crops within lowland regions below 1800
meters, such as Denki, are far less diversified than in nearby highland districts. The
predominant cropping pattern on lowland rain-fed plots in the last decade or so has
been the cultivation of teff, maize, and two distinctive cultivars of sorghum, using
fallowing but no pulse rotation to preserve soil fertility. On lowland farms, the
majority of farmers I interviewed cultivated only three cultigens over a four-to-five-
year period, chick-peas being the only pulse sowed.[59] Of all crops, sorghum has
enjoyed the greatest stability. Each climatological zone has maintained its own set
of distinctive sorghum cultivars: *cheraqit,* an early maturing white; *zangada,* an
early-maturing red; and *mashila,* the long maturing, drought-tolerant white, are the
dominant types in the district.[60]

Though they do not appear as discrete events in historical sources, several historical trends in agriculture emerge from the combined evidence of cropping found
in interviews and nineteenth-century descriptions. The best historians can do to
determine causality is to associate general changes in cropping and the characteristics of particular crops with broader environmental and socioeconomic phenomena. In highland areas, fragmentation of plots, diminishing pasture, and lack of
access to markets have clearly worked against specialization. Risk management
strategies have pushed farmers to diversify crops according to crop conditions and
to maintain tight crop rotation to compensate for the virtual disappearance of fallowing. Declining per-capita ratios of oxen have also promoted increasing reliance
on pulses and on such cereals as oats and maize, which require less oxen and human
labor and/or yield better fodder. This overall process has represented a tightening of
the crop and livestock management regime.[61]

In the lowlands three major trends are evident. First, the cultivation of cotton, a
mainstay of the Argobba household economy in the mid-nineteenth century, has
declined in the past one or two generations. The decline of cotton has paralleled the

58. This trend, evident in my interviews with oxen-poor households in the Sela Dengay
district northwest of Ankober district, requires further research statistically correlating animal
traction availability with cropping patterns.

59. This evidence derives from 1985 interviews with 15 farmers and intense field observations for two periods in residence in Denki May-June 1985 and June-mid July 1986.

60. For information on sorghum in Ethiopia, see E.G. Damon, *The Cultivated Sorghums
of Ethiopia* (Dire Dawa,1962); E. Westphal, *Agricultural Systems in Ethiopia* (Wageningen,
1975). The initial evidence suggests that the diversity of sorghum cultivars in Ankober has
declined since 1840 though this point needs further research by agricultural historians, agronomists, and plant breeders.

61. Ministry of Agriculture figures on household oxen ownership and my own data gathered from Ankober and Mojana Wodera sub-districts is cited in McCann, "Social Impact."

expansion of cereal agriculture in the lowlands' agricultural economy, which in 1878 Cecchi and Chiarini described as consisting solely of sorghum and cotton. This decline may be attributable to post-World War II events, since Forbes in the 1920s reported Argobba cotton cultivation as widespread as well as noting at Ankober that "in front of every hut we saw one or two hand looms where men plied wooden shuttles on a primitive frame." In this craft economy men were the weavers and women provided the thread or cash income by spinning the raw cotton. By the 1980s many lowland farms had shifted to cereal crops such as teff and maize in addition to sorghum; cotton was a minor crop whose cultivation had largely disappeared in favor of annual cereal crops. Drought-affected farmers who sought to supplement household income by resuming weaving had resorted to importing raw cotton from the Awash Valley through Aliyu Amba.[62]

The second event was the expansion of maize as a major lowland crop; though difficult to date precisely, this is clearly a twentieth-century phenomenon, possibly associated with stable moisture availability over the past few decades and improved cultivars introduced in the 1960–80 period. My 1985–86 interviews show that farmers who had planted maize over the previous five years did so on rain-fed plots dependent on reliable rainfall in both the *belg* and *meher* seasons. Overall, the expansion of maize in lowland cropping systems needs to be studied in the context of changing variables of draft power, household labor demands, and overall household income strategies.[63]

Finally, the sophisticated production of irrigated, market-oriented crops such as fruit (bananas, citrons, limes), *chat,* and limited amounts of coffee, reported by nineteenth-century observers in the lowland river valleys such as Denki, has certainly not expanded and has probably declined. This trend away from perennial cash crops and toward annual food crops suggests the fairly recent expansion of the highland ox-plow complex's annual cropping regime to a new ecological zone as a by-product of highland migration in the postwar period. While banana gardens today dot the banks of the Denki River, their fruit is not a major income earner for either households or the district, nor do the handful of coffee gardens still in place affect either local income or the regional economy.[64] In a few cases farmers have

62. Forbes, *Red Sea,* 128, 132. Many Argobba weavers in Denki had shifted from weaving to full-time agriculture in the last decade to increase their household income. In the 1980s drought-affected farmers in Denki were importing raw Awash cotton from Aliyu Amba to resume weaving to supplement household income. By 1986 several Christian farmers were learning the trade. Interviews with Abdullah Sawmawadad, and Zergabachew Menyilu, Denki, June 1985.

63. Weissleder listed maize as a *wayna daga* and lowland crop, though not a major one, in the 1960s. Weissleder, "Political ecology," 34. Maize appeared as part of the cropping cycle alongside teff and sorghum in my interviews in both *wayna daga* and lowland farms.

64. The evidence for irrigation is based on my interviews and observations in four Peasant Associations located along the Denki River.

begun to produce maize on irrigated plots. While this action suggests an adaptation to the moisture-sensitive nature of maize, it also indicates the shift from perennial cash crops to annual subsistence crops, a change which reflects the increasing dominance of the ox-plow annual cropping regime.

The 1984 Drought

The evolution of settlement, cropping, and agronomy over the course of the twentieth century has created an economic and social configuration particularly vulnerable to short-term agricultural shocks. In 1983 and 1984 the weather delivered such a shock to the Ankober lowlands in the failure of two consecutive *belg* (short) rains and main rains which ended in mid-August rather than mid-September. The failure of the spring rains meant loss of a small May harvest but, more important, left the soil too dry to cultivate for the main sowing season. Combined with the foreshortened main rains, the growing season in those years shrank from 100–120 days to 30–60 days, reducing crop yields to near zero.[65]

While the drought affected Ankober District as a whole, its effects were most pronounced in lowland areas such as Denki, where rainfall variability was most severe and new migrant populations depended on rain-fed lands for annual crops. In 1974 a severe drought struck but was short lived: cattle died quickly and one year's crops failed. Rains returned again in 1975 and continued through eight straight growing seasons. Pastureland and livestock recovered quickly, and the 1975 land reform which eliminated tenancy, drew a new wave of highland migrants to the lowlands to take up rain-fed agriculture. Good cereal prices in that period also convinced local weavers to take up annual crop agriculture full time.[66]

The 1983–84 crisis, however, had an entirely different character. Successive droughts and an infestation of stalk borers caused the failure of sorghum and maize crops. The disappearance of livestock forage with lack of rain meant the oxen died from the sustained lack of nourishment. Local residents themselves labeled the crisis *saltateq* (we were unprepared), *sholko wagegn* (it stabbed me), *adaltegn* (I slipped on it), and *agurat* (it pounced), names which evolved while the crisis lengthened and deepened. In early summer 1985, just before the rains, a pandemic of fever swept Denki in the late spring, killing perhaps 10 percent of the population and again delaying sowing for those who survived.[67] In the 1985–86 season excel-

65. Need reference on rainfall.

66. James McCann, "Preliminary Report on Oxfam/ILCA Oxen Seed Project." Unpublished Report to Oxfam America, July 1985. Abdullah Sewmeweded and his brother Yusef, Argobba weavers, are examples of this trend.

67. Fever may have been typhus and malaria; my own case of it was tentatively diagnosed in Boston after several months as malaria, though the epidemic outbreak would also suggest a tick-borne febrile disease as well.

lent rains fell on the highlands above the escarpment. In the lowlands, however, the main rains, which had begun promisingly in June, abruptly ended early in September and virtually eliminated the maize crop, which dried up quickly; some farmers who had planted early were able to consume it or sell it at the green stage to avoid a total loss. The teff, which had germinated in late August, was quickly eaten by an infestation of grasshoppers *(fenta)*. Farmers planted three or four times, with each sowing successively eaten by grasshoppers that had propagated in the extended dry season. The good spring rains of 1986 brought germination of sorghum and chickpeas, later destroyed by stalk borers.[68]

The cumulative effects of drought, food shortage, and loss of livestock, especially oxen, produced disaster but not chaos. The social response and effects within the agricultural system were already fundamentally conditioned by the area's ecological, social, and agricultural history. First, the climate crisis was not universal. In May 1985 during my first walk down from the town of Ankober to Aliyu Amba and Denki, the changes in the landscape from the highlands into the valleys below were shocking. In the highland zones only one to two hours' walk from the Denki Valley fields of light green maturing barley from spring rains were in stark contrast with the dry, drought-stressed fields below 2000 meters. From Denki looking back up toward Ankober—nineteenth-century travelers' favorite image—the alternating green and gold of crops nearing harvest stage were a constant reminder to the Denki residents of their dependence and relative poverty. For the new generation of migrants, the view also represented a glimpse of their own family past.

Within the Denki economy itself, the effects were also mixed. In the valley of the river, bright green patches of irrigated bananas and a few irrigated plots were closely juxtaposed to yellowing fields on rain-fed lower slopes. In May, plantings of maize and even hardy sorghum were yellowed from drought stress; many farmers had already driven livestock onto the fields to feed on the stunted seedlings. This configuration was a graphic illustration of economic and social hierarchy. The green riverine plots belonged to long-standing Muslim Argobba, whereas new migrants depended on rain-fed annual crops. Farmers fortunate enough to rent irrigated, alluvial land paid up to half of their harvest to others for use of the land and oxen rental.

Still other households, those headed by women, had no agricultural land at all, since it has long been male labor which defines a household's claim on plots. The products of a high divorce rate (see chapter 2), death of a spouse, or male labor migration, female-headed households have been fragile and short lived on an individual basis, but as a social category an enduring part of the agricultural community. They represent the bulk of the noncultivating households which always exist within the agricultural community. In the Denki Valley female-headed households in 1985 made up 68 (14.3 percent) of the 474 registered households, a figure which

68. James McCann, "Oxfam Ox/Seed Project Evaluation." Unpublished report to Oxfam America, 1986, 14-15.

probably has tended to increase during economic crisis and constitute a measure of the social stress and mortality of a near-famine situation.[69]

Drought and food shortages are not an aberration or a deviation from environmental and economic norms. Rather they are more extreme forms of the historical stresses and risks inherent in the ox-plow system. The effects bring into sharp relief the historical forms and stresses within the agricultural economy. Droughts, crop pests, and human disease are also conjunctural events which play against specific historical conditions in demography, crops, and conventions of property. In Denki's case drought arrived after a period of sustained immigration of annual cropping farmers, whose presence dominated the irrigated, perennial cropping system and populated the valley with capital-poor households vulnerable to debt.

Conclusion:
The State and Agriculture in Ankober

There are several important underlying parameters of agricultural trends in Ankober. One is the shift of state power and associated economy of the royal court to the south after the mid-nineteenth century. Royal control over regional land and the concentration of resources and specialized demand at the political center had produced intensive land use, coordinated land management, and provided protection of key natural resources such as pasture and forest. Specific political decisions—the relocation of the capital, the new southern-based national economy, the postwar central state's unwillingness to invest in the "low potential" northern peasant agriculture—changed the context of local agriculture in Ankober. The gradual loss of direct contacts between the state and local resources in the late-nineteenth and early twentieth centuries accelerated with the growth of a new political national economy based upon higher productivity in the south and west, as well as with the decline of international trade to the Awash Valley when the Djibouti railroad reached Addis Ababa in 1917.

The nineteenth-century evidence of intensification in resource use and forms of agriculture in Ankober, however, does not suggest a Boserupian model of population increase driving increased productivity per unit of land. Ankober's postwar population increases, in fact, seem to have had the opposite results. Then population growth and local property practice drove disenfranchised individuals to marginal land, where periodic drought and poor market access fostered a recrudescence of subsistence-oriented agriculture and expanded rural poverty.

The evidence of the 1980s suggests that, despite the obvious need to intensify agriculture in the district to meet food shortages, the use of specialized agriculture

69. McCann, "Preliminary Report," 7-8. Figures drawn from three Peasant Associations (Denki, Lalo, and Addis Alam).

in Ankober has declined substantially since the mid-nineteenth century. By the late 1980s virtually no commercial agriculture existed in the district, in fact probably less than in the mid-nineteenth century, when lowland valleys produced irrigated fruits for highland markets and cotton for cloth exports. The canals which intersected the highland regions in the mid-nineteenth century have largely disappeared. In the Denki river valley, only a few irrigated banana and coffee gardens remain; local residents claim that the amount of available water has declined markedly as new land cultivated upstream has choked off their supply.

The decline in state support of labor-intensive agriculture in fact may have begun shortly after Ankober's decline as the center of Shawa's political economy. As early as 1890 Traversi described a royal irrigation scheme abandoned when Menilek moved the political center south.[70] Not only did royal support of irrigation provide state sanction for maintenance, but also the gradual erosion of royal tenure into individual tenures reduced the potential for local cooperation in resource management.

The importation of an annual crop regime has also changed the nature of the lowland economy. The structure of poverty within Denki has been, in the short term, a product of the drought of 1983–84. The conditions of agriculture in the area, however, are not short term but deeply embedded within the historical patterns of settlement and agriculture over the past few generations.

70. Traversi, *Let Marefià*, 62.

5

The Plow in the Forest: Agriculture, Population, and Maize Monoculture in Gera

> "As you gave me perpetual darkness, let your country be
> swallowed up by forest; if you cut a single tree let it
> be multiplied by ten. May you not see light, only darkness."
> —the Curse of Abba Bosso, a Gera oral tradition,
> as told by Yalew Taffese

Though the ox-plow complex had its origins in the relatively open plateau environment of the northern highlands, it also spread across the Abbay [Blue Nile] and onto an entirely different cultural and physical landscape. In the Gibe River basin in the early nineteenth century a group of small, well-organized Oromo kingdoms emerged to take advantage of a burgeoning export trade in exportable extractive trade goods—slaves, civet musk, and aromatic herbs and spices. The region's dense hardwood forests also included an abundance of wild coffee, another extractive export crop, which by the mid-nineteenth century had percolated into northern markets and into the Red Sea trade in small caravans. Far more than export trade, however, the growth of these southern forest state systems rested on a highly diversified agricultural base which took the highland ox-plow as its primary implement and integrated annual crops into a much different, older horticultural tradition.

This chapter treats the evolution of plow agriculture in the remnants of the nineteenth-century kingdom of Gera, an Oromo polity at the heart of the broadleaf for-

I am grateful for the help of Redd Barna (Norwegian Save the Children), Helge Espe, and its project staff at Gera for the logistical help in completing this research. I am especially grateful to the late Wayzero Yeshimebet Asfaw, Ato Dub Gelma, Ato Daniel G. Michael, and Ato Belachew Soboku.

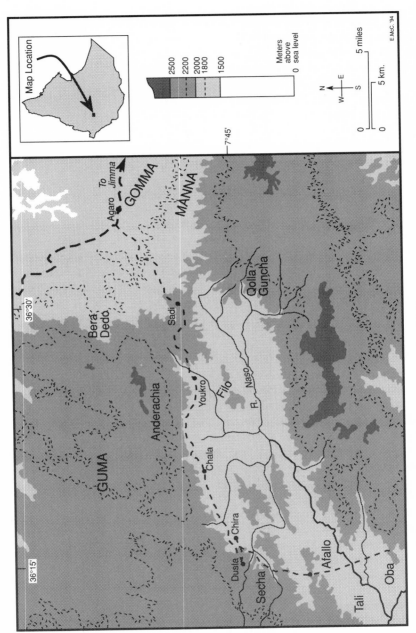

Map 5.1. Gera

est zone of the Gibe River basin. The chapter examines the nineteenth-century evidence and treats the twentieth-century development of a specific agricultural system and a coffee-maize complex as emblematic of the historical evolution of modern agriculture in an important highland ecological system. The long-term effects of shifting patterns of land use and changes in Gera's distinctive forest cover were part of this economic change visible in the historical landscape itself, described by a succession of travelers and by the testimony of Gera's oldest residents. The change was also as profoundly but less visibly evident within individual farmsteads, where diet, forms of labor, and vulnerability to environmental shock shifted in accordance with the exigencies of new crops and the emergence of a centralized national political economy.

Physical Setting

The dense broadleaf montane forests which have historically covered the highlands of the Kaffa and Ilubabor regions of southwest Ethiopia contain dense stands of hardwoods exceeding 35 meters in height and, compared with northern open and often treeless plateaus, this was a diverse ecosystem which supported a staggering variety of natural products. In the primary canopy, Emilio Conforti in 1939 counted 35 species of "tall trunked" trees. In 1991 an afternoon's trek up the forested slopes still revealed 33 locally recognized species of trees, ground cover shrubbery, and broadleaf plants, all with specific uses in the local economy. The forest provides shade and organic matter, which supports an even greater diversity of bush cover at ground level. It is within this moist, shaded "curtain of vegetation" that Ethiopia's rich varieties of *coffea arabica* evolved in a wild state. Although what many call wild coffee may be a naturalized domestic form characteristic of secondary forest growth, the large number of genetic types found in southwest forests argues for its origins there. Whatever the origins of the germ plasm, certainly coffee in this "wild" state predates the southwest's coffee trade itself.[1]

Gera presents an environmental contrast with the northern cradle of ox-plow agriculture. Elevations in the region range moderately between 1800 and 2400 meters; temperatures across the year vary only between 17 and 20 degrees centigrade. Of the Gibe Valley regions, Gera contains the highest elevations of coffee forests (averaging 2200 meters elevation) and uplands, such as Secha and Bera Dedo which historically have probably resembled the more open wooded grasslands of the northern plateaus. In the valleys of the central plains, low-lying lands,

1. Conforti, *Impressioni,* 157. For origins of coffee see Huffnagel, *Agriculture,* 204-5, and Raffaele Ciferri, "Primo rapporto sul caffé nell'Africa Orientale Italiana," *Agricolo Coloniale* 34 (1940), 135-44.

chafe maret, are seasonally waterlogged; marsh grasses and reeds replace the dense forest growth of the higher elevations.

The climate, though bimodal in its seasons, also differs subtly from the northern highlands in both amount and distribution of rainfall. Of the annual average rainfall of 1525 millimeters, about 83 percent falls in the March–September main rains, which come earlier in the humid southwest than in the more northern highlands. More important, those rains are extremely reliable: the probability of at least 60 millimeters in each month of the rains is almost 90 percent.

Soils in this environment have also had a role in shaping the agricultural regime. The reddish brown to dark brown clays are eutric nitosols which range from slightly to strongly acid with a high nitrogen level but little available phosphorus. A rich variety of coffee germ plasm evolved in this acid, nitrogen-rich soil, and, not surprisingly, coffee trees have no need for fertilizer, and indeed show little response to it.[2] The high nitrogen content probably also accounts for the lack of cereal-pulse rotation in the area's agriculture, in sharp contrast with northern systems which depend on pulse rotation to fix nitrogen for primary cereal crops. The reddish brown forest soils are also quite deep compared with the brown-black northern soils, but the southwest's reddish soils have a forest soil's characteristically thin layer of organic matter near the surface. Unlike West African forest soils, however, Gera's fields require only short fallow, two or three years after five years of cereal crop rotation.[3]

Political Foundations of Agricultural Change

Historically, Gera District has been best known for the nineteenth-century kingdom which developed a reputation for exotic trade goods and volatile relations with its neighbors. Among its most visible historical figures was its queen mother, the *ghenne* Gumiti, who imprisoned the Italian geographers Antonio Cecchi and Giuseppe Chiarini in the early 1880s, an event recalled in both European literature and local oral tradition.

Gera's history as an Oromo kingdom rested on a much deeper tradition of social organization and agricultural production that predated the early eighteenth-century Oromo invasion of the region. The Gibe region as a whole first appeared in historical sources when Emperor Sarsa Dengel in 1586 conquered the area's autochthonous Sidama-speaking population, bringing Christianity and perhaps plow

2. Kassahun Seyoum, Steven Franzel, and Tesfaye Kumsa, "Initial Results of Informal Survey Coffee Producing Areas of Manna and Gomma Woredas, Kefa Region,"Working Paper No. 4/88 Addis Ababa, 1988, 6.

3. Daniel Gamachu, *Environment,* 12. Interviews with local Oromo and with northern migrants in three zones of Gera in 1990 and 1991 indicated an absence of the tight cereal-pulse rotation systems found elsewhere in the ox-plow complex.

agriculture itself to the empire's southwestern frontier. Though local memory contains few references to the pre-Oromo past (i.e., before the late-eighteenth century), a number of nineteenth-century observers reported remnants of an early Christian population who had lived four generations in the region without contact with wider Christian communities in the north. Remnants of the earlier Christian community, known locally as Bussase, still live in Afallo and in the Gojeb Valley.[4]

The historical evidence of the plow is more definitive. The ox-plow complex along with a full range of annual cereal crops was well entrenched by the mid-nineteenth century i.e., well before Emperor Menilek's late-nineteenth-century expansion, and probably predated the Oromo invasion.[5] Emperor Menilek's 1886 conquest of Gera and the exile of its royal family were part of imperial expansion to capture the valuable extractive trade of the entire southwestern region, but for many northerners, Menilek's expansion also represented a reincorporation of former cultural and political hegemony.[6]

The genesis of Gera's ruling dynasty derived from the imposition of Oromo settlers into the Gibe basin in the late eighteenth century. Oral tradition from Jimma and from Gera itself also places the foundation of the Oromo kingdom in Gera fairly late in the process of state formation in the Gibe region as a whole.[7] While accounts of its first rulers vary, a local oral tradition concerning the "curse" of Abba Bosso, one of Gera's early kings, links Gera's origins to its neighboring states, to agriculture, and to the forest environment. The tradition recounts the curse laid by Gera's ruler Abba Bosso on the conspirators who deposed him:

> He also called for the chief of Gomma [a neighboring state] and asked him to bring him a red ox; he asked from Jimma a white ox; and asked them also for a third ox. When he got the three oxen, he slaughtered the white ox and turned towards Jimma and blessed the country. He also slaughtered the red ox and facing Gomma blessed that country. Finally, he turned toward Gera

4. J. Spencer Trimingham, *Islam in Ethiopia* (London, 1965), 109-110.

5. Claims that the southwestern plow differed from that of the north are unfounded; see Cecchi, *Da Zeila*, vol. 2, 164; and Mohammed Hassen, *The Oromo of Ethiopia*, 118. The only difference between Gera's local plows and the northern version is the fixed angle between the handle and beam, which is more acute in the Gera version. All other parts and principles are the same.

6. Massaja encountered Orthodox Christian exiles, the Bussase of Afallo, in Gera in 1861. Massaja, *I miei*, vol. 4, 170-71, 204-5; also see Trimingham, *Islam*, 109.

7. Massaja, *I miei*, vol. 4, 221; Cecchi used dates within the gada cycle to estimate 335 years since the Oromo arrival in the Gibe region. Cecchi, *Da Zeila*, vol. 1,283-84. Gera may have been one of the latest royal lineages formed. Abba Bor Abba Magal claims his families arrived in the early nineteenth century. For Jimma oral tradition see Mohammad Hassen, *The Oromo of Ethiopia*, 112-13.

and cursed it. "As you gave me perpetual darkness, let your country be swallowed up by forest; if you cut a single tree let it be multiplied by ten. May you not see light, only darkness. Let others get honor and wealth and let you remain wretched."[8]

The elements of this story—oxen and an encroaching forest—allude strongly to an agrarian, plow-based society for which oxen are an important symbol and an intrusive forest is an obvious curse.

Though dominated by an Oromo ruling class, Gera's population came from diverse sources. In the nineteenth century it included Oromo settlers, Kafficho migrants, a Muslim merchant community, and a group of southern migrants who Cecchi called the Wombari, and Christian Bussase exiles from the Kaffa kingdom to the southwest. Guglielmo Massaja a Capuchin missionary who set up two Roman Catholic mission stations in Gera in 1859, described Gera's ruling class as the product of marriage between Oromo settlers and "Kaffini" women of the local royal house, producing a *mescolanza* lineage which transformed itself by the early nineteenth century into a ruling family.[9]

Gera's political and social practice reflected the exigencies of sedentarism and Oromo concepts of social property. While elements of other Cushitic and Omotic languages persisted, the Oromo language dominated political and social life. Property and social practice, however, reflected long association with settled agriculture. Unlike northern ambilineal inheritance, Gera Oromo practiced firm unilineal primogeniture, with subsequent sons inheriting significantly smaller portions of their father's patrimony; young men paid brideprice (usually in livestock) to obtain a wife. Throughout the period of this study, marriages, unlike the northern model, appear to have been quite stable, though polygyny was common. Though thoroughly sedentary by the mid-nineteenth century, the Oromo of Gera nevertheless retained elements of their former pastoral polity. Cecchi described the celebration of *butta,* "the greatest national celebration" of the Oromo, led every eight years by one of five grades of the *gada* system.[10] Traces of the Oromo political past did not, however, survive the nineteenth century. The last recorded case of butta's celebration in Gera was 1880; though some older Gera natives know of it, none recalls its celebration after 1940.

8. Interview with Yalew Taffese, Yukro (Gera), 24 November 1991. Traditions of origin vary: the last claimant to the royal line traced its origins to an Oromo branch in Raya (eastern Tigray). Another traces the royal line to Tullu Gunjee, a grandfather of Abba Bosso, who was covered with hair from head to toe. Interview with Abba Dura Abba Bora, Yukro (Gera), 25 November 1991.

9. Massaja, *I miei,* vol. 4, 221; interview with Shisay Abba Bor, Chala (Gera), 22 and 23 November 1991.

Islam also provided an important binding element in Gera's diverse population, though it arrived well after the agricultural economic base was already well entrenched. With the resurgence of regional trade in the mid-nineteenth century, Muslim traders penetrated the forest areas and found converts among the royal families. At some point in the 1860s Abba Magal, the most powerful Gera king converted to Shafiite Islam and had converted most of his population by the time of his death in 1870. When Cecchi arrived almost a decade later, Gera had all the elements of a coherent economic and political culture: its population spoke Oromo, practiced Islam, recognized the legitimacy of the centralized authority of the royal lineage, and cultivated with the classic highland ox plow.

Gera's population was not, however, homogeneous. Slaves probably made up the largest single sector of Gera's diverse nineteenth-century population; they were possibly even a majority. Cecchi, who had traveled widely in southern Ethiopia where slavery was an integral part of the social economy, concluded that Gera contained the widest variety of slaves he had seen anywhere. He noted slaves from Kullo, Walisso, and Konta mixed with Kafficho, Janjaro, Gurage, Kambata, and Shankalla (i.e., western Gojjam), and with "Galla" of various branches of Soddo, Botor, and Agallo. Cecchi counted 3000 slaves at the royal residence at Chala, that is, one-fifth of his total population estimate for the kingdom. At the royal capital they served as weavers, iron makers, and arms makers, enhancing the material wealth and prestige of the court.[11]

If the political and social origins of Gera remain open to question, the long-term presence of ox-plow agriculture is fairly clear. When the Oromo invasion overwhelmed the autochthonous Sidama in the early eighteenth century, a sophisticated blend of ox-plow cultivation and horticultural tradition was already in place. The agricultural system blended the limits of the forest environment with the annual cropping regime of the ox-plow complex. Amon Orent has argued for nearby Kaffa that incorporation of annual crops which could be collected and stored by the political elite may indeed have been a response to an Oromo threat, which in fact the region resisted for almost two centuries. Annual crops such as teff, eleusine, and barley also constituted a sumptuary differentiation between social classes, which reinforced elite authority. The presence of such crops in Gera may well have

10. For marriage and property in the nineteenth century, see Soleillet, *Voyages en Ethiopie*, 262-64; Pearce, *Life and Adventures*, vol. 2, 314-20. Among my Oromo interviewees, some of whom had been married for 60 years, divorce was very rare. Cecchi, *Da Zeila*, vol. 2, 283-84; for accounts of the *gada* system, see also Asmarom Legesse, *Gada: Three Approaches to the Study of African Society* (New York, 1973); and Enrico Cerulli, *Etiopia Occidentale (dallo Scioa alla frontiera del Sudan): Note del viaggio 1927-1928* (Rome, 1933), 31-53.

11. Cecchi, *Da Zeila*, vol. 2, 258, 381-82.

accounted for the growth of a political hierarchy which could control wealth in the form of agricultural land and stored food, a base of authority which superseded the pastoral economic base of the Oromo *gada's* eight-year cycle of authority.[12]

It is likely as well that Oromo invaders brought with them a knowledge of annual cultigens, since the royal family claimed Raya (eastern Tigray and Wallo) and a passage through Shawa, Wallaga, Jimma, and neighboring states before settling in Gera. In Gera, names for a number of key crops suggest Cushitic roots rather than northern origins: white sorghum (Oromo: *masinga hadi;* Amharic: *mashila*), red sorghum (Oromo: *masinga dima;* Amharic: *zangada*), maize (Oromo: *boqolo;* Amharic: *YaBahar mashila*) barley (Oromo: *gerbo;* Amharic *gabs*). Other crops, i.e., the primary annual cultigens of the ox plow retain names of northern origin: teff (Oromo: *tafi;* Amharic, *teff*), wheat (Oromo: *gamadi;* Amharic: *sinday*), eleusine (Oromo and Amharic: *dagusa*), chickpeas (Oromo: *shumbura;* Amharic: *shimbra*), field peas (Oromo: *atara* and Amharic: *ater*).[13]

Forest, Ensete, and the Plow

The evolution of agricultural practice in Gera has been intimately shaped by its physical setting in the forest and by the human activity capable of transforming that setting. Gera's agricultural substratum which predated plow cultivation included food crops adapted to a shaded forest environment and horticultural forms of cultivation. This tradition included a bundle of domesticated root crops well adapted to forest soils: taro (*godare*), yam (*boye* or *wachenu*), and a small, potato-like tuber, *Ya Oromo dinnich.* The repertoire also included ensete, a crop widely considered to have predated annual crops in several highland cultivation systems (see chapter 2 and below).

The first clear descriptions of crop mixing, diet, and agricultural practice for Gera appear only in the mid-nineteenth century with the arrival of foreign travel accounts. The first of these accounts belonged to Massaja, who established two mission stations in Gera in 1859. Two decades later Antonio Cecchi, held captive by the queen mother for over a year, provided a much richer description of a diverse system of annual cropping mixed with forest crops. From Cecchi's description, it is clear that the ox-plow complex was firmly in place in Gera by 1880, and possibly even predated the arrival of ensete. He described the principal products of the Gibe states as "different qualities of teff, maize, sorghum, eleusine, wheat, barley, peas,

12. Orent, "From the Hoe to the Plow," 192.

13. The modern Amharic used the Cushitic form for maize, which is also *boqolo;* the historical name *yabahar mashila* (the sea's sorghum) has given way to the Cushitic root in most usage.

Figure 5.1. The forest setting, c. 1880.

Figure 5.2. Clearing the forest for maize, 1991.

fava beans, and haricot beans." He also pointed out specific differences in crop management between Gera and its neighbors: Gera farmers sowed both white sorghum and *dagusa* a full month earlier than neighboring states. For maize, Gera farmers planted in February and harvested in August as well as on a July–September cycle, whereas other regions sowed in April and harvested in August.[14]

Gera's farmers sowed wheat and barley in the limited zones above 2300 meters in July and harvested them in December, the classic highland cropping schedule. Pulses appear to have been a minor crop, sometimes intercropped with sorghum or double-cropped with barley or wheat in upper elevations on the moist, nitrogen-rich, red-brown clay soils. Gera's cultivators, according to Cecchi, double-cropped where soil moisture permitted, perhaps on the boggy bottom lands between Chala and Filo, where Cecchi spent the bulk of his captivity in Gera. The illustrations in Massaja's book show the royal *masera* (compound) surrounded by ensete and lemon trees (see figure 5.5 below). Farmers also cultivated ginger, coriander, and fenugreek as well as collecting aromatic herbs from the forest.[15]

By 1880 maize was apparently an important but still secondary crop, well suited to the relatively moist climate and middle elevation of Gera as well as to its nitrogen-rich soils and bottom lands *(chafe maret)* that permitted double-cropping with teff. Massaja makes only one oblique reference to maize in regional agriculture in his voluminous descriptions of southern economies. Cecchi, however, reported an apparently low-yielding variety of maize, which "like our own quantino *(zea mays praecox)* does not exceed 1.25 meters and has ears of 10 rows of small grains, smooth of a faded yellow." This description suggests a "flint" variety, resistant to insect damage but with more hard starch and probably a lower yield than later, improved varieties. Cecchi speculated that maize had arrived with Muslim merchants from the north.[16]

Cecchi's description and Massaja's brief reference suggest that maize occupied a position of secondary or tertiary importance in a highly diversified cropping system which blended sorghum and maize with the northern "package" of highland cereals. Ensete and root crops (not mentioned in the nineteenth-century accounts) balanced the annual cropping and seasonal diet. Class elements of diet also emerge indirectly from observers' accounts. When Cecchi and Chiarini's status as guests changed to captives early in their residence in Gera, their diet provided by the queen shifted from meat and bread made of white teff and wheat to "a ration of teff or maize of lowest quality, on rare occasions accompanied by a vessel of acidic milk or very bad beer."[17]

14. Cecchi, *Da Zeila*, vol. 2, 278. In 1990 maize in Gera was also doublecropped with teff on bottomlands, that is, it was sown in January and harvested between April and June. Most fields of maize, however, were still on the late-spring sowing schedule.

15. Cecchi, *Da Zeila*, vol. 2, 277-78, 280-81, 279.

16. Cecchi, *Da Zeila*, vol. 2, 278; Massaja, *I miei*, vol. 4, 204-5.

17. Cecchi, *Da Zeila*, vol. 2, 297.

Another key crop derived from forest conditions and horticultural traditions was a part of the ox-plow annual crop complex by the mid-nineteenth century. *Ensete ventricosum* (Oromo: *qocho*). Ensete appears to have been an important food crop across the highlands, perhaps for as long as cereals in some areas. The plant shows two important signs of long-standing domestication: it exists only as a domesticate; its seeds are sterile and thus it can be propagated only vegetatively by using suckers. Chapter 2 describes ensete's the cultivation and preparation of bread from its scraped, pulverized, and fermented pseudostem. Ensete's origins are not clearly understood, but the plant historically must have evolved in its domestic form within a moist forest environment, since its 70 cultivars are well adapted to variations in elevation between 1600 and 3100 meters but require fairly high levels of fertility (organic matter)and soil moisture.[18]

Ensete is a horticultural crop adapted to close management within the domestic compound in contrast with the extensive cultivation of annual crops. It also involved a different seasonal and gendered allocation of labor. Cultivation and transplanting of suckers by men using a hand hoe (Gurage areas use a two-pronged digging stick called a *wonet*) takes place after the main rains to remove weeds. Manuring (and thus integration of livestock) and mulching, or at least incorporation of organic matter, are also an important part of cultivation. In the past, a single holding may have consisted of 300–500 plants, a third of which might be consumed and replaced over the course of a year. In some areas of Gera, such as Oba near the Kaffa border, 3000–4000 plants per household were the norm. Ensete suckers reach maturity in three to nine years depending on elevation and local conditions.

Ensete cultivation need not, however, imply historical density of population; rather, it implies a stable forest system which did not require major land clearing. Both tubers and ensete have thrived under forest cover with only minor clearing required to sustain food supply. These food crops reflect horticultural traditions of vegetative propagation (rhizome plantings rather than seed broadcast) which fundamentally separate the annual-crop cereal traditions of the north from older traditions of forest agriculture. Indeed, even recent migrants from the ecologically similar Kaffa region to the highland Secha area of Gera arrived with no knowledge of plow cultivation.[19]

---

18. In the southwestern forests ensete appears to have retained its primacy, whereas northern cultivation is reported by James Bruce around Lake Tana and in the Simen region. See chapter 2. Alvares described ensete as "so like the Indian fig that they can only be distinguished from very near." Alvares, *Prester John*, vol. 2, 47. Lockhart, *Itinerário*, 245-46; Huffnagel, *Agriculture*, 285-88; field interviews, 1991.

19. Interview with Abba Bulgu Dono Maru, Secha (Gera), 27 November 1991. Abba Bulgu's father had migrated from Kaffa in the late nineteenth century without a knowledge of plow cultivation.

Ensete's position as a dominant food crop was by no means uniform over the forest zone. In 1859 Massaja described the dominance of ensete over cereals in the local diet of the Kaffa region bordering Gera on the south, where he observed that almost one-third of the total land of Kaffa was covered with ensete cultivation.[20] Describing the Kaffa diet, Massaja noted that "only the merchants, the Muslim families, the Galla, and some Christians from outside maintain the custom of eating Abyssinian bread i.e. composed of teff or admixtures of other cereals; all the rest of the population eat bread of *coccio* [ensete]."[21] By 1880 ensete was much less important in Gera. Cecchi estimated that it stood behind sorghum, teff, and maize as the most cultivated crops.

Ensete, which has been critical to local food supply in Kaffa and areas of southern Shawa, may have been of fairly recent origin in Gera. An elderly slave recounted for Massaja a story of ensete's arrival from Kaffa to Gera where it had not been known before. Cecchi elaborated the story to suggest that ensete's introduction had come during a famine.[22] Only in the zones ecologically contiguous with Kaffa was ensete a primary part of the diet. While Massaja was remarkably silent on diet and crops in Gera, he mentions them obliquely, noting that only his less senior Gera acolytes ate it: "This bread is inedible to anyone not accustomed to it. And in fact, only the young men [*giovani*] of Ghera ate it, while all of the others satiated themselves with meat and milk, fortunately abundant." By contrast, Gera's royal family, who I interviewed a century and a half later, admitted eating ensete on occasion, but pointed out that their children did not like it.[23]

None of the nineteenth-century descriptions of Gera or the forest region offer any inkling of livestock shortage as a critical constraint on agricultural production. On the contrary, livestock resources, particularly cattle, were well developed and quite specialized. In Gera and in the southwest as a whole, highland zebu varieties adapted by size and hardiness to plow traction appear to have dominated southern species such as the southwestern dwarf breed still evident around Jimma.[24] According to Cecchi, the large landlords owned "not less than 400–500 head each, while the grand dignitaries such as the Abba Corò [*abba qoro,* regional landlords representing the king] have 2 or 3 thousand." Pasture and green biomass was abundant on fallowed fields and on seasonally flooded lowlands below

20. Massaja, *I miei,* vol. 4, 220-21, 218.

21. Interview with Geriti Abba Magal and Shisay Abba Bor, Chala (Gera), 22 November 1991.

22. Massaja, *I miei,* vol. 4, 221; Cecchi, *Da Zeila,* vol. 2 , 279; interview with Abba Dura Abba Bora, 25 November 1991.

23. Massaja, *I miei,* vol. 4, 193-94, 221. Interview with Shisay Abba Bor, 22, 23 November 1991.

24. These may be remnants of the Sheko, a short-horned humpless breed. See Alberro and Haile Mariam, "Indigenous Cattle," 4.

the forested slopes. Cecchi also described stall feeding of oxen and steers with specialized diet within the homestead: breeders reserved such fattened steers, called *uotafò,* for payments of tribute to the king.[25]

The forest environment also offered natural products which supplemented the cultivation of the soil. Gera, long known for the quality of its honey, produced eight types, both domestic and wild, reflecting the seasonal cycle of flowering plants. *Tufo,* among the best types, derived from the yellow flowers of the Hypericaceae, which grows only on open, fallowed fields. The presence of this honey type, adapted to open postcultivation fields in 1880, thus offers further evidence of extensive cultivation.

By far the most important product of Gera's forest for its twentieth century economy was wild arabica coffee. Unlike neighboring areas of Jimma, Gomma, and Gumma, which produced cultivated coffee in the early twentieth century, Gera's coffee was a natural product of the forest. Wild coffee brought clear advantages in that it was more resistant to disease than the domesticated strains propagated in Jimma, which lacked the wild coffee varieties. Massaja noted in the neighboring Kaffa region, the only other district with wild varieties, that while no house was without a small plot, "the best comes from that which grows spontaneously." The best and freshest fruit was consumed locally, whereas the windfalls and less-ripened crops of February found their way into the regional trade network. By 1859 Massaja in his diary described wild coffee along Gera's southwestern frontier: "29 September 1859—passed forests of wild coffee, loaded with fruit which had already begun to ripen. All of this forest belonged to the king of Ghera who received a rich harvest of coffee without any cultivation; since when it reached maturity and fell little by little, the population was obliged to go and harvest it and bring it to the agents of the royal house."[26] Two decades later Cecchi described coffee growing within reach of the king's quarters: "At a short distance from the village of Filo a hill rises; there rises majestically the royal maserà of Ghèra; at its feet at the end are forests full of coffee, citrus, and bananas, in the midst of these rises the conical roofs of the chief of Cialla, the capital of the king." (See figure 5.3.)[27]

Though coffee had its origins in Ethiopia's southwest, there is no pre-nineteenth-century evidence of coffee trade from the Gibe region. Increasing demand for export into the resurgent Red Sea trade, however, encouraged royal families in Jimma and Gomma (to the east of Gera) by the mid-nineteenth century to respond by planting coffee bushes. By the second half of the nineteenth century, both kingdoms had easily surpassed Gera in export production.

For the local farm economy in this period, coffee was not a major income crop. In the 1850s farmers could trade dried coffee berries [*jenfel*] only for an equal mea-

25. Cecchi, *Da Zeila,* vol. 2, 282-83.
26. Massaja, *I miei,* vol. 4, 184.
27. Cecchi, *Da Zeila,* vol. 2, 248.

sure of grain. Until the late nineteenth century, Ethiopia's northern Christians disdained coffee as a Muslim drink, and merchants of the major export goods of slaves and ivory ignored coffee as a potential export crop. Only small-scale merchants of beeswax and coriander (two other forest products) collected coffee for local consumption and for their caravans destined for the regional market centers. Massaja noted that most wild coffee went unharvested altogether.[28]

There is, however, conflicting evidence on local coffee use in the nineteenth century. While coffee consumption was widely restricted in the Christian north, Massaja argued that freeborn farmers could collect fruit in the forest for their own use.[29] Oral testimony from one of Gera's oldest residents claims, however, that coffee consumption was part of a restrictive sumptuary code. Abba Dura, born at the turn of the century and son of an important *abba qoro,* recalled the political culture of the late nineteenth century:

Q. Was coffee generally drunk at that time?

A. Yes. people drank coffee when I was a child. Prior to that people did not drink coffee, as it was the property of the state. Only the state used it because coffee belonged to the state. Only with permission of the state that people got the right to drink coffee . . . A soldier who had scored brilliant victories and gained a reputation for bravery—such a figure would get permission from a committee which gave him the advantages he deserved as an honorable person, including the right to drink coffee.

Q. Weren't the common people allowed to drink coffee?

A. No, they were not allowed; if they were found drinking coffee without permission, they were sold as slaves for breaking the rule! During the reign of Menilek the forest and the coffee alike belonged to the chief; it was only for the use of the royalty that the coffee was collected.[30]

People, Land, and Forest

An enduring theme of agriculture in Gera has been the subtle, dialectic interaction between human settlement and the forest cover. The forest provided income and the means of sustenance as well as the covering, cooling shade which allowed coffee and ensete to flourish. At the same time, however, annual cereal crops required open

28. Mohammed Hassen, *The Oromo of Ethiopia,* 122-23; Massaja, *I miei,* vol. 4. 227.

29. Massaja, *I miei,* vol. 4 184. Also, interview with Abba Bor Abba Magal, former balabat of Gera, January 1990.

30. Interview with Abba Dura Abba Bora, Yukro (Gera), November 1992.

fields, soil fertility, and protection from the depredations of the forest's wild animals. The forest itself was not primeval and unyielding; when I first saw its vegetation in 1990 it clearly bore the marks of several historical layers of human action in the form of recently fallowed clearings, and secondary growth of specialized, quickly maturing, sun-loving trees, later dominated by the resurgence of tall species which block the forest floor from sunlight.

Local residents can recount in great detail the progression of growth of forest species. Within the forest on a fallowed plot or abandoned homestead site, the first stages of forest, softwood saplings, appear within a year along with bushes of a wild clove-like aromatic *kefo qamalay,* amomum, "monkey peas," baboon's cabbage, and flowering *tufo* plants (in open areas). Within four years mimosa and hardwood saplings provide sufficient shade for spontaneous growth of coffee seedlings. The broadleaf forest's tallest trees also grow surprisingly fast, creating a canopy 20 meters or above within 10–15 years. The climax forest in Gera develops within 40–50 years and includes hardwood trees like the *sombo (Ekebergia capensis),* which reaches 45 meters with a trunk diameter of 2 meters. The 50-year climax forest thus may include 30–40 species of trees, shrubs, and large plants, and a high canopy that might reach 50 meters.[31]

Reconstruction of Gera's demographic and agricultural history must begin with the evidence of human effects on this specific forest environment. The ox-plow complex and its cultigens were well fixed in Gera and had succeeded in transforming forest cover into open landscape supporting diverse annual cropping and horticulture evident in the crops and diet borne out in the snapshots offered by observers from 1859 and 1880 and described in the living memory of local residents. The presence of livestock husbandry—especially skilled in producing oxen—and the diversity of highland crops bear witness to a sophisticated agricultural system. Labor needed to clear the forest must also have been in sufficient supply for the removal of the vegetative cover which impeded the plow.

The forest's primary defense against the plow is the enormous demand for labor in the process of clearing the way for new cultivation or reclaiming a long fallowed plot. For its part, the ox-plow complex has within it, where labor allows, an expansionary imperative which in periods of population growth can and did push back the forest, creating in the forest zones its own contours and clearings near farmsteads rather than continuous frontiers of fields. Our first glimpse of Gera's environment in the mid-nineteenth century, in fact, indicates the dominance of ox-plow open cultivation in areas of Filo, Chala, and Afallo, where in 1900 and even today forest cover prevails. For the Asante state in the West African forest, that transformation from root

31. Names of plant species in Oromo and Amharic cited here derive from a 1991 field visit and interviews with Shamsu Tewfik and Seyid Abba Karo (Yukro) and Abba Fogi Abba Chebsa (Afallo). For identification of Oromo names as Latin species, see Wolde Michael Kelecha, *A Glossary of Ethiopian Plant Names,* 4th ed. (Addis Ababa, 1987).

crops to open cereal fields took place with the arrival of new, unfree labor.[32] For Gera, this change may also have included the presence of new population, but it included as well a new economic system in the form of the ox-plow complex, which required cleared land but did not provide a new technology to cut dense vegetation. Forest clearing required the use of a *gejera* (machete) and firing the base of large trees to prepare fields for cereal crops. The success of annual cereal cultivation in cleared areas in turn supported denser population and state formation.

Given the available technology, there was an intimate link between population density, labor, and the state of Gera's forest cover. The historical evidence of changes in Gera's forest challenges images of a primeval forest pushed back steadily in a linear progression of human effort. Far from a progressive front of cumulative agricultural expansion, the evidence supports an episodic, conjunctural process where open, cultivated land dominated the landscape when labor allowed and a resilient forest cover encroached once again when human labor diminished.

For Gera there is little precise historical information on forest clearance, though the process must have come at a point when population and labor resources had reached a critical mass. Local oral traditions from the Gibe states suggest that forest clearance took place at the beginning of the eighteenth century, though Gera's settlement probably took place much later. Prior to that century, low-density human activity such as root crop and perhaps limited sorghum cultivation as well as hunting and gathering, could have and probably did shape forest vegetation around the Gibe Valley's Sidama-speaking settlements.[33] In any case, not all the region was covered with dense forest. Some highland zones above 2300 meters, such as Bera Dedo in northeast Gera or Secha in the northwest, may have been savannah woodland similar to the northern highlands. Other areas, such as the seasonally waterlogged bottom lands between Filo and Chala, would not have supported hardwood growth so remained relatively open.

Our first sustained panoramic view of the forest and settlement in Gera comes rather late and derives from Italian exploration in 1859 and 1880, which also informs us about crops and local politics. These accounts provide a benchmark against which to measure settlement patterns of the twentieth century and the basis for reconstructing the late-nineteenth-century human and physical landscape. The picture which reaches us from these accounts is of a setting quite different from

32. Ivor Wilks, "Land, Labour, Capital and the Forest Kingdom of Asante: A Model of Early Change," in J. Friedman and M. J. Rowlands, eds., *The Evolution of Social Systems* (Pittsburgh, 1978), 501-2. Wilks' work on the West African forest indicates that clearing one hectare of virgin forest required removal of 1250 tons of vegetation; a 15-year fallow field would require 100 tons.

33. Mohammed Hassen, *The Oromo of Ethiopia*, 116. For a comparative study of North American Indian land use effects on forest, see Silver, *A New Face on the Countryside*, 46-48.

northern Shawa's highlands; rather, it is a highly varied one shaped by a peculiar combination of human action and environmental forces. With no aerial view existing, our images derive from a ground-level perspective: travelers on muleback or foot moving through forest to clearings or, to the royal *masera* along well-marked routes of travel.

Travelers first entered nineteenth-century Gera along a caravan route from the small kingdom of Gomma. Though the Gibe states emerged from roughly similar environmental and political roots, conflicts between them, differences in economic structure, and land use set them apart from one another. The frontier of Gera with Gomma, for example, comprised socially vacant spaces created by political conflict. Cecchi described the frontier zone as "immense and gloomy forest rich in time-honored trees with colossal trunks. Among them were podocarpus, euphorbia, cussi, sycamore, wild olive, acacia, and gardenia." Passing through the "majesty of the superb forest" toward central Gera, the dense vegetation broke: "In the valleys and along the slopes of the hills are the plowed fields with the scattered frequency of the thatched huts and villages, posted in the most pleasant [*ridente*] positions." Farther east approaching the capital, Cecchi described boggy bottom lands near Filo: "The end of this valley presents an extended plain, little cultivated although the soil was very fertile and rich in water which for lack of slope here and there was waterlogged. The dry part forms a luxurious meadowland where beautiful cattle and sheep grazed."[34] Homesteads themselves were set back from these open plains, either isolated or collected in small hamlets of 10 or 15 families. Gera families constructed their houses, "with a certain air of elegance," of bamboo, ensete leaves, and teff straw, revealing the products of their mixed farm and forest economy. Houses of the wealthy also boasted *gombisa* [granaries] of threshed cereals as well as containers for unthreshed teff and *dagusa*.[35]

Abba Dura recalled descriptions of the nineteenth-century open landscape around Filo: "There was no forest. People had to use dung as fuel. The whole area was covered with houses. There was no forest; there was nothing to use for fuel. There was no wood!"[36]

At the capital of Gera at Chala (or Ciala, in the Italian rendering) the royal *masera*, was set—as is the home today of the royal family's descendants—on one of a chain of hills at the western edge of the bottom land meadow. Massaja recalled that in 1859 it was "dressed with plants of citrus, coffee, bananas." Two decades later, Massaja's published engraving of the setting of Cecchi's confinement showed a horizon of open, cultivated fields with the royal compound in the foreground surrounded by ensete, lemon trees, acacias, and bananas as well as the stumped remnant of a large tree. The capital itself was a substantial settlement,

34. Cecchi, *Da Zeila,* vol. 2, 248.
35. Cecchi, *Da Zeila,* vol. 2, 256.
36. Interview with Abba Dura Abba Bora, Yukro (Gera), November 1992.

consisting of the royal compound, animal enclosures, and quarters for 3000 of the king's slaves, who prepared food for the royal household and cultivated extensive royal fields.[37]

The district of Afallo to the south and west of the capital toward the Gojeb river valley was, at least in the nineteenth century, the most densely settled region of Gera. Massaja described the people as herders and farmers whose homesteads were scattered across beautiful and "flowing" cultivation. At Afallo, Massaja established a Capuchin mission station which served the exiled Christian community and local converts. He had built there a compound of five houses, planting the grounds with what must have been Gera's ambient vegetation: ensete, coffee, lemon; in the mission's fields Massaja's followers sowed teff and "ogni sorta di cereali indigeni."[38]

Twenty years later Cecchi retraced Massaja's path from Chala to Afallo. Like Massaja, he was impressed: "The country is the most beautiful and picturesque, and most populated of the kingdom. Everywhere thatched huts separate or together in groups in the midst of ensete plantations, lemon trees, coffee, [kale], a special variety of potato [*Ya Oromo dinnich*] onions, garlic . . ."[39]

The landscape engraving of the mission at Afallo published with Cecchi's book depicts a bucolic but cultivated garden setting: open ground surrounding the mission dwellings, ensete interspersed with small trees. One can imagine at a distance the mission's fields of cereals, cultivated under the mission's ox plow for 20 years, among hundreds of plantings of ensete (300–500 per household) and pasture. The engraving, published variously as a rendering of the mission station or as the tomb of Chiarini (Cecchi's companion who died *in situ*) is idealized, even antiseptic. It shows closely cropped grass and an open horizon beyond Yet, its image of an open landscape is congruent with both observers' textual accounts of Afallo as densely settled and extensively cultivated in annual crops and ensete. Abba Fogi of Afallo, who spoke to me in 1991, recalled descriptions from his grandfather's time (the nineteenth century), that Afallo was "a field without any trace of forest."[40] That landscape (see figure 5.6 below) was the product of concerted and continuous human action to clear and maintain space for the needs of ox-plow cultivation.

Given the density and resilience of the forest cover, how did Gera's agricultural base achieve its nineteenth-century dominance? The accumulated critical mass of labor appears to have been the key ingredient. Cecchi in 1880 estimated Gera's population at between 15,000 and 16,000, an estimate which appears not to have included the huge population of slaves he reported elsewhere in his

37. Cecchi, *Da Zeila,* vol. 2, 292.

38. Massaja, *I miei,* vol. 4, 224.

39. Cecchi, *Da Zeila,* vol. 2, 382.

40. Interview with Abba Fogi Abba Chebso, Afallo (Gera), 25 November 1991.

Figure 5.3. Chala *masera* (royal residence), Gera, c. 1860.

Figure 5.4. Chala royal residence site, 1991.

165

account.[41] Slaves' most important long-term role was as a captive class of agricultural labor. Unlike in much of the north, slaves in Gera had an active role in field labor, cultivating for the royal houses as well as for landed aristocracy. Free farm households, many of whom may have been migrants, also worked royal lands, but they served as tenants and probably bore responsibility for clearing new forest fields as part of their tenancy. Cecchi observed: "In Gera the rich land owners cultivate with their slaves; where they do not have them or do not have enough, he distributes the land to free persons [*coloni*] who cultivate in his name and in his interest."[42]

Given the level of technology of agriculture and implements for cutting trees and vegetation, only a rapid rise of population and labor supply could account for the massive early-nineteenth-century forest clearing implied in the open lands at Afallo and along the central highlands around Filo and Chala which, except for marshy bottom lands, supported forest. While dating this transformation is difficult, it is reasonable to assume that the foundations of the kingdom rested on the maturation of cereal agriculture as a means of accumulating, concentrating, and storing food, a process far more difficult with forest food products.

What we know of historical property systems and land tenure in Gera suggests that it attracted population from surrounding zones. Fertility rates were far less critical to Gera's historical labor supply than the rapid absorption of new population. Local social practice encouraged such movement. Migrants and refugees did not face a rigid ethnic definition of land ownership. Oromo property law transferred land and property to the eldest son, but provided income to younger sons and dowries to daughters, encouraging and capitalizing on the clearing of new plots in the forest. Oromo lineages also apparently allowed households without issue to adopt new members with the property rights of children. Slaves, according to Cecchi, enjoyed no rights of inheritance and thus tended to retain their status over time.[43] Though the flow of slaves into the Red Sea trade drew down Gera's labor force, over time new sources of slaves meant captive labor lay along the southern frontier. Moreover, their use value to the royal family and local yeomanry may well have exceeded their exchange value for imported goods.

The expansion of the ox-plow complex both in Gera and the Gibe region as a whole thus took place simultaneously with the application of concentrated labor power on the forest cover and the formation of small-scale kingdoms. Once cleared, fallowed forest plots with their characteristically shallow nutrient base needed to be fallow three of every five years, and new plots had to be cleared from the forest. Once cleared of virgin cover, however, forest plots required significantly less labor

41. Cecchi's estimates of the total population and of 3000 slaves at the royal capital strongly suggest that his reckoning for Gera's population did not include slaves.

42. Cecchi, *Da Zeila*, vol. 2, 277.

43. Cecchi, *Da Zeila*, vol. 2, 292-93; Soleillet, *Voyages en Ethiopie*, 262-64.

to bring into production again. West African virgin forest required the removal of 1,250 tons of vegetation for first cultivation; subsequent fallow clearance requires only 100 tons.[44] Thus, over time the labor invested in clearing forest plots reached a plateau and leveled off to a point of equilibrium, where the norm was retilling fallowed plots rather than clearing virgin forest.

A primary motive for forest clearance must have been the agronomic imperatives of annual cereal crops preferred in elite diets. Sorghum was probably the earliest forest cereal crop, given the Cushitic root of its Oromo name; it is a crop ideal for forest clearings, since it requires little labor after sowing and its root structure helps break down clods of virgin soil. Teff and eleusine, by contrast, require heavily worked soil and precise sowing and harvesting schedules ill-suited to the early stages of forest clearance. Those crops, and also barley and wheat, probably arrived in the forest crop repertoire somewhat later, after forest clearance opened land to continuous cultivation and with the arrival of northern migrants. Certainly by the mid-nineteenth century several cultivars of teff, wheat, and barley were well integrated into the elite diet.

By the mid-twentieth century farmers clearing forest preferred maize as the initial crop on newly cleared forest plots. Maize was well suited to Gera's moist, nitrogen-rich soil. The single cultivar, yellow flint, reported in the nineteenth-century, was a relatively recent arrival, though its quick maturation and its edibility in the green stage after three months made it attractive as a double crop to meet the needs of the hungry season. Cecchi's experience with it as a food offered to a prisoner, however, suggests its low regard in elite diets.

Thus, the political and economic kingdom first visible in outside accounts by middle of the nineteenth century derived from the conjunctural concentration of labor, the efficacy of a technology of production in the form of the ox plow and livestock husbandry, and the application of a diverse bundle of food crops which complemented both an earlier horticultural tradition and the forest's natural products. Gera's mid- to late-nineteenth century landscape as described by outsiders and the historical memory of its oldest citizens did not resemble, even remotely, the open highland plateaus of the north. The rolling hills above the cultivated plain were permanently green with forest, ensete, and marshy pasture. Homesteads cultivating forest plots appeared here and there, surrounded not by contiguous fields but by clearings in the forest containing both crops and fallowed plots. Where concentrations of population collected, around Afallo or at Chala, open, cultivated expanses appeared, turning brown by December and showing a marked contrast with the forest cultivation in less populated zones.

Cecchi noted the specialized cultivation of these intensively cropped zones, probably those royal lands around Chala, where slave and corvée labor produced cereals for the royal table: "The soil, in difference to that of Abyssinia, is disposed

44. Wilks, "Land, Labour, and Capital," 501-2.

in great plains crossed only by a network of paths to allow access to the fields and for weeding the bad weeds. Everyone who participates in the activity—the most important after seeding—puts them to dry and then burns them. The ashes are dispersed some days before the rains on the ground and this fertilizes the soil."[45] The sophistication and organization of this production suggests a thorough penetration of the ox-plow complex and its capacity to shape the terms of labor and to transform the landscape. The forest itself remained pristine only along the frontier, a status determined by movements in human history rather than by nature itself.

The Plow Retrenches, 1881–1940

The slow multigenerational process of clearing the forest and the consequent expansion of ox-plow cultivation stood in marked contrast with swift movements in the political sphere, which had significant environmental implications. Cecchi and Chiarini, the two inquisitive Italians who had appeared on the scene in June 1879, were prime suspects as a Shawan fifth column. Cecchi's detainment in Gera by the queen mother, the *Ghenne* Gumiti, was a result of the queen's well-founded suspicion that Negus Menilek of Shawa, still nine years from attaining the imperial throne as Emperor Menilek II, intended to expand his tributary hegemony to the resource-rich Gibe region. The most immediate effect beyond detention itself was a shift in diet provided to her guests by the *ghenne*. On 5 October 1880 his companion, Giuseppe Chiarini, died of complications from the poor diet and an endemic but unidentified fever.[46] The *ghenne,* however, had been prophetic; in March 1881 the threat of Shawan invasion forced Gera's submission to Menilek and payment of tribute. Six years later, a small detachment of Menilek's army under Dejazmach Basha Abuye occupied Gera and sent the king Abba Rago and the queen mother, Ghenne Gumiti, into exile in Jimma, where they died in captivity.[47]

There are very few records of the Shawan administration of Gera and fewer still of the effects on agriculture. Oral tradition, only one generation old, suggests the basic outlines. The Shawan conquest meant the arrival of a military governor and appointment of eight *abba qoro,* local agents who organized civil and fiscal admin-

45. Cecchi, *Da Zeila,* vol. 2, 277-78.

46. Chris Prouty Rosenfeld, *A Chronology of Menilek II of Ethiopia* (East Lansing, 1976), 85.

47. Rosenfeld, *Chronology,* 99. One local informant recalled that Abba Rago died shortly after the occupation and that it was his son Abba Magal who had gone into exile. Interview with Abba Dura Abba Bor, Yukro (Gera), 24 November 1991. See also Cerulli, *Ethiopia Occidentale,* 159.

Figure 5.5. Afallo mission, c. 1880.

Figure 5.6. Afallo mission site, c. 1991.

istration on behalf of the Shawans. The presence of the Shawan state, however, was
in main form represented by the *malkanya,* officers to whom the military governor
assigned Gera farmers as *gabbar,* local cultivators who owed them labor and tribute. How many were there? The great grandson of the *ghenne* remembers: "It is not
possible to say one or two; there were many of them. The country was not as *taf*
[barren, impoverished] as it is now, it was *lam* [productive, wealthy] in those days.
It was the *malkanya* who impoverished it and then left it for Jimma, Gomma, and
other places."[48] Abba Dura, born during Menilek's occupation, recalled local conditions: "As a result of Menilek's invasion the people faced a shortage of food; they
had to feed the army of Amharas and their officials. The population decreased. People could not exchange their products in the market . . . . Those who submitted to
the Amhara's pounded the stalk of the bush used for making beer. People had nothing to eat and lost their will and energy. They ran away."[49]

⸱ *The malkanya,* the arm of the occupying army, quickly went about the process
of extracting Gera's resources. The initial phase of occupation did not last long.
Coffee, though a potential source of wealth, was not immediately extractable without surplus labor, nor was it of particular interest to the occupying forces, who
sought resources more immediately convertible to prestige goods. Gera's most
moveable assets were its slaves, whom the Shawans apparently fed into a still burgeoning Red Sea trade. Given the large numbers of slaves reported at the height of
economic growth in the 1850–80 period, the loss of labor to the agricultural economy was devastating.

While other areas of conquest received and retained large numbers of northern
soldier-settlers, few remained in Gera, and most probably continued to the west,
where ivory hunting and further sources of slaves were more attractive.[50] Most
northern settlers had gone by the 1920s. On the imperial scale, conquests of Harar
and Arsi and tribute extracted from the larger and better-located Jimma provided far
more important economic resources than Gera. If it was short-lived, the actual occupation was harsh, driving Gera's residents to neighboring regions such as Jimma,
which Amhara rule had not yet reached. One elder from Afallo recalled: "The
inhabitants were wiped out, finished . . . . A large number of people left Gera and
fled to Kaffa, Jimma, Gomma, Diddu, and the like . . . . My grandfather also
escaped to Jimma."[51]

48. Interview with Abba Bor Abba Magal, Chala (Gera), 21 January 1990. *"Lam"* meant
"fertile," but actually referred to presence of labor on the land.

49. Interview with Abba Dura Abba Bor, Yukro (Gera), 24 November 1991.

50. Gera's lack of extractable wealth, in fact, made enslavement a form of punishment.
Gera's second governor, Dejach Gename, was appointed there as a result of transgressions
in the wealthier Wallaga region. Interview with Abba Dura Abba Bora, Yukro (Gera), 23
November 1991. For an example of where settlers remained, see McClellan, *State Transformation,* 131-46.

51. Interview with Abba Fogi Abba Chebsa, Afallo (Gera), 25 November 1991.

The loss of slave labor and of an unknown number of refugees among the free Gera population and the departure of even the occupying troops emptied Gera. Frank de Halpert, head of Ethiopia's antislavery commission in 1932, estimated that the population of the region "probably decreased by three-quarters," a figure confirmed in local traditions. In 1910 Montandon, passing through the nearby administrative center at Agaro, referred to it as almost abandoned.[52] In fairly short order the absence of labor permitted the forest to recapture what it had yielded only a few decades before to the plow, machete, and short-handled ax. In Gera's humid setting and without the vigilance of human labor, the forest cover regenerated quickly. By the late 1920s virtually all the open cultivation had reverted to climax forest.

Testimony to the forest's regrowth and to the effects of the flight of human labor from Gera comes from Enrico Cerulli, who revisited Gera in 1928, determined to retrace the steps of Cecchi and to locate the tomb of the "martyr" Chiarini. Cerulli arrived at the Gera frontier a month and a day from the 49th anniversary of Cecchi's entry. After four hours' trek through the frontier forest, Cerulli described the effects of 40 years of secondary forest growth on Gera's once thickly settled central highland. Emerging from the primary frontier forest, he observed:

It was possible to discern only—very indirect alas!—traces of human life. We entered the region of Filo, at least this was told to us by the guide. These which were one time the names of villages, and now passed and indicated simply by some tracks in the forest; in some years these will be entirely forgotten. At Filo Cecchi had found a village and from this rose at that time Ciala [Chala]; now there is nothing, not even a hut and the forest has covered all the hills, not leaving anything to see. . . . Of the ancient capital nothing remained but the memory.[53]

Cerulli continued his trek along the old route past Chala, southwest to the site of the mission station at Afallo, which both Massaja and Cecchi had called the most densely settled region in Gera. He found the open, cultivated plain depicted in Cecchi's nineteenth-century engraving and narrative entirely reclaimed by the forest. Chiarini's tomb and the mission station itself were covered in a dense secondary growth of hardwood trees and ground cover. Cerulli's party described the result of the aftermath of depopulation and almost a half century of regrowth: "In Chira from here we have not seen even a village, we have not seen even one man. The forest is most thick and intricate . . . . The mission was cleared forty years ago; and with this equatorial climate and with this humidity an abandoned village situated at the

52. De Halpert cited in Margery Perham, *The Government of Ethiopia* (Oxford, 1959), 321-22; George Montandon, *Au pays Ghimirra: Récit de mon voyage à travers le massif éthiopien (1909-1911)* (Neuchatel, 1913), 350.

53. Cerulli, *Etiopia Occidentale*, 156-57.

Figure 5.7. Yalew Taffese, migrant coffee farmer (Gera).

margins of the forest, disappears, victim to the forest, in still less time."[54] In 1991, when I visited Afallo, the old mission site remained covered by a dense forest canopy; only the oldest residents remembered it as open, cultivated land, from a story two generations old.[55]

The Italian occupation of Ethiopia in early 1936 brought little if any change to Gera. In 1937 they appointed Abba Bor, the son of the last king of Gera, as *balabat* of Gera, and he established himself in a tin-roofed house 200 meters west of Chala Guda (Big Chala), the former site of the royal *masera,* now tumbled down and the stones carried to the royal cemetery. After his appointment as *balabat* by the Italian authorities, he and his family lived on a small salary and the receipts from his family lands in the immediate area, which, despite the diminution of crop land, provided his table with teff, wheat, maize, beans, and peas—crop types which "came from our grandfathers." The 1938 Italian *Guida dell'Africa Orientale Italiana* tourist guide described the site of the former capital for potential visitors as "a few huts menaced by the forest near the site from which rose Ciala, the former capital of Ghéra."[56]

Within Gera, the effect of the conquest by Menilek's army was less occupation than the massive collapse of the political and economic structure that had pushed back the nineteenth-century forest. Little documentary evidence exists either locally or from the outside to chronicle the rapid decline of local labor and agriculture. Local life histories and recollections may be the only means to reconstruct the shape of events during in the interwar years. The mute testimony of the resurgent forests, however, corroborates the massive loss of population between 1880 and 1928 described in oral accounts.

By the late 1920s, imperial government representatives at Agaro were negotiating monopoly trading rights in coffee for the region. In 1933 the imperial government completed its first major road project, an all-weather road connection from Addis Ababa to Jimma, the center of the coffee trade.

The Coffee-Maize Complex, 1941–75

Gera's postwar agricultural economy derived in large measure from the transformation of the role of coffee in the national economy and the local effects on areas which began to produce it. The depletion of the ivory frontier and suppression of the slave trade refocused attention in the national economy on the productive rather than the extractive capacity of the southwestern economy. Postwar growth built upon prewar

54. Cerulli, *Etiopia Occidentale,* 162.

55. Testimony to the mid-nineteenth-century lack of forest in Afallo parallels that for Filo and Chala. Interview with Abba Fogi Abba Chebsa, (Afallo), 25 November 1991.

56. Interview with Abba Bor Abba Magal, Chala (Gera), 21 January 1990; Consociazione Turistica Italiana, *Guida dell'Africa Orientale Italiana* (Milano, 1938), 532.

tendencies. Coffee as an export crop from the southwest began to be economic around 1910, with its major growth after 1917, when the French-built railway finally established its railhead at Addis Ababa, drastically reducing transport costs. By 1925 the exports of western coffee (i.e., that from the Gibe region and Sidamo) had surpassed production from the Harar region (the source of mocha coffee).

In the early years the major leap forward in total coffee exports from the Gibe region resulted from the production from cultivated plantations that had been put in place in Jimma by Abba Jifar and had reached maturity in the 1920s. In the interwar years, however, Gera remained dormant as a coffee-producing region, with its labor crisis apparently preventing increases in exports even from the collection from wild bushes.

Oral evidence from community elders recalls Gera's expansion of coffee production as being linked to an overall resurgence of the coffee trade at the end of the Italian occupation when Italian-built roads promoted the use of motorized transport from coffee collection centers such as Jimma and Agaro, the latter being Gera's major coffee market. Table 5.2 indicates the scale of expansion nationally, a trend which parallels evidence of Gera's own role in coffee production.

Coffee plantations on a large scale had begun in Jimma after the turn of the century and spread to other regions. Cultivation of domestic trees for other areas, including Gera, began primarily in the postwar period, when prices, transport, and new population expanded the prospects. The rise in coffee production overall resulted in and reflected ripple effects in local farm economies. In the postwar era new incentives for transforming rural land title into cash profits resulted in massive purchases of tax-delinquent land for new coffee plantations. By the 1970s in Kaffa as a whole, 59 percent of the rural population cultivated its entire land holding as tenants who paid either rent or labor to landlords, who in turn accumulated wealth in profits from coffee exports.[57]

In Gera, as in the old Kaffa kingdom to the southwest, increased production initially involved the expansion of wild coffee collection. Spontaneous forest coffee, however, was never totally a free good; forest lands often belonged to the royal family or, later, to landlords, who retained rights of collection along with corvée obligations from local farmers. Coffee in unclaimed forest areas could be collected by anyone who chose to invest their own labor. The harvest of the wild fruit, however, imposed its own rigors, since berries began maturing in September, with the bulk of labor required in October and November. Plants had to be "slashed" (weeded) three times per year, though coffee farmers did little pruning.[58]

57. Guluma Gemeda, "Some Notes on Food Crop and Coffee Cultivation in Jimma and Limmu Awrajas, Kafa Administrative Region (1950s-1970s)," in *Proceedings of the Third Annual Seminar of the Department of History* (Addis Ababa, 1986), 93; John Cohen and Dov Weintraub, *Land and Peasants in Imperial Ethiopia: The Social Background to a Revolution* (Assen, 1975), 51.

58. Interview with Abba Bor Abba Magal, Chala (Gera), 21 January 1990.

TABLE 5.1. Coffee exports (in metric tons) via Djibouti, 1910–34.

Year	Harari	Abyssinian[a]
1910–14	15,949	1,160
1915–19	18,803	619
1920–24	22,035	8,991
1925–29	30,186	34,337
1930–34	41,564	42,115

Source: McClellan, *State Transformation,* 122.

[a]Abyssinian refers to coffee from southwestern Ethiopia, including that from Gera.

TABLE 5.2. Coffee exports from Ethiopia, 1943–57.

Year[a]	Exports in English tons	Year[a]	Exports in English tons
1943–44	11,883	1950–51	27,503
1944–45	13,338	1951–52	25,451
1945–46	15,882	1952–53	36,219
1946–47	14,663	1953–54	37,219
1947–48	17,795	1954–55	38,614
1948–49	17,829	1955–56	32,953
1949–50	21,152	1956–57	49,220

Source: Huffnagel, *Agriculture:* 219.

[a]Years are Ethiopian Calendar 1936–49 E.C.

Open access to forested land, favorable markets, and ideally acid, nitrogen-rich soil also stimulated small farmers and outside entrepreneurs to establish new coffee plantations in Gera, albeit somewhat later than in areas closer to markets. Few documents exist, but local living memory accounts detail the process. Gera's first plantation was established in the early 1930s on 20 *fachasa* of land (5 hectares) near Filo near the caravan track to Agaro. It was established by a Gomma entrepreneur, Abba Milkiyas Abba Waji, who introduced transplanted seedlings to Gera. He also hired local day labor to plant, slash, and harvest the coffee.[59]

In the postwar coffee boom, coffee planting and systematic collection of forest coffee increased, drawing outside capital from merchants, salaried town-dwellers,

59. Interviews with Yalew Taffese, Abba Dura Abba Bor, and Abba Bulgu Doma Maru, (Gera), November 1991.

and traditional landlords to purchase forested land. Gera's own landed elite prospered by selling forest land to new migrants, though members of the elite themselves invested few resources directly in coffee. Abba Bor, Gera's *balabat* and later district governor, sold his land rights to entrepreneurs but developed little, if any, of his own holdings.[60] Yalew Taffese, a coffee migrant who came to Gera in 1960, recalled that the coffee boom was infectious:

> The benefit Abba Milkiyas got from his coffee plantation started it, and everybody began planting coffee. Farmers, elders, and young men cleared their forest and prepared their plots for coffee.

> Q. When did the modern farmers who plant coffee in rows come to this area?

> A. This was recently, during the reign of Haile Selassie. They came (in the name of the father, the son, and the holy spirit) around 1948 (A.D.1955–56) and afterwards . . . . It was Gitcho Yadeta who paved the way. People like Berhane Gelata, Shone, and Yihune [urban entrepreneurs] followed.[61]

Beyond a steady increase in coffee prices and new forms of transport, which enhanced profitability for the low-value-per-unit commodity, perhaps the most important spur to coffee's local expansion was an increase of population through migration in the post-1941 period. New immigrants came from a variety of sources. Unlike in the north, land titles in Gera were for sale as freehold. After the Italian expulsion in 1941, cashiered soldiers could easily buy land from local landlords willing to sell empty forest tracts at low prices. Segaye Sidu, a native of Jirru in northern Shawa, for example, arrived in Gera in 1943 with his uncle and father, who purchased 80 hectares of land from an Oromo landlord. Ato Segaye recalled that Gera cultivated little coffee at that time and that most land, with the exception of bottom lands, was forested. Only one Orthodox church existed (Chira Maryam) to serve the needs of Gera's Christian population arriving from Gojjam, Salale (northern Shawa) and Gonder. However, Segaye also recalled arriving to find an ox-plow system not strikingly different from what he had left behind in Shawa. Some of the crops and the diet-sorghum, teff, and maize—were familiar but also included strange root crops such as yams and taro, which were unfamiliar to him.[62]

Abba Karo and Adda Reda, husband and wife, were in the first wave of new Oromo migrants who arrived in the region to participate in Gera's agricultural transformation, arriving there in 1936, six months into the Italian occupation, with 30 other migrants from Wallaga to the north. They purchased 5 *fechasa* of land on the forested lower slopes above the Filo bottom lands. Though forested with "big trees"

60. Kassahun, Franzel, and Tesfaye, "Initial Results," 7.

61. Interview with Yalew Taffese, Gera, 29 November 1991.

62. Interview with Segaye Sidu, Yukro (Gera), 13 January 1990.

in 1936, they learned that fifty years before (i.e., prior to the mass exodus) local farmers had cultivated what would become Abba Karo's and Adda Reda's homestead site.

Though mature adults, Abba Karo and Adda Reda were newly married and had no children to provide family labor. Nevertheless, they quickly set about cutting a farm out of the forest, drawing upon *dabo* (exchange labor) to clear plots and plant ensete. In their first season they also planted maize on the newly cleared plots and followed two years' sowings of maize with teff, according to the local practice. They also planted ensete, common locally, not as a primary diet crop but as a hedge against failed annual crops, also a Gera pattern. Breaking from local models, however, Abba Karo began to plant and cultivate coffee seedlings in the forest above his homestead, with knowledge brought from their native Wallaga. Within three years the plantings yielded coffee, which they sun-dried as *jenfel* (dried beans) and carried by mule to the Agaro market. Within a decade, their farm harvested coffee from almost 3000 trees, producing as much as 24 *farasula* [1 *farasula*=17 kg] of *jenfel* in the fall harvest. The farm also produced maize and teff as well as fruit, *chat,* and root crops from the homestead's compound.

Abba Karo and Adda Reda were pioneers in resettling Gera's forests, but their model was replicated many times over by new waves of migrants after the Italian occupation. New arrivals were seasonal migrants who arrived annually during the prime coffee-picking season between October and December to supplement local labor in picking coffee on new plantations and areas of wild coffee on lands controlled by local landlords. These laborers, drawn from eastern Gojjam, southern Begamder, and the Gurage area of western Shawa, sought off-farm wages to supplement income from their own production. For the southwest's coffee districts as a whole, this annual influx, which began in the 1940s, had reached 50,000 persons annually by 1970, increasing the population in some coffee growing areas by 5–10 percent.[63] The seasonal migration also brought permanent population as many laborers used wages from the coffee harvests to purchase land or settle in new markets as petty merchants. Gera's portion of this influx may not be possible to determine, but the presence of new forms of labor stimulated the clearing and cultivation of new coffee lands, a process which over time transformed the agricultural economy and the relationship between annual ox-plow cropping and allocations of seasonal labor.

In the postwar rural economy access to coffee income was a critical factor in the organization of farm labor and small-farm economies. Locally, two strategies for expanding coffee production dominated. Residents with land title could recruit labor locally from tenants in exchange for waiving rental fees, or they could recruit corvée labor. Coffee pickers recruited locally sometimes also received half the product, which they could convert to cash in local markets. Collection of wild coffee on a

63. Adrian Wood, "The Decline of Seasonal Labour Migration to the Coffee Forests of South-west Ethiopia," *Geography* 68 (1983), 53-56.

landlord's estate was, in fact, an important strategy for new farmers to supplement farm income to purchase oxen while also planting trees on their own land.[64]

Establishing coffee plantations in Gera did not even require time delays for seedlings to develop, since it was possible to clear the forest floor around existing wild trees, thin them, and replant young bushes. The forest itself provided the required shade and soil moisture. No fertilizer was needed—only labor to pick berries, prune back old growth, transplant seedlings, and slash weeds. Most important was the control over land and the ability to mobilize labor for the harvest and cultivation of the crop from local tenants or from seasonal northern migrants, many from the landlords' own home areas.[65] Absentee landlords often required tenants to develop coffee plantations from uncleared forest land in lieu of rent paid in kind.

Other farmers were less fortunate. If coffee production was an effective source of income, it was not a universal option for farmers engaged primarily in annual crop production. The labor calendar for coffee plantations, or even for the picking of the wild forest crop, involved substantial conflicts with the labor needs of annual crops, especially teff, eleusine, and wheat. Coffee cultivation coincided with important field activities of annual crops. For oxen-poor farmers, the conflict was an absolute one: land preparation for the main sowing period for the major cereal crops required borrowing oxen, which were available only late in the planting season, a key time for tending coffee plantations. By contrast, oxen-rich households could either prepare their land early and then tend coffee trees or exchange oxen usage for labor.[66] For teff and eleusine, the main land preparation and planting season conflicted with four of the major coffee cultivation activities. The harvest in particular was critical, since teff and coffee had to be harvested in overlapping periods. If left in the field, teff was susceptible to major mold damage from late fall rains.

Changing world-market processing requirements in recent years also skewed the labor equation, affecting farms which attempted to produce food crops and manage a coffee plantation. Small farmers preferred the dry-processing method, which allowed collection of coffee berries in slack times in the food crop cycle, drying them on the ground, and marketing the dried berries in bulk. The "wet" method introduced, in the 1960s, required continuous harvesting of only ripe berries thus shifting small farm labor priorities, and requiring continuous harvesting over the September–December period, otherwise devoted to weeding and harvest of annual crops. Only *cashira* (wet-processed coffee) reached the world market, drawing three times the price of *jenfel* (dry-processed) coffee.[67]

---

64. Such farmers included those who arrived with Abba Karo in 1936, but also many who arrived up to the 1970s. Interviews with Abba Karo and Abba Fogi Abba Chebsa, (Gera), November 1991.

65. Wood, "Seasonal Labour," 53.

66. Interview with Galbi Senbeto, Wanja (Gera), 14 January 1990.

67. Interview with Galbi Senbeto, Wanja (Gera), 14 January 1990; Kassahun, Franzel, and Tesfaye, "Initial Results," 16.

The conflicts between the agricultural calendars of coffee and annual crops over time brought about a specific historical adjustment in cropping patterns, what might be called the maize-coffee complex. Maize, already present in forest cropping systems by the late nineteenth century, proved a superb complement to the demands of coffee labor needs for both tenants and small farms invested in coffee. Maize's primary advantages as a complementary crop were its short maturation cycle, relatively high yield, and comparatively low labor cost for sowing and processing. Moreover, maize performed particularly well in the nitrogen-rich soils of Gera and on bottom lands allowing a second, dry-season crop. The original "Oromo" maize variety, with its ears' distinctively bare tips, was well suited to uplands, was planted just before the reliable main rains, and matured in five to six months, providing food before coffee income in the fall. Its relatively low yield probably restricted its spread in the pre-Italian years. More recently, early maturing, open-pollinated, improved varieties—what Gera farmers refer to as "Kenya"—have offered a higher yield and an earlier crop. The improved cultivars' short cycle also suits them to double-cropping with teff on bottom lands, where soil moisture permits an early-spring teff harvest.[68]

Perhaps maize's most important quality as a partner to coffee cultivation was the flexibility within its harvest schedule. Maize can provide food at its green stage after only three months, arriving early in the traditional hungry season, that is, the summer months before coffee harvests, when farmers acquired debt from food purchases. Once mature, maize can remain in the field without moisture damage, awaiting harvest until after coffee harvests have ended. Farmers collected ears and stored them unshelled in open cribs for a few months. Maize's liabilities as a subsistence crop (see below) were not much of a factor when it complemented coffee income. The coffee-maize complex, however, economically differentiated resource-rich farms, which derived income from coffee cultivation and also produced maize and some teff, from low-resource farms, which increasingly concentrated on low-labor, subsistence, annual crops supplemented by wage labor.[69]

The rise of the coffee-maize complex across the southwest was a by-product of increasing coffee production in Ethiopia as a whole. Kaffa's percentage of Ethiopia's total production rose along with the total increase. While Gera's total production was below that of its neighboring districts, Gomma, Manna, and Jimma, it nonetheless was a high production region. By 1971–72 (i.e., just prior to the revolution), Kaffa as a whole produced 40,884 metric tons, over 40 percent of Ethiopia's total production. The regional economic boom which stimulated in-

68. Improved varieties of composite maize introduced by the Institute of Agricultural Research and the Ministry of Agriculture since 1967 include open-pollinated types that have mixed with bare-tipped local varieties over time. Though farmers sometimes differentiate characteristics of Oromo and Kenya varieties, such distinctions refer only to phenotypes visible in a given field, not to genuine varieties. Interview with Tassew Gabayehu, Institute of Agricultural Research Station (Jimma), 28 November 1991.

69. Guluma Gemeda, "Some Notes," 95-97.

migration to Gera also extended well beyond the elevations amenable to coffee production. Although Gera produced less coffee than some of its neighbors, the role of coffee income in the food crop economy linked annual crop agriculture to the forest cultivation of coffee. High prices for teff, driven by a lively cash market for prestige food crops, encouraged farms above coffee elevations to produce teff, barley, and small amounts of wheat for the market. In areas above Gera's central valley, migrants reoccupied highland areas above 2300, meters ideal for annual crop production. These new migrants, who came as tenants or as purchasers of land titles, came from a wide variety of agricultural traditions, bringing with them crop production techniques, dietary preferences, and conceptions of environmental resource use. In those resettled areas, highland farming systems and environmental use differed widely, reflecting imported variations of the ox-plow system.

The new coffee economy, which brought both economic expansion and population growth, also brought changes in the forest landscape. New migrants reoccupied the lower slopes with dispersed homestead sites. Ensete plantations reappeared within two to three years after houses and forest plots for maize and teff emerged, scattered near the housing sites. The new population, however, did not push back the forest entirely, since soil fertility permitted short-fallow cultivation. Moreover, coffee plantations promoted the preservation of forest cover because they required at least 30 percent shade.[70] When new migrants began to populate the landscape it took new forms: at coffee elevations (1800–2200 meters) forest cover remained; at higher elevations conducive to annual cropping northern migrants began forest clearing to make way for teff, wheat, and barley.

The human imprint on Gera's highland landscape differed widely in those zones concentrating in annual crops. Changes in the physical appearance of the land were a direct result of the extensive cultivation tradition imported from the northern Shawan highlands in Gera's northeast district of Bera Dedo. There, migrants from Manz in northern Shawa in the mid-twentieth century steadily and consciously reduced forest cover to make way for the extensive cultivation of highland cereals and horsebeans in a five-year rotation followed by three years' fallow. Grain from the area found an easy market at Bashasha, a market town on Gera's border with coffee-rich Gomma. The evolution of the coffee-maize complex and its corollary food markets also stimulated a parallel expansion of transport, communication, and capital markets, which linked Gera's local economy to regional and national markets. The small market towns of Bera, Chira, Dusta, Bashasha, and Sadi served as points of exchange between the coffee and food crop economies.

As a product of this exchange, a class of small-scale merchants also emerged in Gera's small postwar market towns, entrepreneurs who moved small quantities of coffee and grain from Gera's market to other markets and accumulated small pools

70. Kassahun Seyoum, *Initial Results,* 18.

of capital, which they lent to coffee-income households at 100 percent interest for three months. Yalew Taffese was a coffee migrant-turned-merchant who quickly learned the emerging annual economic cycle of coffee farmers:

> In the past people planted coffee only and neglected the growing of teff and grains. They work on coffee and cover their necessities such as clothing with the money they earn from selling coffee. When it is finished they borrow money, which they could repay from the next season's income.
>
> If they borrow coffee they pay back in coffee. If they promise to pay in cash they pay it in cash . . . . If a person borrows a quintal [100 kg] he will pay two quintals; 100 birr [Ethiopian dollars] was to be paid with 200. . . . The payment is due in January, mid-January.[71]

On a regional level Agaro (in Gomma), 50 kilometers from Gera's old capital, grew quickly in the 1960s from an Amhara garrison town to a major coffee market and administrative center. The growth of food markets there permitted farmers to diversify household diet by purchasing preferred food crops such as teff, eleusine, and pulses, which were produced by nearby noncoffee highland zones and generated local income for farms concentrating on food crops. Within the region Gera was a major source of food crops to areas such as Gomma with a higher concentration of coffee production. Agaro also housed entrepreneurs who invested government salaries or mercantile profits in new coffee plantations along the new market roads which reached into the forest.

Gera's postwar economy bore little resemblance to its nineteenth century antecedents, except in its relative prosperity. Were Antonio Cecchi to have retraced his steps 90 years after his 1880 visit he would have found an interesting set of transformations in the rural landscape. In 1970 the forest itself would have looked familiar to Cecchi, opened as it was by new cultivation of farmsteads both in the forest and in open areas. Farms around Filo and Chala would have looked familiar to Cecchi as well in their tools and in their methods of cultivation and storage. For women living within an hour's walk of a market town, however, diesel flour mills, which chug away into the evening hours, have transformed their work and expanded their social worlds away from their homesteads' grinding stones. Towns also gave expression to women's new economic roles as hotelkeepers, brewers, and prostitutes. The route connecting Gera to world coffee markets in 1970 would most likely have presented the most startling contrast to a late-nineteenth-century traveler. Instead of food moving along caravan tracks to royal storehouses, a twentieth-century traveler would see donkey loads of grain arriving at roadside markets for loading onto private trucks bound for Agaro's open market to feed cash-rich coffee farmers and new urban populations.

71. Interview with Yalew Taffese, Filo (Gera), November 1991.

As a curious traveler Cecchi would have been fascinated by the expanding, almost industrial plantations of coffee and the trucks plying back and forth on all-weather gravel roads between local markets and Agaro. With a closer look at the fields cleared in the forest and on the bottom lands, he would have been surprised by the domination of maize, in 1880 a poor farmer's crop, the diversity of crops which had delighted him in 1880 much diminished. At table in most farmers' homes, however, Cecchi in 1970 would have eaten *enjera* made of teff or eleusine grown in the highlands but purchased in the local market with coffee income. Masked by the veneer of modernity and prosperity of coffee-producing households, however, oxen-poor families, female-headed households, and tenants have had a diet consisting primarily of maize, root crops, and kale; teff and food crops from the market have been well beyond their means.

Economic expansion in the coffee regions not only fueled local economies driven increasingly by cash income, but also transferred considerable wealth to towns and cities, where absentee landlords converted coffee income to urban investments in housing, transportation, and education for their children. New Christian migrants, who experienced both economic and political success, founded at least three new Christian churches in Gera in the postwar period to serve the needs of baptizing their children, burying their dead, and marking religious festivals. In the market towns such as Chira, Bashasha, and Sadi merchants built storehouses and brought in scales to aid in bulk measurement and to transfer both food and coffee from producers to consumers or larger markets.

Agaro, almost abandoned in 1913, became a boomtown, sporting corrugated-iron-roofed houses, bars, and brothels owned and staffed primarily by northern migrants, and a burgeoning mercantile community.[72] The linguistic dominance of Amharic and the political hierarchy of migrants from Ethiopia's north and center reflected the increasing national economy and political culture. Bars with a few beds to rent or a few cases of Pepsi-Cola or St. George beer to sell bore the names of the owners' places of origins: the Blue Nile Bar, the Selale Hotel, the Jimma Tea House. Young women, also largely from the north, worked in those establishments offering drinks and sexual services; some parlayed their cash income into enterprises of their own. Beyond Agaro and Gera's own market centers, Jimma and Addis Ababa were only a few hours away by tarmac road.

Chala, Gera's nineteenth-century thriving capital, had no postwar role as an administrative center or point of tribute collection, though Abba Bor, the descendant of Cecchi's captor, the *ghenne* Gumiti, still served as local district governor until the 1974 revolution. He built his residence two hundred meters from the spot of Cecchi's captivity, secluded among ornamental trees. Abba Bor enjoyed the trappings of modernity in a coffee economy—a corrugated-iron roof, tiled floor, glass windows—and his interior decor displayed evidence of past authority in the forms

---

72. Ato Belete Chekole, who owns a hotel-bar in Chira and was born in Chira of northern migrant parents (Gojjam), is typical. His three daughters live and work in his hotel.

of family photographs, threadbare Oriental carpets, and memorabilia of a prosperous past. In 1991 the nearby site of the old royal *masera* still had evidence of forest growth, but was overplanted with a field of horsebeans, the stones of the old royal house having been removed to build the grave sites in the royal cemetery. Only a huge millstone used by royal household slaves and too large to move remained to mark the site of the royal enclosure.

1971–1990:
A Tyranny of Maize, 1971–90

Gera's overall increase in coffee production through the 1950s and 1960s was followed by decreases in both total production and percentage of the national total after the 1974 revolution. By 1980–81 the Kaffa region as a whole saw a decline in both its gross output of coffee and its percentage of production, which fell to less than 30 percent of the national total. While Gera's production was only a small portion of this figure, perhaps one-fifteenth, it reflected overall trends. Table 5.3 indicates Gera's scale of decline in coffee production over the last decade.

The decline in Gera's coffee and consequent transformation of its agricultural economy were the result of two independent events, one environmental and the other political. The environmental shock came in the form of coffee berry disease (CBD), a fungal disease which first appeared in Kenyan plantations in 1971 but

TABLE 5.3. Gera annual coffee production, 1980–90.

Years	Production (in metric tons)	% Change
1980–81	1055	—
1981–82	651	-38
1982–83	400	-38
1983–84	775	+93
1984–85	420	-45
1986–87	226	-46
1987–88	300	+32
1988–89	350	+16
1989–90	308	-12
Total 1980–90		-71

Source: Ministry of Coffee and Tea Development, Chira (Gera).

spread to Gera's trees that same year. Though the disease, called cholera locally, does not kill the trees themselves, it results in black, dried, and empty berries useless to commercial production. Gera's blessings of altitude and rainfall, which contributed to its agricultural potential, proved deadly to its coffee; CBD had its greatest effect in moist forests above 2000 meters. Thus, Gera suffered the steepest drop in production of any region in Ethiopia. Not even Gera's disease-resistant wild strains escaped. CBD spores spread through the air, but were dispersed even more effectively through the local practice of picking berries by hand-stripping branches. By 1975 farmers had abandoned coffee plantations in central Gera; only those coffee forests around Afallo and Oba, at lower altitudes, continued to produce commercial quantities. A walk through Gera's coffee forests in 1991 revealed thousands of trees bearing blackened, empty berries.

The second and more complex set of changes in the agricultural economy emanated from Addis Ababa in the form of rural development policies in the wake of the 1974 revolution. The first of these was the 1975 land reform, which nationalized all land, turning tenants into landholders. Local peasant associations took charge of land allocations and distributions, transforming traditional Oromo property systems into equal, partible shares available to all male claimants.[73]

For Gera, the effect of land reform on the coffee economy was minimal, since large holdings of coffee had already lost their economic value, their forest cover restricting their use as annual crop land. Although some absentee landlords lost their coffee holdings, coffee laborers living locally as tenants and newcomers from the north were able to make land claims on annual cropping land. The entire class of rural tenants became landholders, and large holdings transmogrified into small farms, which, because of the loss of coffee income, had little choice but to shift to annual cropping. On those farms where land was cleared for the first time, maize was the crop of choice, given its low labor requirements, early maturation, and relatively high yield. Those households which had oxen or could borrow them and had sufficient labor for timely weeding and harvest could also rotate teff and *zangada* (red sorghum), a northern import, along with maize as a subsistence crop. For most households, however, the loss of coffee income left them with a single crop, maize, as the basis for diet, exchange, and savings.

The revolution and agricultural marketing policies formed by technocrats in Addis Ababa had, however, an important indirect effect on Gera's agricultural economy. Strict control of the marketing and pricing of coffee as an attempt to maximize export earnings in fact reversed the relative profitability of coffee and food crops. Farmers thus received a fixed price, which did not reflect the high world market prices in the 1975–80 period. The decline in coffee income and restrictions on movement of trade transformed the terms of trade between coffee and food. Limit-

---

73. Despite the best intentions of drafters of the land reform proclamation, women in Gera derived no direct benefit from the legislation, since the national guidelines failed to overcome strong Oromo and Islamic male dominance of property rights.

ed road networks permitted effective control of coffee trading by inspections at key choke points along the roads around Agaro and Jimma, which restricted interregional movement of food as well as the free flow of coffee. By 1980 steep rises in food prices, brought on by declining production per capita, reduced incentives to plant new coffee or maintain old plantings. From 1975 to 1990 food prices quadrupled while coffee prices remained static. Gera farmers who still had viable coffee trees, like others in neighboring Gomma, shifted quickly to food production, sometimes removing coffee trees from prime land.[74]

The shift to food production by coffee farmers and annual cropping farms continued a longer-term trend. Labor demands by the state this time, rather than landlords, placed a premium on early yields, low labor, and crops which would escape the attention of the Agricultural Marketing Corporation, the parastatal structure which set prices and quotas on marketable crops such as teff, wheat, and pulses. Maize was the primary answer, since it was already well integrated into the coffee production cycle; its early maturity addressed both food shortage and the problems of security of tenure. The arrival of new, early-maturing cultivars distributed in the 1960s and 1970s by the Ministry of Agriculture permitted drained bottom lands to be double-cropped in maize during the main rains and teff to be sowed on residual moisture. By the 1988–89 cropping year maize made up virtually half of all crop production in Gera (see table 5.4). The cropping patterns indicated in table 5.4 rep-

TABLE 5.4. Percentage distribution of total annual crop production, 1986–91.[a]

Year	Maize	Teff	Wheat	Barley	Sorghum	Horse beans	Field Peas	Millet	Haricot Beans
1986–87	31.8	31.9	15.4	1.4	1.5	16.1	1.1	.01	.02
1987–88	34.8	31.9	15.0	1.4	1.5	13.6	1.1	0.1	.27
1988–89	48.9	21.6	9.9	6.2	1.4	10.8	0.8	—	—
1989–90	46.6	21.5	10.0	6.1	3.9	10.7	0.8	.12	.18
1990–91	47.5	22.7	10.9	.68	4.3	12.1	1.1	.24	.22
Average 1986–91	41.9	25.9	12.2	3.1	3.1	12.6	1.0	.01	.02

Source: Ministry of Coffee and Tea Development, Chira (Gera).

Note: Percentages shown are estimates as reported by the ministry.

[a]Percentages are calculated on basis of hectares planted

74. Kassahun Seyoum, Hailu Tafesse, and Steven Franzel, *The Profitability of Coffee and Maize among Smallholders: A Case Study of Limu Awraja, Ilubabor Region.* Research Report No. 10 (Addis Ababa, 1990), 10-22. Also see Kasahun, Franzel, and Tesfaye, "Initial Results," 30-44.

resent an extraordinary concentration in a single crop, particularly given the inclusion of production from some areas above 2300 meters, that is, beyond maize's environmental range.

Two further interventions in the rural economy implemented by Addis Ababa contributed to the movement toward both the reduction in perennial cropping and the expansion of maize as the dominant food crop. In 1984–85 the central government began a program to resettle farmers from the drought-stricken north on southern lands. In the first wave, farm families from eastern Wallo and Tigray arrived in Gera in nine resettlement sites. Ethiopia's Relief and Rehabilitation Commission placed Amharic-speaking Oromo from Wallo at Filo on that area's prime bottom lands. Tigrayan resettlers were less fortunate and found themselves placed in a remote lowland zone, where only food allocations from international agencies sustained their food supply. In both cases, like their northern migrant forebears, these new farmers were dedicated to annual crops. Though maize was relatively new to their cropping systems, the white sorghum they were accustomed to was too slowly maturing, and they quickly adapted to maize's quick yield to ensure their food supply.

Gera's new northern farmers found a new environment in which to adapt their agronomic methods and cropping schemes. Interviews with and observation of migrants importing northern farming systems into Gera's forest ecology reveal important elements of the ox-plow system itself. Unlike coffee areas, which required preservation of forest cover, the imported northern land management system promoted systematic expansion of fields in annual crops and therefore required the routine clearing of forest ground cover and burning of large shade-giving trees.

Northern farming systems, however, were not imported in their entirety. In fact, they were transformed by both local agricultural practice and the eccentricities of the local environment. Farmers from northern Shawa, where over time land shortages fostered careful livestock-crop management such as tethering and hand feedings of oxen on crop residue, have given up their former practices in favor of open pasture. Gera's northern migrants have also largely abandoned tight cereal-pulse rotation practice on restricted holdings in favor of expansion to new virgin forest land.[75] This practice has emphasized the use of maize for two years of the five-year cereal rotation and accounts for the expansion of maize even within dedicated annual cropping cycles. Northern farmers have also dropped the practice of rotating pulses with cereals, a method which fixes nitrogen and limits the need to fallow on intensively cultivated plots. Gojjami and Gondere farmers claim that pulses, with the exception of horsebeans, will not grow in Gera's red and brown clay soils.

Though most farming practiced by northern migrants to Gera has emphasized extensive rather than intensive crop and livestock management, there is evidence of

75. Interview with Mekeday Ketema, Bera Dedo (Gera), November 1991. Mekeday was a migrant resident who arrived in Gera from Manz in 1967.

adoption of local agronomic practice. Gera's moist climate promotes rapid, virulent weed growth, which requires three labor-intensive weedings for most annual crops, particularly teff. Northern farmers who complain bitterly of the weed problem have learned from Oromo neighbors to use seedbed flooding to control weeds. By building temporary channels and dikes and flooding a field after its first plowing for teff, farmers suffocate weeds, and then follow with three more plowings, breaking of clods, and sowing. The practice, though labor intensive, shifts the weight of labor from weeding in the early fall to the flooding practice in early summer and increases yield in weed-sensitive teff cultivation.[76]

A more profound transformation in Gera's agricultural base has been the 1987 villagization program by which government officials have relocated farm homesteads away from their historically dispersed pattern into rigidly planned village sites. Villagization is a policy conceived in Addis Ababa by the Ministry of Planning, academicians, and bureaucrats and it has been executed by agents from the regional planning offices and the Ministry of Agriculture, who selected sites, forced movement of reluctant farmers, and often countermanded earlier decisions about village sites. Had Cecchi returned, he would have watched the distinctive and attractive houses nestled within the forest being replaced by uniform, regimented, rectangular units built hurriedly by farmers under duress.

Villagization has simultaneously reshaped human settlement, destroyed the remaining vestiges of perennial crops, and exacerbated insecurity of tenure. By 1990 the forest had already reclaimed the former homestead sites, including plantings of coffee, fruit trees, ensete, and root crops. Three years after the onset of villagization, homestead sites in the forest had already reverted to pasture and the first stages of forest regeneration. By contrast, new village sites were slow to regenerate household plantations of ensete, taro, and kale. One farmer in Bera Dedo told me that the newly allocated compounds were too small for ensete cultivation and "besides, they broke our hearts."[77]

In Gera the effect on individual farm economies was almost immediate. Farmsteads located within the forest on lower slopes were moved down onto cultivated areas and away from ensete, coffee, *chat,* and household gardens of yam and taro. Abba Karo claimed to have earned 1000 birr per year on sales of his fruit, *chat,* and vegetables alone. Those plantings, investments of decades, quickly deteriorated, leaving only animal-damaged remnants of his former house site.[78] In the short term, villagization thus has contributed to vegetative regrowth in the forest. In the longer

76. This practice was described to me by two separate communities of northern migrants as learned from the local practice of Oromo residents.

77. This farmer who I met in Bera Dedo did not give his name.

78. A visit to the former site confirms Abba Karo's claim. In 1991 he lived in the new village, where he had replanted only kale and a few cuttings of ensete. Interview with Abba Karo Abba Bor, (Yukro), November 1991.

term, villagized farmers have invested heavily in maize to ensure a quick supply of food and to hedge against a further reallocation of land.

Across Gera, villagization has nearly completed the movement to annual crops and to maize in particular. In Gera the loss of perennial crops in six peasant associations was about 80 percent. Table 5.5 indicates the nature of those losses. Farms in the Oba area, which traditionally maintained 3000–4000 ensete plants, lost virtually all.

By 1989, the cumulative effects of coffee berry disease, villagization, and disincentives for coffee production meant that 94 percent of Gera households surveyed by the Norwegian Redd Barna nutritional survey were dependent upon maize for their daily diet.[79] Maize as a primary agricultural product and as a principal food source presents specific problems, particularly when juxtaposed with alternative crops and a mixed repertoire. Maize is the most drought vulnerable of the major crops, In Gera, drought is rare, but maize's need for moisture, particularly at the silking stage, means it grows best during the main rains and thus occupies land which might otherwise produce other crops. Moreover, maize, is high in starch, and less nutritious than other cereal crops. Both food consumption and marketing patterns were heavily driven by a further characteristic of maize: it is the least storable of all the major annual crops in the historical Gera repertoire.

TABLE 5.5. Gera Warada: effects of villagization on perennial and bi-annual food crops.

Peasant Association	Before villagization			After villagization			Effect (%)
	Ensete	Taro	Others	Ensete	Taro	Others	
Tolly	3920	3400	4250	830	830	460	-81
Oba	1300	1100	1000	330	200	140	-81
Kersa. Geba	1500	1350	1150	560	201	—	-83
Qolla.Selaja	450	100	710	40	10	30	-92
Qwacha. Tulla	1160	650	850	—	—	—	100
Wanja. Kersa	1040	610	1420	200	130	150	-87
Total	9370	7210	9380	1960	1371	780	-83

Source: Adapted from Redd Barna–Ethiopia, "Food Availability Survey in Gera: Report to Redd Barna—Ethiopia," Gera Integrated Rural Development Project P.-4008, November 1989, Appendix.

79. Redd Barna-Ethiopia, "Food Availability Survey in Gera: Report to Redd Barna-Ethiopia," Gera Integrated Rural Development Project P-4008, November 1989, 8.

Conclusion:
Maize and the Forest

Far more important than examining the causes for the fairly rapid decline of coffee production in Gera over a period of 15 years is the question of why a virtual maize monoculture has replaced it. Why was there not a return to the diverse patterns of agricultural production evident in the mid-nineteenth century? After all, the historical evidence suggests that diversity and balance within the cropping repertoire were well entrenched in Gera's agricultural history.

The transformation of Gera's forest landscape over the 1850–1990 period was a direct result of human activity and adaptation within the ox-plow complex. Given the limits of technology and the characteristics of crops, the removal of primary forest cover and, at certain points, secondary forest regeneration resulted directly from shifting population and labor resources in the region. Though the population doubtless grew over the period of study, from Cecchi's estimate of 15,000 (plus a large slave population) to current census figures of 43,000, growth was not linear or dependent on rates of fertility or mortality. Concentrations of population within that trend of general growth appeared as a result of conjunctural factors: the massive influx of slaves, out-migration after the Shawan conquest, and immigration in the postwar coffee economy.

Moreover, the precise effect of population depended on both the nature of the labor supply and the dominant agricultural system. The annual cropping regime, which required open fields and a short-fallow system necessarily reduced forest cover when population expanded. Ensete and root crops were compatible with shaded forest, since they were less extensive in their need for both cropland and pasture. In the long term, the indications are that the ox plow's annual cropping, livestock-dependent regime dominated and pushed back both perennial cropping and the forests themselves in the central areas of Gera to a low ebb in the late nineteenth century. The economic growth of the Gera kingdom in the early nineteenth century appears to have taken place as a result of a rapidly increasing labor supply and increasing production of foodstuffs based on an annual cropping regime in which eleusine, teff, maize, and sorghum replaced an earlier regime of root crops and sorghum based, perhaps, on hand-hoe cultivation. The massive forest regeneration between 1880 and 1940 was something of a historical aberration resulting from a political shock and a massive labor loss. The subsequent rise of the coffee-maize complex was a product of economic change in the regional economy and world demand for a local product, which drew new population and cemented the relations between Gera and urban economic centers in the region and in Addis Ababa.

The forests themselves and their effect on agriculture form an enduring theme through the 1800–1990 period. Though much of the forest remains on the steeper slopes of the hills, the annual cropping regime and its implied strategies of land management promise to push back the forest as population increases. Even with their gradual diminution under population pressure, the broadleaf forests of the southwest still account for 65 percent of Ethiopia's total remaining forest.[80] The coffee-maize regime itself had been compatible with maintaining secondary forest cover, since coffee plantations require shaded sites and maize can be sown on bottom lands or in small forest clearings near the homestead. The transformation of the coffee-maize complex to a full-blown maize monoculture resulted partly from an environmental factor (CBD) but more from policies in the political arena—fixed coffee prices, land reform, and villagization—which projected state power and urban priorities onto the rural landscape.

Maize as a subsistence crop in the Gera agricultural economy also represents a meeting point between two conceptions of agriculture in a forest environment. The coffee-maize complex combined use of perennial cash crops that use forest cover and annual food crops (maize and teff) grown on small plots. The extensive annual cropping regime brought by northern migrants, however, has equated population growth directly with forest clearance, a process which emphasizes maize as a necessary part of cropland expansion. Potential conflict between these opposing ideas of environmental management will be an arena in which central government policy and the regional economy may be decisive, and the resolution will shape the long-term future of Gera's landscape.

80. Conforti, *Impressioni,* 157-58; Daniel Gamachu, *Environment,* 12.

6

Addis Ababa's Kitchen: Food, Forage, and Intensification in a Closed Ox-Plow Economy, Ada, 1800–1990

> What a splendid panorama! An expansive plain spreads itself at the
> foot of the mountain, the plain slowly undulates, divided
> like a chessboard by so many verdant paths, bounded by so many
> cultivations of barley, wheat, and tieff. . . . What fertility!
> —Leopoldo Traversi, "Da Entoto al Zuquala" (1887)

> Family size and population density is higher for Ada than for any area
> of Shoa or for Ethiopia as a whole. Despite the high percentage
> of land in production, the agricultural population density is higher than
> for any area of Ethiopia. Average per capita gross value of crop
> production is 45 percent higher and the family gross value of crop
> production is 80 percent higher than for Ethiopia in general.
> —Borton, et. al. *A Development Program or the Ada District . . .* (1975)

By the second half of the twentieth century, population growth and, more important, concentrations of population around urban centers on Ethiopia's highlands intensified demand for food supply and increased land values. The influx of migrants and investment to agricultural zones around cities paralleled the immigration and capital growth in the cities themselves. Land prices as well as strong markets for food crops and meat promoted the conversion of open land to cropping and pushed microagronomic economies toward "closed" agricultural systems (see chapter 3). In such settings, the economics of crop-livestock management came into conflict with market and extramarket incentives for farmers to expand cropped land in particular crops,

191

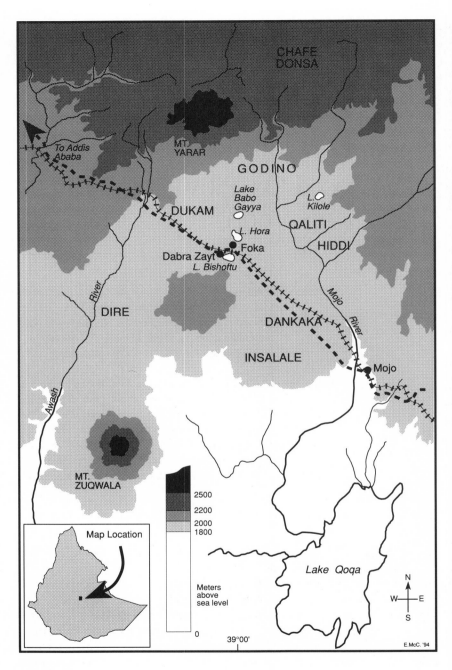

Map 6.1. Ada

192

especially cereals. The agricultural history of Ada District is wrapped up in the intensive engagement of an ox-plow economy with a burgeoning capital city and the transformation of a national political economy.

Like Ankober, Ada's agriculture evolved in the shadow of an imperial town. Unlike its nineteenth-century counterpart, however, Addis Ababa presented its periphery with an urban center possessing a permanent and expanding infrastructure, an economic engine that drew power, people, and a food supply on a scale which Ankober never even remotely achieved. Addis Ababa's economic and political power was national and international and transformed its hinterland as it grew, first higgledy-piggledy as Menilek's capital and then as a modern metropolis under Haile Selassie. For Ada's farmers, the urban leviathan next door meant a powerful market, a state and elite which sought control of land, and especially pressure on their agricultural system to adapt to an increasing population and high land values.

In empiricist terms, the relationship between population increase and agricultural practice in the ox-plow complex is a function of the interaction between crop and livestock production, the gradual expansion of cropped land at the expense of pasture. As an agricultural system, the ox-plow complex has been dependent on the ability to sustain both its livestock population and the expansion of cropped land, a relationship fundamental to its movement over time and its expansion across the highland landscape (see chapter 2). Agricultural economists have seen this process as an agricultural econometric model:

> Evolution of crop-livestock interaction is an inverted U-shape with integration being very weak, increasing, and then decreasing in response to changes in population density. Associated with closer interactions are growing constraints on obtaining inputs in markets and contracts. Such constraints create cost advantages in providing inputs directly on farm, thus encouraging crop-livestock interaction.[1]

To a large degree, this generative description of crop-livestock integration and population parallels farmers' own subjective accounts of agricultural change over their lifetime.

The goal of this chapter is to reconstruct the agricultural history of Ada District as an example of agricultural intensification and crop-livestock integration through a narrative which describes change in the language of both social science empiricism and farmers' own, sometimes dissonant, recollection of the process. This conjuncture of factors, including a revolution and land reform in the 1970s, has pushed ox-plow cultivation toward intensification of land use and specialization in a single crop. How farmers have adapted their agricultural practice toward these ends is the topic of this chapter.

1. McIntire, Bourzat, and Pingali, *Crop Livestock Interactions*, 3-1.

Figure 6.1A. Oromo farmers, c. 1888.

Ada: A Closed Microagronomic Economy

Ada District, 50 kilometers south of Addis Ababa, is the best example of an agricultural zone which has approached, perhaps exceeded, the status of a "closed" microagronomic economy in the second half of the twentieth century. Its story is special, given its proximity to Addis Ababa's burgeoning urban market, the Djibouti railway, and Ethiopia's Highway 4, which links the capital to its southern and eastern hinterlands.

Ada district is a trapezoid of 1750 square kilometers, which lies at the northwestern edge of the Rift Valley. Ada's landscape includes a range of elevations from about 2300 meters on its northern edge to about 1600 meters at its lowest along the arid and eroded country in the south near the Awash River. The bulk of Ada's land (57 percent) lies in the 1800–2000-meter zones on an undulating plain which includes isolated volcanic cones and six lakes, which Leopoldo Traversi in 1887 observed were "formed of so many spent volcanoes, bizarre, without a tree, disposed in a straight line from East to West."[2]

From perhaps 1960 on, Ada has comprised Ethiopia's most intensive and specialized smallholder, ox-plow, microagronomic economy. From the mid-1960s as well, Ada's farmers have cultivated in the midst of Ethiopia's highest concentration

---

2. Traversi, "Da Entoto a Zuquala. Lettere del dott. Leopoldo Traversi al sig. conte Buturlin," *Bollettino delle Società Geografica Italiana* 24 (1887), 583.

Figure 6.1B. Oromo farmers, c. 1888.

of agricultural empiricism: research and extension activities involving crop and livestock field and on-farm testing, privileged access to inputs (especially improved cultivars and fertilizer), extension advice, and the attention of a central state apparatus and national bourgeoisie headquartered only 50 kilometers away.

Although Ada is doubtless Ethiopia's best documented microagronomic economy over the course of the 1960–90 period, the historical context of its agricultural development is poorly documented in both secondary and primary sources. Key foreign observers such as W. Cornwallis Harris, Johan Krapf, Antonio Cecchi, and Father Giuglielmo Massaja reported little on Ada's political context and even less on the state of agriculture. Then as now, foreign visitors were smitten by Ada's physical beauty and particularly the geology of its six crater lakes. In the middle nineteenth century Ada lay outside political events which focused on Tigray and northern Shawa. Only Leopoldo Traversi's 1886 and Luigi Cappucci's 1887 accounts provide glimpses of Ada's economic and agricultural world. Written sources in Ethiopian languages on Ada are thinner still; of 40-odd churches in Ada, even the oldest has no substantial, verifiable written account of its foundation, local history, or even land grants.[3]

Oral sources therefore play an important role in providing a skeletal structure of Ada's settlement, politics, and economic history. Life histories of the oldest Ada residents in 1992 reached back to the first decade of the twentieth century; accounts of their parents' lives can push the lens back a generation further, to the last quarter of the nineteenth century, on issues such as migration, social property, and settlement. Histories of settlements in local oral tradition often include oblique evidence of livestock management, agriculture, and diet. Oral traditions of church foundations also help in piecing together history of land use and, ultimately, the evolution of the relationship between livestock and agriculture in the periods before empirical baseline survey data.

Ada's climate reflects both its dominant altitude range between 1800 and 2000 meters and its proximity to semiarid zones on its southern frontier with the Rift Valley. Twenty-two years of rainfall data (1965–85) indicate a strong unimodal rainfall pattern, with an annual average of 845 millimeters (decreasing through the 1970s), 70 percent falling between June and September, accounting for Ada's single growing season. Ada is warm, dry, and very dusty in the dry season.[4]

Along with the seasonal pattern of rainfall, the most important environmental element shaping Ada's crop and livestock patterns of use has been its soils. Ada's soils range between light, red loam soils on its sloped, upland areas to its dominant, heavy, black clay vertisols, which farmers call *koticha* or *walka*. Ada's black soils share the vertic properties of "cotton soils": deep cracks in dry conditions and a tendency to smear and cake in wet conditions. For Ada, this soil's most salient agronomic characteristic has been its propensity to seal its surface layer on saturation and hold standing water by late July and early August, making ox cultivation diffi-

3. Leopldo Traversi, "Da Entoto," 581-95; Luigi Cappucci, "Condizioni dell'agricoltura nello Scioa," *Bollettino della Società Africana d'Italia* 6 (1887), 281. Donald Crummey has indicated from his exhaustive survey of church documents in northern Ethiopia that Shawan churches in general tend to be relatively bereft of historical documentation compared with more northern counterparts in the Gojjam, Tigray, and Gondar regions. At Ude Emmanual, the only church in Ada I visited which claimed to have *tsehuf* (writing) on its foundation, the single half-page of church history declared its foundation by Wayzero Bafana (young Menilek's first consort) in 1812, an obvious error. She was not born until at least two decades after that date.

4. Its annual mean temperature is 16 degrees (Celcius), ranging between 10 and 22 degrees. Gryseels and Anderson, *Research*, 9; Meseret Showamare, *Climatic Records for ILCA Research Sites, 1985* (Addis Ababa, 1986), 19-20.

cult and mechanized cultivation in the rainy season virtually impossible. In some areas such as Lake Chialanca, the seasonal flooding of low-lying vertisol plains with several meters of water prevents main-season cultivation, providing only dry-season forage for local livestock and a wet-season haven for an astonishing variety of water fowl, including thousands of migrating lesser flamingos.

Ada on the Shawan Periphery, 1800–89

Ada's geography and its proximity to the Ethiopia's expanding twentieth-century state formed a strong current in its agricultural history. The region's position on the edge of the Rift Valley and to the east of the Awash River determined that it would be a point of engagement for the expansion of central Ethiopia's two dominant forces. The Oromo and Amhara populations of Shawa engaged in warfare and cultural exchange along a fluid frontier in northeast Shawa since the sixteenth century; various Oromo-speaking groups occupied parts of southern and western Shawa perhaps as early as the fourteenth century. Ada seems initially not to have been a focal point for intense settlement or conflict, occupying a place on the hinterland of more concentrated settlements of agropastoralists farther north and west.

The earliest Oromo traditions of settlement extend five generations prior to the first Shawan Amhara hegemony in the second quarter of the nineteenth century. The oral narrative tradition of the three Ada brothers who migrated in search of pasture and purgative waters for their cattle suggests that the first arrivals in the early eighteenth century (roughly contemporary with Oromo expansion into Gera) were predominantly pastoralists. Badada Kilole, an octogenarian native of Ada, recounted Ada's narrative of origin to me while seated beneath a *warka* tree outside an Oromo cemetery a few hundred meters from Lake Kilole, the site of the purgative waters:

> Oh, that has a long history of its own. It was a long time ago. The origin or the birthplace of this Ada was not here. He was born at Muger [northwest Shawa]; he set out from Muger with his herd of cattle in search of *hora* [purgative mineral water]. He arrived at a place called Nacho [near Dukam in Ada] and temporarily camped there. From there Ada led his cattle to the present Dabra Zayt [Ada's chief town] area to get *hora* [the cattle smelled it and led him there]. This *hora* is found around Arsadii. After letting his cattle drink, he returned to Nacho. There were three brothers who followed him to this *hora*. Following the three brothers other relatives also came to the Nacho area. They took for granted that the region that was good for their cattle would also be good for them!
>
> The eldest brother was known as Boorii, the second one was called Adii, and the third was called Walato. These were the three brothers who came to the area. These form the clan of the three Adas.

In the beginning there was only cattle raising aided by a bit of agriculture; it was subsistence. It was gradually that growing crops came into existence among the Oromo. Milk and meat were the staples. While the Oromo were leading this life, the Amhara came and took control.[5]

Ada's settlement by the three clans apparently was not an occupation of empty lands, but a conquest. The Ada tradition also recounts that a certain Botora, son of Adii (one of the founding brothers), had to fight Liben Oromo to expand his holdings within the area.

Ada's settlement history and its effect on Ada's agricultural history are less an issue of ethnicity as an arena of conflict than the evolution of a relationship between Ada's agricultural and livestock economies and their engagement with the political economy of the Shawan state and, later, the emerging modern national economy. In the initial phase, the local Oromo tradition describes the arrival of the Amhara as a product of rivalries between two clans of Oromo settlers, suggesting a preexisting tradition of cultural and presumably economic contact across the central highlands. Luigi Cappucci confirmed the basic thrust of this tradition from his visit of 1887 by noting that Ada had "first belonged almost all to only two Galla," presumably Botora and Bunee, son of the eldest brother, Boorii, who led the two major clans. Thus the initial "conquest" of Ada by the expanding Amhara state must have taken place at some point in the late eighteenth century, its Oromo leadership under no more than a loose tributary relationship between the Amhara and their southern frontier.[6]

This relationship seems to have continued into the second quarter of the nineteenth century when Bullo Jille, a Christian and a great-great-grandson of Boorii—one of the three Ada brothers—and the son of an Amhara mother from Bulga in northern Shawa, established a working relationship with Negus Sahla Sellase, then expanding his domain's frontier south from Ankober. On one expedition in 1841 Sahla Sellase founded and endowed Ada's first church, at Yarar Sellase; Bullo Jille himself followed suit two years later, founding Kajima Giyorgis Church in the 1843. Bullo had the arks (sg. *tabot*) for those churches brought directly from Bulga, reflecting cooperation rather than competition with Sahle Sellase's expansion.[7]

5. Interview with Badada Kilole, Hiddi (Ada), 25 May 1992. For the original Liben settlement, see interview with Dabtara Haile Masqal Wandemu (Dabra Zayt), 26 May 1992. All accounts agree on the Ada clan's origin in the Ambo (Mecha) area.

6. Cappucci, "Condizioni," 281. The reference to Gonderine connections is interesting because this is the time when Gonderine migrations to northern Shawa also formed the bulk of the troops used in the reconquest of the areas south and east of Manz. See chapter 4.

7. Local Oromo traditions consider these churches as part of their local patrimony and not inteventions. See Table 7-1; also Badada Kilole, 25 May 1992 and Dabtara Haile Masqal Wandimu, Kajima (Ada), 26 May 1992.

These church foundations provide some insight into the nature of early settlement and Amhara-Oromo relations. Neither of these churches is located on the central vertisol plain; Yarar Sellase is near the top of Mt. Yarar, overlooking the Ada plain, a spot from where Traversi (see above) surveyed the "open" agriculture of southern Shawa in 1887. Kajima Giyorgis is located on upland red loam soil on the northwestern approach to the vertisol plains. Both are "frontier" churches, not placed to serve a densely settled agricultural community, but rather to function as outposts to draw local converts and project northern Shawa's plans for expansion. Their placement suggests that Ada was an area still awaiting both Christianity and incorporation into the ox-plow-based political economy of the dominant Shawa.[8]

There is convincing evidence that Ada underwent a substantial economic and demographic transformation in the 1870s and early 1880s, reflecting not only changes in the local agricultural economy but also a wider a transformation in the Shawan state itself. The best evidence of changes in local settlement patterns is the spate of church foundations of the 1870s, endowments by local aristocracy (the Bullo Jille family) and also by Bafana, Negus Menilek's consort. These were not frontier churches, but institutions placed on and around the central vertisol plains—often near sites associated with Oromo religious rites—to serve a growing population of settled agriculturalists, that is, northern settlers and Oromo Christians. The arks for these churches came almost exclusively from northern Shawa. Unlike other regions of the south and west where churches served a conquering soldiery-turned-landlord, these churches served an ethnically mixed local population, Oromo and early migrants from northern and central Shawa (Minjar, Fantale, Rogge).[9]

Cappucci, who described Menilek's relations with Oromo regions of southern Shawa in 1887, argued that the changes in relations in the late 1870s resulted from Menilek's own changing political status and ambitions:

Eight or ten years ago, when the king was at Ancober or at Debra Berhan and made his military expeditions or zemecha against the Abiccu, against the Galan, against the Ada, against the Guleli, against the Gumbicco, in sum when he had still not acquired in a stable way the provinces north of the Awash, he had many fewer soldiers and fewer arms, and his lands in [northern] Scioa were in most part cultivated on his account and his supervision, as

8. Interview with Dabtara Haile Masqal Wandemu (Dabra Zayt), 26 May and 31 May 1992. Haile Masqal stated that local conversions among the Oromo began only after the establishment of the two original churches. For a full list of Ada churches, foundation dates, and land endowments prior to 1939, see Commissariato di Governo di Biscioftu, "La chiesa etiopica nel R. Commissariato di Governo di Biscioftu," 1939, IAO fasc. 791.

9. Interview with Dabtara Haile Masqal Wandemu, Kajima Giyorgis church, 26 May 1992. Haile Masqal, born in 1925 at Kajima (Ada) of parents born in Ada, is the informal historian of Kajima Giyorgis Church. For a comparative case on church expansion see McClellan, *State Transformation*, 85-87.

such mild, likewise for maintaining him in state. But when the king distanced himself from the old provinces to establish himself at Entoto, the situation changed. He began then to possess many firearms, and therefore extended the conquest. . . .[10]

Thus, as Cappucci implies, two factors affected the nature of Ada's absorption into Menilek's conquest state. First, it was incorporated in the late 1870s when Menilek's power was insufficient to causes a direct alienation of lands from local cooperating elites. Second, the relationship of Menilek's state with southern periphery populations and their local economies shifted dramatically when in 1886 he moved the site of his capital to Addis Ababa, four to six days' ride southwest of his former royal town of northern Shawa. Given its earlier absorption under relatively mild terms, Ada avoided the violent conquest which befell populations in Kaffa, Arsi, and Harar.

More important, Ada's proximity to the new capital meant it received a new status within the imperial domain, that of *madebet* (kitchen or pantry), an area whose land tenure and tributary obligations focus on providing food directly in support of the palace. While the Shawan capital had been seasonally moved between the royal towns of Ankober, Liche, Dabra Berhan, and later Enewari, Menilek had relied on his chief of the *madebet,* Walda Giyorgis, who Massaja stated "had custody of the royal granaries and the stores of comestibles and provided all the necessities of the king's table, of the court, and of those maintained by the government."[11] While the emperor rotated between his northern Shawan royal towns, he transformed local tenure to include in-kind payments of food and services directly to the palace rather than to resident military officers or the church.

Invocation of a territory as *madebet* or as a royal town was a prerogative of the *negus,* who also exercised the right to restore income rights to local officials and aristocracy when the court no longer needed support from that region.[12] Ada's oldest residents offered differing interpretations of the meaning of *madebet.* Jamanah Bayu, an Ada native born in 1916, associated the designation with Ada's main crop:

> It produces good quality teff and is so close to the city. Soldiers were scrambling over it. It is a teff producing area. That is why it is known as the *gwaro* [household garden] kitchen of the state.

Alamu Falaqa, born in Ada in 1924, understood *madebet* in terms of land tenure:

10. Cappucci, "Condizioni," 280.

11. Massaja, *I miei,* vol. 8, 204-5.

12. These tenure prerogatives were used in establishing the royal towns at Angolala (by Sahle Sellase), Enewari (by Menilek), and at the abortive fortress at Feqre Gemb (by Menilek). See Massaja, *I miei,* vol. 8, 204-5; Woobeshet Shibeshi, "Regional Study," 15-16.

The physical closeness of Ada to the palace made Ada the place from where the rations of the soldiers were drawn. *"Madebet"* means a nearby vassal area to the state and palace. The land of Ada was divided into different categories such as *gabbar* land, loaders' land [*chagni*], land for the artillerymen [*madf maret*], land for the servers of honey wine [taj maret], and tailors' land [*safi maret*], land for the church [*samon maret*], etc.

Sahlu Berhane, a second-generation Ada septuagenarian, described Ada's *madebet* role in more economic, utilitarian terms:

It [*madebet*] means an area situated near the state and the palace. Ada is located near Addis Ababa and that is why it was *madebet,* the kitchen of the imperial city. The state used to get service from the city: large sums of grain [and] honey were collected from Ada. Ada fed the court and the palace because it had good harvests of grain and honey. Arsi used to provide the cattle and Ada provided the grain.[13]

Dabtara Haile Masqal, unofficial historian of the Kajima Giyorgis Church, described *madebet,* and within that designation, the role of *samon maret* [church land] and *yamengist hudad* [the state's farm]:

The name *madebet* was given to Ada by Emperor Menilek. Ada was a rich agricultural area. So Emperor Menilek selected it as a supplier of food to the palace and court. The portion of the income went to the church and its servants. There were [also] large farms managed by the state. It was a granary of state and hence the name *madebet.* The state granary was right over there on top of that hill.[14]

Menilek's economic and geopolitical decision to shift his seat of power south to Entoto and thence to Addis Ababa transformed the context in which Ada agriculture would develop.

The earliest documentary reference to Ada's new status as royal kitchen appears in Jules Borelli's 1888 encounter at Menilek's court with "Ayto Bethsabe [Bezabeh] chief of the party of the Mad biet of the house of the king and also governor of Adda."[15] The dual role of chief of the *madebet* and governor of Ada in the years leading up to the foundation of Addis Ababa indicates the directions of Menilek's ambitions as well as his anticipation of Ada's future status as an imperial pantry.

13. Interviews with Jamanah Bayu (Hiddi), 25 May 1992; Alamu Falaqa, Dankaka (Ada) 23 May 1992; Sahlu Berhane (Hiddi), 28 May 1992.

14. This point was clarified by Dabtara Haile Masqal Wandemu, interview (Dabra Zayt), 26 May 1992.

15. Borelli, *Ethiopie méridionale,* 206.

Conversely, the Ankober region's loss of its status as *madebet*, as much as the movement of the political center of Shawa may account for the decline in its agricultural productivity, described in chapter 4.

Notwithstanding Ada's aetiological origins in its founders' search for purgative waters for their cattle, by the mid-nineteenth century it had an agriculturally based economy, one shaped by its peculiar environmental niche and shifting terms of crop-livestock integration. While the available sources for the early nineteenth century offer few glimpses of Ada's agriculture, evidence of settlement and benchmark observations in the 1880s provide a few clues. Above all, Ada acquired its reputation for fertility and agricultural potential early on. Gustavo Bianchi, in his habitual hyperbole, compared Ada with Italy's Po Valley, perhaps less an allusion to physical properties of the area than an allusion to its potential to support Italian economic ambitions. Cappucci praised the *"fertilissimo"* land of Ada as superior to other Oromo lands; he estimated that farmers, with little effort, could produce 100 *dawla* [c. one metric ton] on one-third of a *gasha* [i.e., about 10 quintals per hectare].[16]

Nineteenth-century narrative descriptions, as well as the seasonal rhythms dictated by Ada's vertisol plains and rainfall patterns, suggest a range of problems facing local ox-plow farmers a century and a half ago. In higher vertisol areas the waterlogging of black soils shaped but did not forbid cultivation. Most cultigens in the ox-plow repertoire fare poorly on waterlogged vertisols. Wheat and pulses, for example, cannot tolerate waterlogging and must be planted late in the rainy season or on residual moisture after the rains end. Most cultivars of barley, the historical staple of highland farms, grow poorly at low altitudes, and virtually no types of barley tolerate vertisols' fickle personality. By the mid-nineteenth century, however, Ada farmers had accumulated a battery of agronomic strategies to cope with specialized local conditions. They learned, for example, to plant wheat and specialized barley cultivars on upland, well-drained red soils and to rotate field peas and red teff in a 5–10 year fallow system. Ada's local wheat varieties had relatively early maturity, yielding less than the improved varieties to come in the 1970s but avoiding the risk of early cessation of the rains.

Teff, however, occupied first rank in Ada farmers' repertoire, especially Ada's trademark white variety, *manya* preferred on Menilek's table. Teff resists the effects of waterlogging and prefers sowing in early July after its characteristic labor-intensive, elaborate seedbed preparation. By rotating two seasons of teff (or one of teff and a second of wheat) with chick-peas, Ada farmers achieved over time a healthy balance of soil nutrients and good yields. Moreover, Ada farmers perfected a method of following the slow drainage of vertisol plots with staggered planting, thus maximizing human and oxen labor during the harvest and threshing season, when labor bottlenecks occurred in the agricultural cycle. Tedla Desta, an early

16. Bianchi, *Alle terra*, 423-24; Cappucci, "Condizioni," 281.

entrepreneur who first viewed Ada as a child in the 1920s, recalled the method for planting pulses on receding floods on black soils:

> When the land dries they just plant. They follow the land, follow the water. The water dries eventually. It stays a long time and then starts drying. As it dries, it recesses, then they follow that recession and plant. And it used to produce a very good crop. But not wheat, not teff, but *shimbra*, chick-peas, a lot of chick-peas. That land was even known as chickpeas land, *shimbra maret*.[17]

In January 1887 Leopoldo Traversi remarked on the results of this strategy for cereals from his vantage point on Mt. Yarar overlooking the fields of Akaki and Ada: "It is curious. Beside the fields where grain is already dry, another where it is scarcely germinated, another where it is scarcely grown its stalk. What fertility!"[18]

If agriculture was well established by the third quarter of the nineteenth century, it coexisted with a powerful and still largely autonomous livestock economy, still only weakly integrated with ox-plow farm economies. Ada's extensive livestock herds grazed on fallowed fields and along cultivated borders. In the dry season, pasture existed in relative abundance, much of it lower-lying, virgin, vertisol plains. The dry-season pasture on dried, black soil *chafe maret* (seasonally marshy lands)—provided abundant forage on lands too labor intensive to attract cultivation. These open, yet fertile, pastures which still existed during Menilek's time and early in the reign of Haile Selassie composed the primary arena of competition and innovation for cropland in the late twentieth century.

By the last quarter of the nineteenth century Ada comprised an open agricultural livestock system with only weak integration between its livestock and crop economies. Ada's markets as late as 1886 showed the absence of a forage market, only a minimal exchange of cereals, and domination of livestock exchange between the Boran stock from south of the Awash and the growing money economy of central Shawa. Traversi offered a useful account of Ada's Tuesday market (Addo Gabaya), the area's major market center, set between two volcanic outcroppings amidst a vast vertisol plain:

> Relative to the country, the market is rich, but not because of its people who gather there: for a sack of berbere or for a handful of dry chickpeas would 20 people come from 20 kilometers distance? Likewise would they come from

17. Interview with Tedla Desta, 17 April 1991 (Alexandria, Va), Tedla Desta, born in Harar in 1919, had served as a train conductor in the 1930s and later established a tractor rental scheme in Ada in the early 1960s.

18. Traversi, "Da Entoto," 583. Gene Ellis reported the same practice in 1968. Gene Ellis, "Man or Machine, Beast or Burden: A Case Study of the Economics of Agricultural Mechanization in Ada District, Ethiopia," Ph.D. dissertation, University of Tennessee, 1972, 61.

Gurage to carry [back] a cloth of black cabbage? What impressed me was the cattle of Arussi, which truly surprised in a country like this, where the herds are generally of skin and bones. . . . With eight or nine talers one can have two superb ones. There comes in second line [i.e., after cattle] cloth from Gojjam, which here is quite esteemed, then skins, horses, mules, pepper, and grains.[19]

Thus, grain counted for little in Ada's exchange economy; the Ada diet apparently reflected this orientation toward livestock. Both Amhara and Oromo accounts, albeit twentieth century informants, argue that a good portion of the nineteenth century diet in Ada consisted of meat and milk, supplemented by cereals.[20]

The trend toward crop production, forage markets, and the integration of the two agrarian sectors awaited the growth of Addis Ababa in the twentieth century. Cappucci in 1887 observed the lack of integration and implied little population pressure on land resources: "Even so, it is difficult to put more land in production because of their not having the custom of cutting the straw to give to the livestock in the dry season, in which vegetation is scarce; they oblige themselves to take a large piece of land to maintain livestock there in which they live together as pasture."[21] His comment was prescient in that it was precisely the increase of cropped land and its straw production which supported Ada's livestock population in the post-1960 period, when pasture declined.

Addis Ababa's Kitchen: Ada Land, Population, and Agriculture, 1889–1935

Ada's conversion to *madebet* status in the early 1880s put in train a number of processes which would provide a context for its agricultural development in the twentieth century. In the first instance the status of *madebet* meant the emperor quartered no soldiers on Ada's farms, nor did *shalaqa* (military leaders) receive rights of income and corvée labor over Ada's farmers as they did on lands elsewhere in southern areas. Menilek, in effect, retained these rights for the palace (i.e., direct control over the land itself), setting Ada apart from newly conquered zones where the military tribute system *(neftenyna-gabbar)* dominated.[22] In Ada, land granted by

19. Traversi, "Da Entoto," 587.

20. Interviews with Badada Kilole (Hiddi), 25 May 1992 and Sahlu Berhane (Hiddi), 28 May 1992 pointed to changes in Oromo diet from the late ninteenth to the mid-twentieth century.

21. Cappucci, "Condizioni," 282.

22. For *neftenya-gabbar* system, see Charles McClellan, "Perspectives on the Neftenya-Gabbar System: The Darasa Ethiopia," *Africa* (Rome) (1978), 33, 426-40.

Menilek to members of the palace was freehold, a system Haile Selassie would implement elsewhere in southern Ethiopia beginning in the 1920s.

Second, almost simultaneously with claiming Ada as royal larder, Menilek executed a land measurement program *(qalad)* which divided Ada into units of measured in *gasha*. As a land measurement unit, *"gasha"* had a wide range of meanings, anything from 6–7 hectares in northern Shawa to 80 hectares in southern areas; in Ada, however, it meant precisely 40 hectares. This land measurement in fact transformed the meaning of *"madebet"* to be "freehold" status. By 1886, the official year of Addis Ababa's foundation, the imperial government had measured and auctioned off substantial portions of Ada's land as freehold. Cappucci, who reported on the land auction, claimed that a *gasha* could be had for the price of a mule.[23]

In addition to land sales, Menilek himself made Ada land grants to churches in Addis Ababa, to new churches established in Ada, and also to individuals in his service. Kajima Giyorgis Church had received a 480-hectare patrimony (actually reduced from Sahla Sellase's original grant); other churches received between 400 and 500 hectares. Church land grants could be cultivated by priests directly or given out to tenants who paid their tithe directly to the church rather than to the state granary. Other servants of the palace or ex-soldiers also received Ada freehold lands. Ato Sahlu Berhane, a second-generation Ada native born near Hiddi in 1922, recalled his father's land grant obtained from Menilek in the early 1880s:

> My father had a thorough church education in Gojjam and Gonder. On the completion of his studies he came to Atse Menilek, who tested him and gave him the rank of teacher of the Diqua. Later on he established the Urael church [in Addis Ababa]. After long years of service he was given one *gasha* of land from this area and he settled here. . . . His uncle took care of my father's belongings while my father continued with his preaching and writing prayer books. During those days there were many Oromo here so my father did not want to come here; he preferred to stay in Addis Ababa. The peasants who did not like the intrusion of the Amhara abandoned the land and ran away. They did not want to cooperate with Menilek's campaigns and wars. The Oromo did not farm in those days. They lived on milk and they were ignorant of farming. My father was a chief priest of Urael Church in Addis Ababa. He stayed in Addis Ababa and continued to administer his land in Ada as an absentee landlord. He used to cover the expenses of the land, the taxes and the like, by the money he used to get from his services.[24]

23. Cappucci, "Condizioni," 281-82.

24. Interview with Sahlu Berhane (Hiddi), 28 May 1992. Ato Sahlu served for 15 years as a local *atbya dagna* (district judge) at Hiddi.

Ato Sahlu's recollection of his father's need to cover land taxes with personal income suggests both an initial decline in the value of Ada land and also an early depopulation, the latter a phenomenon already seen in Gera (see chapter 5).[25] Sahlu himself disagreed with the evidence of land sales that I presented to him from Cappucci; he argued that "people [during the time of Menilek] were not ready to pay five birr. . . . A mule cost too much and land was very cheap."

Menilek's victory at the Battle of Adwa required him to reward soldiers who had fought with him, and it provided him with an opportunity to fix a loyal population on Ada's *madebet* land. Jamaneh Bayu, a third-generation Ada landowner, recalled his grandfather's arrival: "He came at the time of the Battle of Adwa. Some came from Gonder, others from Gojjam, and the rest from Manz. When they came back from the battle they were given plots of land in this area, and they settled here permanently. My grandfather and others came here as soldiers of Ras Nadew."[26]

It was not until the first decades of the twentieth century, however, that tenancy on Ada land began to take shape, drawing new migrants to Ada. Low demand for the land forced landowners who wished to put their land in production to offer concessionary terms to attract migrants and Oromo residents. Sahlu recounted the change:

> Gradually more and more tenants began coming to farm on my father's land. Even Oromo started settling on the land as tenants. To bring the virgin land under cultivation, the tenants were given two years of grace, during which they were totally exempted from any kind of payment except the tithe of the government. During the third year the tenant farmer began paying *erbo* [onefourth] payment, and after the third year, the tenant began to pay *siso* [one third] to the landlord.[27]

Most of these new migrants arrived from northern Shawa (Bulga, Tagulet, Minjar), the parents of those former tenants I was able to interview in the early 1990s. In other cases the soldiers or, later, civil servants who received Ada land from the imperial court settled servants or even rural relations on their lands as tenants. Unlike *gabbar* land in newly conquered zones, where the state transferred its income rights to the local governor and resident soldiers, Ada income accrued directly to members of the imperial court.

25. Cappucci observed a deflation of land prices in Ada in the mid-1880s through a decline in the quality of mule accepted as payment. Cappucci, "Condizioni," 281.

26. Interview with Jamaneh Bayu (Hiddi), 25 May 1992.

27. Interview with Sahlu Berhane (Hiddi), 8 May 1992; and Habta Mariam Wandamagagnahu, Hiddi, 25 May 1992. These concessionary terms were also reported in the historical section of a 1969 baseline survey. See Borton et al., *A Development Program for the Ada District Based on a Socio-Economic Survey* (Stanford, 1969), 97.

In Ada no single person held more than about 200 hectares. The government itself directly held only about 100 *gashas* in the district, that is, less than 1 percent of Ada. This land included 2400 hectares in the area of the Kajima Giyorgis Church, mainly light red soils, and 1600 hectares at Foka, a flat vertisol plain between the village of Bishoftu and the Addo (Tuesday) market. From these lands the imperial court received products directly, using corvée labor from Ada farmers, who owed one day's labor a week as part of their tenancy.

Prior to 1935 grain flowed to Addis Ababa through political channels as taxation or rent rather than through a mercantile sector. The arrival of the Franco-Ethiopian rail line from Djibouti and Dire Dawa in 1917 and the reforms of the regent Ras Tafari [later renamed Haile Selassie I] provided important impetus in fixing Addis Ababa and its southern periphery as a focus for both political power and population growth in the capital and along the railway route. Ada had two railway stations, both chosen in the absence of any existing towns, adjacent to market sites. One, called Les Addas was on the vertisol plain near the Addo market, and the other was a few kilometers north at the smaller market site at Bishoftu on high ground between two crater lakes, Bishoftu and Hora. Up to 1935 there was only a station building on what was to become the site of the town of Dabra Zayt/Bishoftu: "At that time Dabra Zayt did not have two huts! There was no trace of a town."[28] Mojo, in the district just south of Ada, grew earliest, since it was the encampment of the Somali railway workers. Within Ada itself, town development awaited the Italian occupation and the postwar economic growth.

Local markets, as in the late nineteenth century, remained primarily in service of the livestock economy. Leading up to the Italian occupation Ada maintained the aspect of, if not a pastoralist, at least a livestock-rich economy based on extensive dry-season pasture on seasonally flooded vertisols and the free movement of animals between grazing sites and the lakes, especially Lake Kilole and its purgative waters. Sahlu Berhane recalled his father's keeping 10 milk cows at a time. Having been a 14-year-old conductor on the Djibouti-Addis Ababa railway, Tedla Desta, who would later become one of Ada's leading agricultural entrepreneurs, recalled trains of the early 1930s having to stop in Ada frequently to whistle large herds from the tracks. Farmers therefore had a large pool from which to draw young bulls and oxen. In addition to open, dry-season, marsh pasture, farmers, even tenants, held pasture in their own land holdings either as fallow or as pasture land rented cheaply from a neighboring landowner.[29] Livestock forage thus consisted primarily of pasture grasses.

In the accounts of land holdings from interviews covering the pre-1960 period, most farm holdings were small, that is, two hectares or less. Most farmers inter-

28. Interview with Sahlu Berhane (Hiddi), 28 May 1992; and Badada Kilole (Hiddi), 25 May 1992.

29. Interviews with Alamu Falaqa, Dankaka (Ada), 23 May 1992; Sahalu Barhane, 28 May 1992, and Tedla Desta, Alexandria, Va, 17 April 1991. See below for the Tedla Desta system.

viewed recalled cultivating plots of two hectares or less, which served to provide the cereal needs of the household. Alamu Falaqa, born in Qaliti (Ada) in 1927, the grandson of a soldier who had migrated to Ada from Tagulet (northern Shawa), recalled his father's farm:

> He had two pairs of oxen at that time. He plowed with one pair while a hired laborer used the other pair. During those days people did not farm large tracts of land. Farmers got sufficient yield from smaller plots of land, a maximum of two hectares (eight *kirt*). The yield from chick-peas is not at all satisfactory these days, though the labor spent on it has increased tremendously. In the past the work was light and the yield was high.[30]

The farm size in this period reflected less pressure on land rather than the lack of incentives to expand area under crops or to integrate crop and livestock sectors of the economy.

Ada's proximity to the growing capital and the rail line also prompted early examples of commercial development in Haile Selassie's Ethiopia. At Godino, in the southeastern foothills of Mt. Yarar, Qagnazmach Babicheff, a White Russian confidant of Menilek, founded and operated a horticultural enterprise similar to one Haile Selassie himself developed at Erer, farther east on the rail line. Babicheff imported workers from Walayta to serve as labor on his irrigated cultivations of sugar cane, citrus, coffee, and potatoes. Babicheff distributed parcels of land to his workers in exchange for their labor and sent his products with Addis Ababa traders to Addis Ababa's increasingly sophisticated market.[31]

The emperor also envisioned Ada and its lakes as an ideal spot for repose for Addis Ababa residents and its growing foreign community. For his 1930 coronation he assigned Makonnan Habta Wold (a future prime minister) and a team of engineers to drain a marshy zone north of Dankaka (near the present air force base) to rid the area of mosquitoes, which tormented overnight visitors. When the engineers drained water from the seasonally waterlogged black soils into Lake Hora, they inadvertently transformed that dry-season pasture into 25,000 hectares of new agricultural land.[32] It was not until the 1960s land boom, however, that the public health and agricultural impact of the action had full effect.

30. Interview with Alamu Falaqa, 24 May 1992 (Dankaka). Haile Mariam Wandama-gagnahu, born in 1907 to a tenant, cultivated a farm of 1.5 hectares in the mid-1920s. Though Alamu reflected a farmer's nostalgia for days of fertility past, there was a consensus among farmers who cultivated small areas of cultivation in the pre-1960 period.

31. Interview with Wayzero Baqalach Gassasa (Godino), 31 May 1992. Wayzaro Baqalach had worked as a personal servant to the Babicheff family. Also interviews with Sahlu Berhane (Hiddi), 28 May 1992; and Balda Azaj (Hiddi), 28 May 1992.

32. Interview with Tedla Desta, 17 April 1991. Tedla later rented land there for his contract tractor operation.

Land productivity was the result of a strategy of rotation well adapted to local conditions and the local preference for teff over other cereals. Ada farmers practiced three rotational schemes, their choice among them depending on soil type and drainage. Alamu Falaqa, born in Ada, recollected rotation strategies which avoided the need to break virgin soil:

> In the past we were ignorant of the uses of fertilizers. If the farmland was reaching exhaustion, we changed the crop. Chick-peas are what the farmers sowed to regenerate the land. Once chick-pea is sowed on a certain plot of land it would be very good for teff at least for two years. The first year white teff and the second year the dark teff is sown, with a good harvest usually. On the third year the plot is used for beans, and the following year that would be good for teff [again]. There was also a variety of wheat known as *galano* which was sown in August, etc. Such a rotation of crops prepared the land for teff and other types of grain.[33]

Rotation had a number of benefits beyond preserving fertility. Crop rotation broke down the life cycle of host-specific pests and weed seed populations, increased water infiltration, and improved microbial biomass (an essential element in fertility). Tedla Desta noted the high value farmers placed on land under pulses such as field peas and chick-peas:

> . . . when they have a fallow land, to turn it into a cultivated land they start with special crops. They just till the top of the fallow land, just the top. We call it *chifleq,* that is done immediately after the rain because at that moment it is soft. And they plant peas, green peas, *ater.* That crop, that place, the next year is a very nicely prepared land—as if it was really plowed land—because the peas fertilize it with the leaves, and the roots penetrate and water penetrates inside. And the next year when the farmer comes it is very easily plowed. They do the same with chickpeas . . . followed by any crop, teff, wheat, any. As a matter of fact, a chick-pea field is considered like a fertilized field.[34]

Farmers in Ada's past also recalled the practice of intercropping on household plots, especially for oil seeds needed for household consumption in an era prior to large grain markets. These included teff or wheat with safflower, teff with rape seed, and maize with sunflower on the Rift Valley southern frontier.[35] Ada farmers also practiced fallowing, though with important local variations. In the red soil areas around Zukwala, land was fallowed 3–5 years; in south central

33. Interview with Alamu Falaqa (Dankaka), 23 May 1992. Also see Borton et al. *Development Program*, 82 and Bekele Shiferaw, "Crop-Livestock Interactions."

34. Interview with Tedla Desta (Alexandria, Va), 17 April 1992.

35. Borton et al., *Development Program*, 82.

Ada, 5–10 years (again, on red soils); and on upland soils around Mt. Yarar, 4–5 years. On black soils fallowing was rarely practiced, despite the availability of virgin plots. By the late 1960s, however, fallowing in most areas of Ada was a punishable offense enforced by landowners trying to push up crop production.[36]

Ada's Colonial Experience:
Agriculture and Empiricism

The 1936–41 Italian occupation marked an important watershed for the national economy and political structure of Ethiopia. For Ada it had important repercussions in terms of town development and ripple effects from the continued growth of Addis Ababa as a national economic magnet. Italy's imperial plans included ambitious schemes for agricultural autarky, the *"paneficazione"* of the rural economy, which they envisioned would not only provision Ethiopia's cities with wheat, but also feed the metropolitan Italian empire as well.[37] These plans had a strong empirical thrust, including national cropping surveys, experiment stations for wheat, poultry, and livestock production, building on three decades of crop experimentation in Eritrea.

The primary impact of the Italian presence on Ada was to transform the handful of houses around the French-built Bishoftu railway station into an urban center from which an Italian resident agent administered the Reale Commissariato di Biscioftu, which oversaw a handful of development schemes, attempted to suppress local resistance, and—the grand scheme—tried to transplant war veterans and entire peasant villages to planned village sites as part of Mussolini's *"colonizzazione demografica."*[38] This last project focused on sites in Chercher, Dabat, which is north of Gonder, and, for the settling of ex-soldiers, Holeta, 30 kilometers west of Addis Ababa on the Ambo road, and Ada. These projects were not only to provide gainful employment to Italy's peasant ex-soldiers, but also to be models for the expansion of modern agriculture in the highlands by Italian farmers.

In Ada the projects of the Reale Commissariato di Biscioftu focused on capital-intensive horticultural, poultry, and livestock schemes, including taking over the Babicheff enterprise by Ente Romagna d'Etiopia, a state-sponsored effort which hoped to settle northern Italian peasants on the 220-hectare irrigation scheme. On a

36. Borton et al., *Development Program*, 83-84.

37. For an excellent overview of Italian agricultural plans, see Haile Mariam Larebo, "Myth and Reality." For examples of Italy's plans see Ufficio Agrario della Colonia Eritrea,"Progamma per l'incremento e il miglioramento della produzione avicola della Scioa," IAO Fasc. 1803, which describes plans for a poultry-production program.

38. For an overview of the Opera Nazionale per l'Combattenti projects at Ada and Holeta, see Ferdinando Quaranta, *Ethiopia: An Empire in the Making* (London, 1939), 40-51.

larger scale, the Opera Nazionale per l'Combattenti (O. N .C.), was a project to house war veterans on a 2500-hectare agricultural scheme at Foka, a vertisol area northwest of the town of Bishoftu, land which had belonged to the empress. It was there that colonial authorities concentrated their efforts at mechanization, seedbed drainage, fertilizer, and use of improved wheat varieties.

In purely agricultural terms, Italian policy had little interest in transforming smallholder production, beyond its organization as a tax base. There are few records of these efforts other than planning documents, and most Ada residents from that time recall few lasting effects. Sahlu Berhane, then in his early 20s, however, recalled in some detail Italian rural agrarian policy in Ada:

> The Italians never interfered in the possession of the landowners. They even did not collect the tithe for two years. The Oromo refused to pay the tithe to the landowner. Later on the Italian government began exacting four quintals of teff per *gasha* of land. This applied to all kinds of grain . . . . With the payment of the tax the landowners' possession rights were guaranteed. The farmer's share was two-thirds, and that of the landlord, one-third of the total yield. The old legacy continued without any change. The state property came under control of the Italians; the property of the poor stayed in their hands.

He also described Italian innovations on the confiscated imperial land at Foka:

> There was a large tract of land that was the property of Empress Manan around Qaliti. This was farmed with tractors, and there were large grazing lands as well all around the valley. The Italians were good farmers and the harvest was immense during that time.
>
> They used tractors to farm large areas of land. The whole forest area was changed into farm lands. They used to sow seeds with machines. They did use fertilizer, but they never gave it to the local people. Yes, the fertilizer was extremely effective; even marshy areas were changed into agricultural lands. They had machines designed for the purpose of drainage and irrigation. Both wheat and teff were widely cultivated in the area. There was a special Italian wheat that gave a good yield. They brought fallow and virgin lands into cultivation.[39]

The most significant change brought by the Italian occupation of Ada was the development of the commercial and physical infrastructure of the town of Bishof-

---

39. Interview with Sahlu Berhane (Hiddi), 1 June 1992; for a list of Italian-planned agricultural schemes in Ethiopia, see Ufficio Studi del Ministero Africa Italiana, "La valorizzazione agraria," 273.

Figure 6.2. Italian tractor cultivation at Foka (Ada), c. 1939.

tu itself. By 1941 Bishoftu was a market and administrative center with a core of buildings, services, and a rural road network which connected Italian schemes to the center. According to the 1938 *Guida dell'Africa Orientale Italiana* (the official Italian tourist guide), Bishoftu boasted a 60-bed hotel (later renamed the Hora Ras Hotel), a post office, telegraph, telephone, infirmary, airport, and official residence. On the former imperial farm at Foka was the Azienda Agraria di Biscioftu dell' O. N. C., where, the *Guida* noted,

> there are under construction 20 colonial houses to be assigned to the ex-combatants; the placing of the first stones took place on 9 December 1937 during the ceremony of the threshing. There prospers there in the Azienda various kinds of grains (wheat, barley, teff, sorghum, maize), legumes (beans, chickpeas, horse beans, peas), beets, sugar cane, various oil seeds, vegetables and gardens, fruit trees, etc.[40]

Few Italian plans for agricultural development in the region, however, panned out. The O. N. C. project, which generated massive documentation in Italian archives in the end built 20 houses and managed two or three harvests. The planned poultry experiment never distributed the Rhode Island red, Plymouth, and Lang-

40. Consociazione Turistica Italiana, *Guida*, 426-27.

shan pullets set out in the plan. These Italian investments at Bishoftu, however, became the foundation upon which Haile Selassie and Addis Ababa landowners built Dabra Zayt, the postwar boom town which fed on the rising value of land and a specialized form of landowner-tenant agriculture.

Ada Agriculture, Freehold Owners, and Tenants, 1941–63

The end of the Italian five-year occupation and the return of Haile Selassie's government brought an era of reconstruction, not in repairing wartime damage, but in recasting prewar institutions into new forms. For the rural economy this meant the expansion of freehold tenure at the expense of earlier *gabbar* tenure and the creation of land markets, a process which greatly strengthened the hand of Addis Ababa as a center of power. In 1941 the emperor restored all previous land rights to former owners in Ada (as elsewhere). Moreover, he also created a new class of *arbegna* (patriots), those who claimed to have served in the resistance and expected compensation in land. The growth in Ethiopia's bureaucracy also created a new class of those worthy of land grants and capable of using salaries and commercial profits to buy rural lands close at hand (i.e., in Ada or other former *madebet* areas).[41] For Ada, the new policies meant the arrival of a new generation of landowners and tenants and a new context for agricultural change.

Haile Selassie's government was well aware of the importance of Ada as a source of income for its upper-level bureaucrats, as a potential source of food for a growing capital, and as a tourist spa. Shortly after the Italians' departure, the government organized a new assessment of land holdings and taxation, adding a fixed land tax of five quintals per *gasha* above the tithe paid by the tenant prior to landowner's claim.[42] At the same time, the emperor granted the former Babicheff plantation to Tsehafi Taezaz (Minister of the Pen) Walda Giyorgis, who expanded the scheme from fruit to include livestock (even pigs) for the increasingly cosmopolitan tastes of the Addis Ababa market. The government also took steps to establish an urban center for administration and commerce, renaming the town of Bishoftu with its Amharic name Dabra Zayt and designating it the district capital. In order to concentrate grain trade in the town Haile Selassie's appointee as district governor Dajazmach Sahlu Difaye, ordered the consolidation of the new Bishoftu market with the long-standing Tuesday market in Addo in 1942. Dabra Zayt's close

41. Interviews with Sahlu Berhane (Hiddi), 1 June; Badada Kilole (Hiddi), 1 June; and Taffara Gabra Hiwot, Dukam (Ada), 17 May 1992.

42. Sahlu Berhane participated in that assessment and later served as an *atbiya dagne* (district judge). Interview with Sahlu Berhane (Hiddi), 1 June 1992. By the late 1960s this payment had been converted to cash (E$15).

proximity to the capital, its Italian-built hotel, and its attractive lakes made it a weekend recreational center for both Ethiopians and foreigners. Also in the 1960s Canadian Jesuits and the Sudan Interior Mission built weekend retreats on Lake Babo Gaya. Dabra Zayt's proximity to the capital also made it the site for the Swedish-trained Ethiopian airforce, which established training facilities southwest of town in the late 1950s.

The expansion of landowning necessarily created a need for new tenants to put land in cultivation. By the mid-1940s, Ada's basic division of land between free-hold land (misleadingly called *gabbar* in Ada) and *samon maret* (church patrimony) was in place.[43] Those with family connections in northern Shawa brought relatives, especially younger sons, who had few prospects for inheriting land (see below and chapter 4).

As a generation earlier, landowners who had bought or been granted virgin land had to offer concessionary rental terms or compete with other landowners for the most hardworking tenants. Taffara Gabra Hiwot, who arrived at Ada in 1953 from Jirru (northern Shawa), began as a tenant with neither livestock nor cash:

> I left my home village willingly without borrowing any property or money. I had food for a few days only. As a good farmer I was popular among the landowners. Landowners were trying to attract me as their tenant and I started my own farm. The first year I got oxen from the landowner on a *menda* [contract], and the next year I was a full-fledged farmer with my own oxen. Later on I hired my own tenants. I used to plow with my own pair while my own tenant plowed with the other pair.
>
> [I had] about 20 *kirt* [five hectares] of land. . . . It was a half and half share, I used to take half and give the remainder to the landowner. There were places where the payment was only *siso* [1/3]. It varied from place to place. I was living a good life, having good relations with my hired farmers and the landowners.[44]

The open economy of Ada rewarded those who could organize their own labor to expand into open areas. As in Taffara's case, the acquisition of oxen provided no obstacles to rapid expansion into open land. Taffara's circumstance of being a tenant who hired tenants was not unusual; it was common for a tenant to become a landowner or to serve as a tenant for several landowners at once. Through most of the 1941–63 period tenant rates remained a third share, after the tenant had

43. *Gabbar* land was eligible for sale, lease or mortgage. Taxes (land and education taxes plus tithe paid in kind) were paid directly to the government. *Samon* land required that cultivators use it to pay taxes and tithe directly to the church; they also had to provide certain services: building, repairing, and support for particular holidays.

44. Interview with Taffara Gabra Hiwot (Dukam), 17 May, 1992.

paid the government tax which was collected at the threshing ground by agents of the landowner.

Former tenants I interviewed identified only a few major landowners in any one area. In the two decades following the Italian occupation Ada's agriculture evolved as an open forage system in which land prices rose but were restricted by the continued availability of virgin and fallow lands, on which tenants maintained their livestock. In Ada's open economy of the 1950s and early 1960s tenancy itself could be a vehicle for accumulation. As in the cases of Taffara and Habta Mariam, successful tenants could expand their own holdings in land and livestock and recruit their own tenants to open new cultivation.

The open economy of land and labor had agronomic implications because some landowners searched for ways to raise productivity of their land. The labor, or rather tenant, shortage in the early 1960s accounts for landowners' early interest in tractor cultivation. Tedla Desta, an entrepreneur and the Ethiopian sales representative for International Harvester, started a tractor contract service to plow and disk virgin fields in Ada. I asked him to recall the state of Ada agriculture in the early 1960s:

> There was a lot of fallow land. That's one of the reasons I went there to offer my services. There were many landlords I met that came and gave us a small contract, not large contracts—two hectares, three hectares. We did the work for them. . . . We would plow for them and they would pay; particularly a few of them that came repeatedly every year and enlarged their field. For the landlords, we just plowed and we got paid and that's it.[45]

As Tedla indicated, landowners, particularly those with virgin, uncropped land showed early interest in mechanization. Ultimately, however, the small, fragmented nature of land holding made tractor contract farming unfeasible on fragmented smallholder plots, and landowners with more than 80 hectares of land accounted for only 7 percent of Ada's total area.[46] Tractors such as Tedla Desta's could break up virgin or fallow soil, but could not carry out the more labor-intensive third, fourth, or fifth plowing pass (the last one for seed cover) or the weeding necessary for teff cultivation; these operations required tenant labor and ox plows.

Although Ada's population had increased dramatically in the postwar years, crop-livestock integration was, by 1963, still weak with few indicators of intensification and integration. With the early tractor contract schemes it was hoped that crop residues could be sold to raise their profitability, but no market was found—

45. Interview with Tedla Desta (Alexandria, Va), 17 April 1992.

46. Average holdings by the late 1960s were still above five hectares per household. A 1961 Ministry of Agriculture survey indicating average holdings of 3.4 hectares may reflect the presence of a still-viable livestock economy. Borton et al., *Development Program*, 96-98.

an indication that dry-season grazing was still sufficient to sustain Ada's livestock economy without further livestock-crop integration.[47] Few landowners were in a position to impose terms on their tenants fearing they would shift their services to another landowner's area.

Ada's Malthusian Scissors, 1963–75

The evolution of Ada agriculture through the twentieth century until the 1960s had proceeded slowly, a product of its own endowments in soil and climate as well the steady growth of population and economy at Ethiopia's political center in Addis Ababa. By the early 1960s the conjuncture of population, market, and agricultural empiricism accelerated and transformed the basic elements of Ada's ox-plow cultivation. By the late 1960s Ada's family and population density was higher than for any other ox-plow region of Ethiopia; its average per-capita gross value of crop production was 45 percent higher than for any area of Ethiopia, and the family gross value of crop production was 80 percent higher than the average for Ethiopia as a whole. By the early 1970s the mean number of landowners per tenant farm was 1.6; individual landowners with 40 hectares of land might have had as many as 10 tenants on their holdings, though not all with homesteads there. In other cases farmers owned small, inherited, fragmented plots and took additional land as tenants from an absentee landowner.[48]

Most critical to this change, as both an indicator of the change and its driving force, was the intensification of crop-livestock integration, the merging of farm-level management in what had been two largely autonomous economies. The result has been a new historical phenomenon in the agronomy and economy of ox-plow agriculture, where growing constraints on obtaining livestock forage have created cost advantages in providing inputs directly on the farm.[49]

While population statistics for the early 1960s are vague at best, there is a widespread perception among Ada's farmers that the population increase which began in

47. Ellis, "Man or Machine," 70.

48. Charles Humphreys, "An Empirical Investigation of Factors Affecting Peasant Crop Production (Based on a Survey of Ada Wereda, Ethiopia)," Ph.D. dissertation, Fletcher School of Law and Diplomacy, Tufts University, 1975, 12. For landowner-tenant data see Vincent, "Economic Conditions," 22-23. For taxation and mixed ownership-tenant holding, see Getachew Tecle Medhin and Telahun Makonnen, "Socio-Economic Characteristics of Peasant Families in The Central Highland of Ethiopia—Ada Woreda," unpublished report, Ministry of Agriculture, 1974. Also see interview with Habta Mariam Wandamagagnahu, 25 May 1992, an Ada native, who remained a tenant his whole farming life (until the 1975 land reform) but inherited his father's tenants and owned four pairs of oxen. Neither he nor his father ever owned land in Ada.

49. McIntire et al., *Crop-Livestock Interactions,* 3-1.

the postwar years accelerated in the 1960s. By 1968 Ada's population had reached 80,000, not including 20,000 residents of the town of Dabra Zayt; by 1974 the rural population had grown to 108,000 and Dabra Zayt to 32,000. In rural agricultural centers such as Hiddi and Godino, small towns grew directly on agricultural land.[50]

The population of Ada grew as a result of both in-migration—due to the attractions of rising agricultural incomes and landowner recruitment—and real improvements in public health in the 1950s and early 1960s, both by-products of Ada's proximity to Addis Ababa and its influential land owning class. A focus on public health, especially an effective USAID-funded malaria eradication campaign which began in Ada in 1960, dramatically reduced infant mortality and adult morbidity. The effect of this campaign and the establishment of health centers in the town of Dabra Zayt not only brought about a natural population increase in family size evident by the mid-1960s but also attracted more migrants to improved social services. Sahlu Berhane, who became a local governor in 1962, argued strongly for improved public health as a primary cause of population pressure on land:

> During the time of Menilek more and more Amharas came and settled here on the land given to them by the state. The local Oromo and the Amhara increased in number as time went on. The people who have come from other parts of the country were hired farmers [tenants], but they were not the cause for the shortage of farm land. Emperor Haile Selassie introduced more and more doctors. Children used to die in large numbers from jiggers and malaria. There was no cure for them; malaria was incurable. DDT was introduced in the last 40 and 35 years. Malaria was eradicated. . . . In 1953 [A.D. 1960] the malaria eradication team used to live in my home.[51]

Notwithstanding the effects of lower infant mortality on family size and population, population movement has been demography's dominant effect on agricultural change, at least within the scale of "social time." Many of the same social factors which drove highland farmers into the Ankober lowlands, toward Gera's coffee harvest and onto other development sites pushed others toward tenancy in Ada.

The principal effect of population growth on local agriculture was the inflation of the price of land relative to that of labor, with downstream effects on agricultural practice itself. Land prices by 1969 had reached E$80,000 per *gasha*, (i.e., E$2000 per hectare). New investments in these circumstances required the expan-

50. Borton et al., *Development Program*, 2; Vincent, "Economic Conditions," 6. For town of Hiddi, established in 1962, see interview with Sahlu Berhane (Hiddi), 28 May 1992.

51. Interview with Sahlu Berhane (Hiddi), 28 May 1992. Badada Kilole agreed on the role of public health investments in population growth, listing typhus, malaria, and jigger-borne disease (all controllable with DDT) as the major constraints. Interview with Badada Kilole (Hiddi), 25 May 1992.

sion and/or introduction of high-value crops.[52] Ada landowners responded to increasing prices of teff and consequent pressure on land prices around Addis Ababa. As a means of increasing cropped land within their holdings (see below) landowners also recruited family and distant relatives as tenants likely to be trustworthy and hardworking. Kuri Walde arrived in Ada in 1970 from Gurage Soddo (western Shawa) to serve as a tenant on the land of her father-in-law, who also owned a grain mill, a Land Rover (a rural taxi) and a large cattle herd. Kuri and her husband provided a reliable source of farm labor, which allowed their patron to expand his commercial interests.[53]

In another case Dasta Haile Mariam arrived in 1971 from Ankober:

> There was a shortage of land back at Ankober. I came here to farm on my uncle's land here in Ada. . . . He brought me here and provided me with the necessities of life including a house. He was living in Addis Ababa but he had land here in Ada. . . about one *gasha*. He was a judge living in Addis Ababa but he owned land here in Ada. He had one ox and I had one. We shared the produce equally. There was no fertilizer at that time. The owner supplied the seed and I took the responsibility of the work.[54]

In addition to landowners' recruitment of relatives as tenants, there is some evidence that tenant households themselves recruited non-nuclear family relatives. The Ministry of Agriculture's 1961 survey indicated that between 1961 and 1973 the average number of persons per household in Ada increased from 5.5 to 6.5.[55]

The movement of new population onto Ada's land and growing family size from new public health conditions increased cropped land at the expense of the livestock economy. In 1960 Ada's cattle population was 92,641 whereas the annual average for the period 1979–90 declined to 80,278 (49 percent of which were oxen or young bulls).[56] The decline in overall livestock numbers mask an even steeper decline in the per-farm holdings, given the increase in population during 1961–90 (at least double). Table 6.1 indicates a measure of the farm-level impact on per-farm oxen holdings.

The pressure on landowners to increase cropped land and to concentrate increasingly in a single crop, teff, a high market value but high-labor crop, was a driving force in the evolution of landowner-tenant relations by the late 1960s. Teff had long been Ada's most important crop, though its proportion of the total cultivated area was necessarily constrained by both its intense labor requirements

52. Tedla Desta paid E$15,300 for half a *gasha*. McIntire et al., *Crop Livestock Interaction*, 3-6, 3-7.

53. Interview with Kuri Walde, Ensalale (Ada), 17 June 1992.

54. Interview with Dasta Haile Mariam (Godino), 11 June 1992.

55. Borton et al., *Development Program*, 96-98; Vincent, "Economic Conditions," 34-37.

TABLE 6.1. Distribution of per-farm cattle holdings, 1969–80.

Type	1969	1970	1971	1980
Oxen	3.21	2.5	2.8	1.86
Cows	1.40	1.1	1.0	.93
Calves	1.50	1.3	.70	.98
Heifers	.31	.30	.20	.33
Young bulls	.80	.80	.2	.48

Source: Data drawn from Getachew and Telahun, "Socio-
economic Characteristics," 25; and Gryseels and Anderson,
Research, 14. Bekele Shiferaw, "Crop-Livestock Interactions,"
gives aggregate figures but not per-farm holdings.

and the need for rotation. The growing demand for Ada's high-quality white teff through the early 1960s while Ethiopia's economy and Addis Ababa's population expanded, pushed Ada agriculture past a threshold toward intensification and concentration.

Pressure on land relative to labor showed in the changes in landowner-tenant relations. By 1969 the vast majority of Ada landowners were absentee, a land owning elite living in Addis Ababa. The Stanford Research Institute study of that year estimated that one-half of Ada's tenants were farming under a *siso* agreement, that is, tenants paid a third of the harvest to the landowner, paid a tenth as tithe, and kept the rest. The other half of the tenants farmed on the half, paying half of the harvest to the landowner plus another tenth as tithe to the government. The Stanford survey also argued that the half-and-half tenancy was concentrated on the best land and that such agreements were increasing.[57]

Another study of 10 Ada farms in 1969–70 indicated a somewhat different balance of the tenancy arrangements in that year. The emergence of fixed-share tenancy indicates a further step beyond the proportional half-and-half tenancy, which at least shared risk between landowner and tenant. Fixed-share rental was the form adopted in most tractor contracts. Though these farms were large by highland ox-plow standards, they reflected the mix of cropped land versus pasture (i.e., the heavy pressure on cropping in Ada). The total average holding of 5.42 hectares (for

56. My owns calculations are based on data drawn from the Ministry of Agriculture's, "Ada Wereda Sample Survey," 1961 cited in Borton et al., *Development Program*, 86 and Bekele Shiferaw, "Crop-Livestock Interactions," Appendix 8.

57. Borton et al., *Development Program*, 9, 98. The group also argued that the new agricultural income tax of 1968 had precipitated the change from third- to half-share tenancy in many areas.

1969) was 91 percent under crops, 4.58 percent under pasture, 2.59 percent in homesteads, and 1.82 percent fallow and in roadways.[58]

Tenancy arrangements were complex, varying by individual farm and between each farmer and one or several landowners. Landowner-tenant relations could be very personal or conducted entirely through a local agent who negotiated and collected harvests for the landowner. None of the farmers in the 1969–70 study had written agreements with their landlords, and most tenants expected to perform other uncompensated labor services, such as building the landowner's granary. On the other hand, the formal division of crop shares hid more informal, but normative, arrangements which benefited the tenant, especially where some rapport had built up over several years. Former tenants argued to me that landowners were a reliable source of short-term credit. Moreover, the division of crops in proportional agreements did not include the *gert*, the portion of grain which remains with the chaff after winnowing. Tedla Desta described these subtle relations:

> Collection comes. The farmer advises the landlord that he is going to thresh on this date; on such and such date. The landlord usually comes late in the day after he has finished [threshing]. He seldom comes and watches the threshing; he would never do that. So when he comes; the *mert* and the *gert* is separated [winnowed]. *Mert* is the clean and *gert* is the seed that contains chaff. So the landlord comes. They start counting, they measure it and the one part is set aside. Every time they have three on one side they put one on another. Then he brings his servant to collect his part, and he tells the farmer, "God bless your *gert*." That is a blessing he must give as a rule, as a tradition. So if he doesn't say this the tenant thinks he is not considered a good sharecropper. It is not the landlord which does the sharing; it is the farmer. You know he takes this thing and he throws it in the air and the part of the seed which is taken in the air is called *gert*. He takes from this [the grain of the first winnowing; he doesn't share from the *gert*. Sometimes he takes as much as he has paid to the landlord. It is known as *gert* when it is threshed [winnowed] again. It is a good thing. It is a taboo even for the landlord to look at the *gert*.[59]

By the late 1960s the pressure of land prices to expand cropped land and the pressure to intensify labor and inputs to improve yield changed these relations, resulting in an increase in tenant evictions. Eviction appears to have been widespread, but few data are available on its rate of incidence. Of the 10 farms studied in the 1969 survey, 3 were evicted on some of part of their holdings in the next year. Sahlu Berhane recalled that by 1962,

58. Getachew and Telahun, "Socio-economic Charcteristics," 20.

59. Interview with Tedla Desta (Alexandria, Va), 17 April 1991.

TABLE 6.2. Averaged percentage of crops on cultivated land, Ada, 1969–71.

Crop	% of land cultivated	% of holding	% of income
Teff	55.5	50.5	57.5
Chick-peas	13.7	12.4	6.5
Wheat	12.9	11.8	7.6
Horsebeans	0.3	4.4	1.3
Barley	0.2	2.5	4.7
Maize	1.3	1.2	1.7
Sorghum	3.4	3.1	1.7
Field peas	3.3	3.0	2.8
Vetch	1.7	1.6	—
Others	0.6	0.5	*10.5[a]

Source: Getachew and Telahun, "Socio-economic Characteristics," 21.
[a]nonfarm income.

. . . land became scarce, the landowners permanently settled in Addis Ababa. A single *gasha* [of land] had to support 20–30 peasants. Leasing land to individual farmers on a contract basis came to be common practice. As a result, the farmers were evicted from their plots by the new contractors. Peasant evictions became the order of the day. . . . Intelligent peasants got advantages out of the system. They [the landowners] evicted the poor farmers from the land.[60]

The pressure to increase cropped land in Ada's rapidly closing microagronomic economy had more than the socioeconomic effects of eviction and decreasing tenants' crop shares. There were fundamental effects on cropping and agronomic practice. By 1969 landowners had begun to impose new practices designed to intensify labor and increase cropped land at the expense of pasture and fallow. In Godino District near the former Babicheff plantation farmers could fallow a plot only with permission of the landowner. At Kajima fallowing by a tenant was a legitimate cause for eviction. In most places in Ada landowners forbade the cultivation of sorghum altogether, since it decreased the amount of land planted in teff. Farm livestock was, after all, exclusively the property of the tenant, not the landowner who therefore had little interest in improving or maintaining forage, providing pasture, or allowing live-

60. Interview with Sahlu Berhane (Hiddi), 1 June 1992. Also see, interview with Badada Kilole (Hiddi), 25 May 1992.

stock rights of passage to pasture and water. Landowners also restricted tenants to only one cow to avoid conflicts over pasture versus cropped land.[61]

Table 6.2 shows a single snapshot of cropping among the ten farmers surveyed in the 1969–71 crop season, in which the move toward concentration on teff is clear. Economic pressure to produce teff had direct effects on pasture, forage, and fallow, since the opportunity costs for these land uses were far too high compared with the advantages of intensifying the production of teff.

Crop-Livestock Integration:
Empiricism, Intensification, and Investment in a Closed Agronomic System, 1975–90

The evidence from the late 1960s and early 1970s indicates an agronomic system approaching closure, that is, with little or no open resources in land or forage. The paradox of the farmers' need to intensify the percentage of cropped land per holding is the consequent reduction of pasture to sustain the oxen needed for cultivation. This conundrum, the need for forage *(meno)*, was the most consistent complaint of farmers I interviewed in Ada and a leitmotif of their recollections of agricultural change. A corollary theme omnipresent in farmers' recollections was the consistent decline in soil fertility and the need for larger holdings to sustain the farm family. As fallow has declined, farmers have searched for a means to maintain both fertility and a high concentration in a single crop (i.e. with reduction in rotation).

Table 6.3 indicates the process of change during 1960–90, the end point of a longer process of concentration in a single crop. The trend indicated there suggests clearly that the concentration in teff has been at the expense of both rotation (a decline in pulse crops) and in other cereals which occupy the same soils as teff. Bekele Shiferaw's 1989 survey of the expansion of cropped land in Ada indicates a strong correlation (95 percent) between the expansion of new crop land for teff and the loss of both fallow and former pasture, but without the increase in other crops (maize, sorghum, chick-peas, and lentils) that has often been associated with newly cropped lands in other ox-plow areas.[62] Ada teff concentration in particular areas which specialize in teff, such as Hiddi, have reached rates as high as 73 percent of cropped land.

How has this process of closure in land and livestock resources affected agricultural practice in Ada? To what extent has the adjustment been an internal agro-

---

61. Borton et al., *Development Program,* 8, 83-84, 92. For sorghum see interview with Badada Kilole (Hiddi), 1 June 1992. For cow restriction see interviews with Kuri Walde (Ensalale), 17 June 1992; Dasta Haile Mariam (Godino), 11 June 1992; and Badada Kilole (Hiddi), 25 May and 1 June 1992.

62. Bekele Shiferaw, "Crop-Livestock Interactions," 76-77, 81.

TABLE 6.3. Crop trends in Ada: Percentage distribution of crops on cultivated land, 1960–89.

Crop	1960	1973	1975	1978	1979	1980	1985–89
Teff	37	37	44	49	56	50	55
Wheat	18	15	15	6	7	7	23
Other cereals	17	23	10	5	5	4	6
Pulses	28	24	30	37	31	36	15
Others	—	1	1	3	1	3	1

Source: Bekele Shiferaw, "Crop-Livestock Interactions," 77. For 1989 in Hiddi, one of Ada's most concentrated teff zones, two study groups of farmers cultivated 72.9 percent and 656 percent teff.

nomic one, and to what extent has it depended on external interventions (research, state policy, land reform)? Ada's experience with the introduction of fertilizer and forage management, mechanization, and agricultural research provides a fairly well-documented view of intensification within an ox-plow system under stress.

Despite their access to food production, availability credit, and markets, Ada farmers began widespread use of fertilizer (DAP and Urea) only in the early 1970s, after its success in the Swedish Chilalo Agricultural Development Unit (see chapter 7). Farmers disagree about who first brought it; some argue the first were innovative landowners such as two sons of Qagnazmach Wandafrash (grandson of Bullo Jille) around Kajima, Zawde Badada at Hiddi, and Grazmach Bisrat at Dankaka. Others argued that tenants and resident landowners who saw Ministry of Agriculture demonstration plots at Dabra Zayt began its use and later organized a buyers' cooperative.[63] Most agree that nonresident landowners tended to be more conservative in agronomic innovation than tenants or resident landowners, the nonresidents fearing cheating by tenants and their local agents in purchase of inputs. Both Ellis and Borton, who conducted research on landlord-tenant relations in the late 1960s, indicated that absentee landowners were reluctant to invest in fertilizers.[64]

Fertilizer's primary effect in Ada's agronomic system was to break down a major constraint to further concentration in teff by eliminating the need for the rotation of pulse crops. Table 6.3 indicates the dramatic effect on cropping strategies between 1973 and 1989, that is, the primary years of fertilizer adoption. In the initial phase of fertilizer use, teff also replaced wheat, which in its improved varieties has a better overall response to fertilizer but had a less ready market in the 1970s and 1980s and cannot tolerate early planting on flooded vertisols.

63. Makonnan Balata (interview, Hiddi, 23 May 1992) refers to the Dabra Zayt association initiated by Ato Kifataw and Taddese Ambaw. Interview with Alamu Falaqa (Dankaka), 23 May 1992.

64. Ellis, "Man or Machine," 121-22; Borton et al., *Development Program*, 85-86.

The growing cash market for teff in Ada and the land tenure system in general provided relatively easy credit for Ada farmers. In most cases loans were of short duration (four to seven months) and were taken to cover shortfalls in the summer months before harvest (65 percent in the July–September period). More than three-quarters (78.7 percent) of all credit were short-term seasonal loans for food, fertilizer, or seed rather than land or capital in the form of livestock.[65]

In a seeming irony, the concentration in teff and the pressure to increase cropped land in the closing system also placed a premium on oxen ownership to maintain productivity on cropped land. As indicated in chapter 3, oxen are the critical capital investment of ox-plow farms; in 1968 they constituted 86 percent of total Ada farm capital.[66] There are both yield and area effects of oxen ownership control demonstrated by research at Dabra Zayt (see chapter 3). This same research indicates that at Ada two-oxen farms produced 81 percent more than farms which owned no oxen and 18 percent more than those with only one. Moreover, there is also a "cropping" effect: two-oxen farms at Ada produced 63 percent more cereal than farms which owned no oxen and 19 percent more than one-ox farms.[67] The issue here is not access to oxen (farms can borrow, rent, or exchange oxen labor) but a farm's ownership and ability to maintain its own pair. The ownership of oxen seems particularly linked to teff production. Getachew and Telahun in their 1969–71 study found that farms which practiced oxen rental *(menda)*, that is, capital-poor farms, accounted for 11.23 percent of maize production but only 1.61 percent of teff.[68]

Although Ada farmers recall the lost soil fertility of Ada's past with considerable regret, their most abiding memory is of the era of open pastures, which disappeared in the postwar land rush:

> In the past grass for cattle was in abundance. There were vast tracts of land reserved for cattle grazing. Now every plot has become agricultural land. There is hardly any pasture for cattle. . . this is the problem. In the past cattle used to graze over large areas of pasture. They bred and interbred in large numbers. There was ample grass. The straw from the crops, grain, and cereals was abundant. After the Battle of Maychaw [1935] the problem of land has become so pressing that everyone is feeling the burdens of life.

65. Though annual rates could be exorbitant (100-350 percent) none of the farmers I interviewed saw loan agreements with landowners as exploitative. Teshome Mulat, "Ada Baseline Survey, Part 4: Credit and Indebtedness in Ada Wereda," IDR Research Doument No. 16, 1974, 6-7, 10; also see Vincent, "Economic Conditions," 32-33; and Borton et al., *Development Program,* 109-13.

66. Humphreys, "Empirical Investigation," 328.

67. Bekele Shiferaw, "Crop Livestock Interactions."

68. Getachew and Telahun, "Socio-economic Characteristics," 22.

During those days there were specially reserved areas for animal grazing beside your plots. People who did not own their own grazing lands used to buy such plots from the landowners—the grass only. Nowadays such grass does not exist.[69]

It is surprising, therefore, that Ada farmers have larger per-farm oxen holdings than the national average (79 percent have two oxen versus 37 percent nationally), despite having the ox-plow complex's densest farm population (see chapter 2).

Moreover, Ada has maintained its high oxen-ownership levels through the population growth of the 1970s and 1980s. The answer to this paradox lies in the dynamics of crop-livestock integration. A World Bank-ILCA study of crop-livestock integration describes in general terms what the Ada case indicates:

Land competition does not limit the stocking rate until relatively high population densities [occur] and hence does not prevent farmers from reaping the benefits of integrating crops and livestock. The simple reason for this is that the expansion of cultivation stimulates livestock production. The apparent paradox is explained by the fact that cropping changes the seasonal pattern of fodder supply. Crop residues provide animal fodder in the post-harvest and dry seasons when pastures are least abundant, thus sustaining greater livestock production.[70]

By the 1969, a baseline survey of Ada pasture showed that it was already in short supply, forcing a market in pasture rental and in forage sales. Several sources of livestock feed had evolved within the local economy. In northern parts of Ada, with the greatest population and cropped land density, farmers coped with dry-season pasture shortages by stubble grazing their animals for two to four months on harvested fields and then providing crop residues. In southern areas, where open land still existed in the late 1960s, landowners rented pasture on a seasonal cash basis, Zukwala and Liben being the least intensively cultivated areas. In 1969 farms along Ada's southern frontier with the Rift Valley also had access to "migration," that is, sending their livestock seasonally to open pasture to the south and west.[71]

Ultimately, however, the primary source of forage within the closed agronomic system of the 1970s and 1980s was crop residues, especially teff straw, which, ironically, become more available as cropping intensified and systems for cutting, storing, and feeding evolved. Farmers had an added incentive to incorporate crop residues into their feeding system, since these were direct farm prod-

69. Interview with Badada Kilole (Hiddi), 1 June 1992; interviews with Dasta Haile Mariam (Godino), 11 June 1992; Alamu Falaqa (Dankaka), 23 May 1992.

70. McIntire et al., *Crop-Livestock Interactions,* 8-1.

71. Borton et al., *Development Program,* 89.

Figure 6.3. Ada landscape: teff straw forage, 1990.

ucts subject to their own management, and not landowner control (as opposed to rented pasture). In some semi-arid, intensive, agronomic systems in Africa, crop residues account for 20–30 percent of annual forage. In highland Ethiopia crop residues average as much as 50 percent of forage in many areas. At Hiddi, Ada's most intensive, teff-concentrated area, residue in 1990 accounted for 71.5 percent of livestock forage, reflecting not only the rate of change in the 1969–1990 period but also the extent of crop-livestock integration within the closed Ada agronomic system.[72]

Ada's ability to concentrate in a single crop, intensify land use, and maintain its high per-farm oxen population resulted to an important degree from the peculiarities of teff. Teff combines qualities ideal for intensification because it combines high prices for its grain, waterlogging resistance on vertisols, storability, and the highest straw yield of any highland crop (see table 6.4).[73]

72. Bekele Shiferaw, "Crop-Livestock Interactions," 94.
73. Bekele Shiferaw, "Crop-Livestock Interactions," 93. Teff's straw yield per hectare is slightly less than sorghum, but is much more palatable as forage than sorghum stalks. Also, sorghum is a minor crop in Ada. Teff straw is also an important building material.

TABLE 6.4. Crop residue yield in Ada, 1985.

Crop	Total area (in hectares)	Crop Yield* (metric tons per hectare)	Residue Yield* (metric tons per hectare)	**Total Residue (metric tons)	% of total residue
Teff	28,554	1.05	2.33	66,530	64.5
Wheat	12,629	1.2	1.78	22,479	21.7
Barley	2,221	1.1	1.53	3,398	3.3
Maize	460	1.5	2.01	924	.8
Sorghum	396	1.1	2.56	1,013	1.1
Chick peas	2,321	0.95	1.46	3,388	3.3
F. peas	2,120	0.70	1.02	2,162	2.1
Horse beans	2,024	0.80	.92	1,862	1.8
Lentils	1,351	0.50	.78	1,053	1.0
Others	296	0.9	1.39	411	.4

Source: Bekele Shiferaw, "Crop-Livestock Interactions," 93; based on data drawn from Ada Agricultural Development Office, Gryseels and Anderson, Research, and the Central Statistical Authority. Percentage calculations of total are my own.

The importance of crop residue by the late 1980s created a localized straw and hay market in which neighboring districts now provide grass, and forage-surplus farms sell teff straw *(chid)* to a burgeoning cattle fattening industry (see below). By 1992 prices had reached E$10 per *chenet* (donkey load).

Mechanization

In 1970 the Ethiopian Ministry of Agriculture submitted a proposal to the United States Agency for International Development to fund 10 tractor-hire project sites in Ada District. The project's designers envisioned that within 10 years 25,000 hectares, or 42.8 percent of Ada, would be under mechanized cultivation. A year earlier, the Stanford Research Institute's study of Ada had already concluded that Ada's shrinking resource base could not sustain sufficient oxen to support its cropping system. Yet, by 1980 Ada's concentration in teff had increased by 35 percent with fewer tractors involved in smallholder agriculture than had been a decade before.[74] The story of Ada's mechanization, however, is less one of the failure of

74. Ellis, "Man or Machine," 1; Borton et al., *Development Program*, 93-94.

mechanization than the resilience of the economics and agronomic persistence of ox-plow cultivation.

The mechanization of smallholder agriculture at Ada began not with the Italians, who used tractors only for their own settlement schemes, but in the early 1960s. Tedla Desta was a former manager of an imperial farm in Chercher (Harerge) and Ethiopian agent for International Harvester. He came to Ada in 1962 to launch a tractor-rental scheme:

> When I started the operation, that was when I bought some agricultural equipment, and then I went down to Dabra Zayt to work for the farmers—for the landlords and the farmers. I had invited, as I told you, many people to open the operation to show them what we could do, and that day we plowed lots of fields just to demonstrate our operation And we did not have much business, to my surprise. We had set the organization and hired an Alemaya graduate to manage the business [laughs], and the business we were getting was a trickle.
>
> No, the farmers would rent our service. We would go there and plow the land for them, harrow disk it and hand it over to them. So the amount of work was very limited, and I decided to rent an area and use our full strength, to put our full strength to work. So I rented about four *gasha* first and 2 *gashas* later on. I set up a price for rental of a *gasha* for [E]$1000 plus 10 quintals of white teff. That was the amount everybody collected for me.

Tedla's system initially provided mechanization of the first plowing and seed cleaning at harvest, but no fertilizer. Labor for the two or three additional plowings, weeding, and harvest came from tenants.[75]

The system was attractive to the tractor hire scheme itself and to the landowner, but less so to the tenant. For his investment, Tedla received the cash payment and a fixed payment of 10 quintals of teff per 40 hectares. While the landowner paid for the cost of the mechanization, tenants paid half the cost of the fertilizer, gasoline, and threshing service. Gene Ellis, who evaluated the system in 1970, estimated that of the increased yield from fertilizer, which averaged 5.25 quintals per hectare, the landowner received 4.94 quintals and the tenant 31 kilograms.[76]

In 1992 I found no farmers at Ada who recalled the Tedla Desta system. Farmers at Hiddi, however, have bitter memories of a certain General Balata, a landowner who brought his own similar tractor scheme. Badada Kilole remembered:

> A certain General Balata come to this area and farmed a large tract of land with tractors. He also used fertilizers and gave us each a hectare of this farm plot.
>
> Q: Was it given freely?

75. Interview with Tedla Desta (Alexandria, Va), 17 April 1991.
76. Ellis, "Man or Machine," 126.

A: What do you mean! We are the ones who went home without a penny. We had to suffer because of the dealings we had with this general. We labored, weeded, harvested, and got the grain ready. But later he began calculating the price of the fertilizer, the transport fee . . . and as a result we could not even get 1 quintal from the 20 produced per hectare!

General Balata also apparently used his control over his tenants to force them to buy fertilizer through him rather than through their own association.[77]

The advantages which accrued to owner-operator landowners such as General Balata encouraged new landowners to evict tenants, buy fertilizer, and hire seasonal labor for the post plowing tasks. Ellis argued from his 1969–70 fieldwork that owner-operators could boost yields from 7.5 quintals per hectare to 12.75, getting a return of nearly E\$130 per hectare over what an ox-plow tenant would pay. Absentee landowners also found advantages in using a tractor-contract farmer who paid a lower rent than tenants but did not require the elaborate patron-client obligations of tenants. Most landowners, however, apparently preferred to continue their higher rents from tenant farmers, whose rent increased to a 50–50 share basis, particularly since they needed tenant labor to complete the entire farm operation through harvest. Evictions of nonresident tenants (i.e., those without homesteads on a holding) proceeded, but for the most part in favor of more productive ox-plow tenants. By 1970, Ada's handful of tractors (26), limited number of users (82), and few cultivated hectares (2484) indicated the tractors' failure to compete.[78]

Empiricism and the Science of Agriculture in Ada

The context of Ada's agricultural change was not merely a reflection of the urban growth 50 kilometers along Highway 4 to the north. It was also part of a postwar Ethiopia which sought to join the world economy, including the fledgling programs of international development aid that were spin-offs of the Marshall Plan. Though founded on geopolitics, such bilateral aid also carried a strong ideology of empiricism and idealism. In 1952, as part of a major agreement with the U.S. Point Four program in Ethiopia, the Ministry of Agriculture and Oklahoma State University opened what was then called the Bishoftu Agricultural Research Station. In 1963 the facility, which had become part of the new university system, used Rockefeller Foundation funds to replace its Quonset huts with offices and a laboratory for the four resident Ethiopian scientists trained at Oklahoma State. Ada thus joined a

77. Interview with Badada Kilole (Hiddi), 25 May 1992; See also interview with Habta Mariam Wandamagegnahu (Hiddi), 25 May 1992.

78. Ellis, "Man or Machine," 122, 126. Ellis describes evictions as a result of mechanization but offers no numbers. The General Balata system, unlike Tedla Desta's, appears to have been one based on hired labor rather than tenancy.

handful of other sites—Asmara, Jimma, Alamaya, Holeta, and later Chilalo—
which received the blessings of agricultural science and an approach dominated by
an American land-grant philosophy of research and extension.[79]

In its first decade, the research station put little emphasis on extension. In its
most active year, 1963, the station distributed 250 quintals of improved white teff
(manya), 70 quintals of improved Kenya wheat, and 30 quintals of chick-peas. The
station also claimed to have 1150 crop varieties under study. By 1969, however,
Ada still had only one extension agent assigned by the Ministry of Agriculture.

The Bishoftu station, renamed the Dabra Zayt station in the early 1960s, had
programs in agronomy, horticulture, and livestock. The agronomy department was
to develop and test "better varieties of economical crops" for use on the Ethiopian
highlands. In fact, the first successful cultivars distributed in Ada, called Kenya 1
and Kenya 5, were from neither Kenya nor Oklahoma, they were bread wheat varieties brought from the Eritrean "Paradiso" research center near Asmara and multiplied at the Ministry of Agriculture's seed multiplication center at Simba (Wallaga)
inherited from the Italians. Used primarily in Arsi, these varieties improved yield
from 6–8 quintals per hectare to 20–25.[80] Such improved varieties were critical to
raising productivity, since local wheat varieties tended to lodge (i.e., collapse) when
treated with fertilizer. The station also introduced three improved varieties of teff,
though with less dramatic yield results.

The Bishoftu station attempted work in poultry development as well. In 1957
alone the program imported by air 1000 day-old New Hampshire hatchlings (Rhode
Island-leghorn crosses) from New York for the breeding program. In later years,
white leghorns arrived to upgrade the level of Ada's local poultry stock. Habta Mariam, a successful tenant with tenants of his own, recalled that poultry program:

> There was a time when we bought what were known as American chickens
> or foreign chickens to rear them on our farms. We were also given the eggs
> of such hens to be hatched by our own hens. We could see the advantages of
> such poultry, though half the eggs perished. Now all of them are gone; peo-
> ple lost hope and stopped breeding them.[81]

79. Oklahoma State University, *The Agriculture of Ethiopia* (Stillwater: Oklahoma State
University, 1964), 50. The facility was officially called the Central Research Station of the
Imperial College of Agricultural and Mechanical Arts. A number of development programs
in Africa paired U.S. land-grant universities with African counterparts (e.g., Michigan State
with Nsukka, Nigeria).

80. Interview with Dr. Tesfaye Tesemma (Dabra Zayt), 17 June 1992. Dr. Tesfaye joined
the Dabra Zayt station in 1961 as a wheat breeder and continues his work there. The wheat
varieties were called Kenya after their breeder Vittorio Nastazi forgot their original names
and just assigned them numbers. Farmers ignored it all and called the variety Israel.

81. Interview with Habta Mariam Wandamagagnahu (Hiddi), 25 May 1992.

Badada Kilole witnessed the dilution of the improved poultry's genetic pool:

In the past we used to have chickens of greater size. The cross-bred were enormous; nowadays such chickens can't be found. The cross-breds are not as big as the previous ones. The cross-bred of the foreign and habasha chickens are the smallest size. The physical change of the American chickens is similar to that of the Karayu cattle. As they got mixed up with the local cattle they gradually decreased in size. The same with the American chicken.[82]

The station's greatest success, however, was not in poultry or teff but with Laqach, the Ethiopian version of a Mexican semidwarf, bread wheat cultivar which was the primary variety used in the Swedish Arsi project and also popular in Ada. Together with minimum package programs' fertilizer inputs and a new approach to extension work (see chapter 7), Laqach was on the verge of making Ethiopia a net exporter of wheat when, in 1974, stripe rust destroyed virtually the entire crop. Wheat production never recovered. Other, more resistant, varieties have been available, but none has had the singular effect of Laqach. Badada Kilole remembered Laqach (though under its local nickname), and its sudden disappearance in 1975, conflating political and agronomic spheres: "Nowadays there are many kinds of wheat. In the past there was only one variety. There was what was known as Haile Selassie's wheat. That type disappeared with the emperor himself."[83] The shock in Ada would seem to be partly responsible for the collapse of wheat after 1975 (indicated in Table 6.3 in the previous section).

Beyond crop science, empiricism also brought development projects designed to raise agricultural productivity and rural incomes. The model for the large "integrated rural development" was the joint Swedish-Ethiopia project in Arsi, the Chilalo (CADU) begun in 1967 (see chapter 7). Ada's version of that program began with an application to USAID. in 1970 to "develop the knowledge and experience necessary to increase agricultural production in the many areas of which Ada . . . is representative." Ada was to be a test for a "package program" of technology and innovation to bring the rural subsistence sector into the money economy as outlined in Ethiopia's third five-year development plan for 1968–73.[84]

The first step toward that goal was the Stanford Research Institute survey in 1969. The next was a major rural integrated development project along the lines of CADU funded by USAID and the Ministry of Agriculture. The goals for the proposed Ada District Development Project (ADDP) were: "1) to increase net farm income; 2) to assist in improving the rights of tenants; 3) to develop the insti-

82. Interview with Badada Kilole (Hiddi), 25 May 1992. See also Sylvia Pankhurst, "Bishoftu Agricultural Research Station," *Ethiopia Observer* 10 (1957), 330.

83. Interview with Badada Kilole (Hiddi), 1 June 1992.

84. Vincent, "Economic Conditions," 3-4.

tutions necessary to change the economy of the area from subsistence to market orientation; and 4) to develop the knowledge and experience necessary for replication in other areas that will substantially increase agricultural production in Ethiopia's highlands.[85] In the end, however, the winds of ideology in agricultural development had shifted, and the ADDP as integrated rural development was too costly. Instead, development planners gave the task of delivering inputs to the Ministry of Agriculture's new Extension and Project Implementation Department (EPID, see chapter 7). Instead of making a major investment in new technology and providing inputs, USAID put its faith in neoclassical economics. In 1973 a survey team headed by a Ph.D. candidate in economics conducted a baseline survey, and in 1974 Haile Sellassie I University's Institute of Development Research and Michigan State University signed an agreement to conduct microeconomic research on rural development in Ada.

Like the Stanford research, the Michigan State project provides the historian with invaluable benchmark data on the state of agriculture in the mid-1970s. The Michigan State group conducted its research during the 1975–76 agricultural year, a period which saw revolutionary changes in land tenure, the disappearance of the land owning class in Ada, and the deepening of Ada's agronomic adjustments to its closed system. Farmers themselves rarely saw results from it and found the efforts curious. Sahlu Berhane recalled:

> Yes, researchers did come from the university with a letter requesting cooperation. They asked all sorts of questions including what we ate and drank. They stayed here for eight days. They usually produced letters bidding our cooperation, etc. Once there were researchers who came to this area to conduct a soil survey. Different researchers came to our area at different times. Sometimes foreigners who did not know the languages of the area came and could not get help or cooperation.[86]

EPID's efforts, though short lived, were instrumental in the early acceptance of fertilizer by tenants. Makonnan Balata, an Oromo tenant, remembered:

> The development agents from the *Geberna* [Ministry of Agriculture] organized tours for us so we could have first-hand knowledge of the application and uses of fertilizer. We visited the demonstration fields and we were impressed by what we saw. We elected Grazmach Bisrat as the chairman of our voluntary committee [to acquire fertilizer].[87]

85. Vincent, "Economic Conditions," 4-5.

86. Interview with Sahlu Berhane (Hiddi), 1 June 1992.

87. Interview with Alamu Falaqa (Dankaka), 23 May 1992; Interview with Makonnan Balata (Hiddi), 23 May 1992.

"Land to the Tiller":
Ada Smallholders and the 1974 Revolution

It is likely that the young military officers and urban intellectuals who led the 1974 revolution and framed the 1975 land reform proclamation took Ada as a model for the abuses of tenant farmers perpetrated by Ethiopia's aristocratic economic class. It was, in fact, precisely the power and influence of the aristocratic class of landowners, which controlled land in Ada, that the young revolutionaries wished to appropriate and place in the hands of a bureaucratic state. Their tool was the 1975 Proclamation to Provide for the Nationalization of Rural Land, which limited farm holdings to a quarter gasha (10 hectares) and also nationalized—in theory with compensation—moveable property such as tractors, threshing equipment, and diesel irrigation pumps.

Though the revolution and land reform aimed primarily at issues of political economy rather than directly at agronomic systems, it brought a number of changes that affected the conjuncture which pushed agronomic change. First, the nationalization of moveable property deflected the movement toward mechanization at Ada. Tedla Desta, for example, had all his tractors and equipment nationalized and sent to new state farms. Others lost their ability to pay off tractor loans and sold off their property quickly at a loss, tossing Ada's mixed rural capital back to a thoroughgoing ox-plow economy. Many tenants, such as Dasta Haile Mariam, whose uncle was a landowner, also sold off oxen, fearing a general confiscation of all rural capital. Kuri Walde's father-in-law lost his flour mill, land, and rural taxi service and moved to Addis Ababa while Kuri and her husband remained as members of the newly formed peasants association.[88]

For tenants, there was an important change: they now held their land not as sharecroppers or as freehold owners but through claims to residence in a peasants association. The new rules provided for tenure through claims based on residence, and thus encouraged new claims. Thus the increase in farm household size through absorption of "relatives" noted in the early 1970s transmogrified into new land claims and increasing pressure on cropped land (see Table 6.5 below).

Ada's closeness to the city and its markets also meant it was accessible to an activist state which sought to reorganize and provide new resources for smallholders. By 1975 78 percent of its farmers could reach an all-weather road within three hours and 29 percent within half an hour.[89] The military government's new Agricultural Marketing Corporation controlled grain marketing and fixed the price of marketable crops. If in some areas this policy pushed farmers toward noncontrolled crops such as oats or maize; in Ada it pushed farmers further

88. Interview with Tedla Desta (Alexandria, Va), 17 April 1992; interview with Dasta Haile Mariam, Godino, 11 June 1992. interview with Kuri Walde (Ensalale), 17 June 1992.
89. Vincent, "Economic Conditions," 100.

toward concentration on teff, whose price still was highest and by-products the most useful.

Between 1985 and 1990 the Ministry of Agriculture's agents had organized most of Ada's farmers into new villages. For some at Hiddi, Dankaka, and Godino, peasant associations and villagization also meant pressure to join an *amrach* (producer cooperative), in which farmers under direction from a new, more activist Ministry of Agriculture (which had disposed of the more "populist" EPID, see chapter 7), pooled land and capital and cultivated in teams. Membership in a cooperative gave these farmers special access to fertilizer and extension advice but removed them from direct control over agronomic decisions. As residents of a "surplus production, high potential" zone, all Ada farmers, not just those in cooperatives, received preferential access to fertilizer, thus allowing the dramatic increase in teff's share of crop production through the 1980s.

Testimony from farmers indicates that, for better or worse, the activity of the Ministry of Agriculture in Ada increased dramatically after 1974. Rather than being an additional source of farm inputs, however, it quickly assumed a monopoly on such resources. The increased availability of fertilizer and easy credit through the Ministry of Agriculture was not a blessing for all. Sahlu Berhane recalled problems:

> The use of fertilizers expanded and became widespread toward the eruption of the revolution. There were times when peasants unknowningly fell into debt by taking surplus fertilizer. The misuse of fertilizers has become a source of misery for farmers after the 1974 people's uprising. Before that the use of fertilizers was rewarding. It had proper management. After 1974 the amount of fertilizer was arbitrarily decided by the peasants association. Its distribution was not based on an assessment of the needs of the farmers and their capacities.[90]

In many ways, however, the changes brought by the revolution simply accelerated processes already in place. Population growth in particular continued unabated. Ada's rural population, excluding town dwellers, went from 115,090 in 1980 to 143,700 in 1990. The residence-based land reform system in place after 1975 meant that new population growth and in-migration translated directly into new pressure to increase cropped land. In the 1979–90 period therefore, the area under cultivation increased by 2.5 percent a year and the grazing land declined by 5.2 percent per annum. Table 6.5 indicates the expansion of cropped land in the 1980s, a process already in evidence by the late 1960s.

90. Interview with Sahlu Berhane (Hiddi), 1 June 1992. The problem of debt accumulation by producers cooperatives who had inputs provided on easy credit terms became clear after 1992. Ada Warada (including two neighboring districts joined to it in 1990) in 1992 had an accumulated farm debt of E$9,377,378.80 of which only E$684,194 had been paid (data from the Ada Warada Ministry of Agriculture Office, June 1992).

TABLE 6.5. Land use in Ada, 1979–90 (in hectares).

Crop Year	Grazing	Cropped	Fallow	Forest	Other
1979–80	20077	42479	452	10212	49080
1980–81	18792	43957	395	10212	48944
1981–82	18322	44615	365	10212	48786
1982–83	16636	46797	337	9499	49031
1983–84	16253	47690	315	9499	48543
1984/85	15034	49539	250	9499	47978
1985/86	14256	50383	210	8910	48545
1986/87	13382	51273	158	8320	49167
1987/88	12833	52451	104	8320	48592
1988/89	12204	53250	87	8320	48439
1989/90	11862	54503	67	8320	47548
1979–90% change	-41	+28	-85	-19	-3

Source: Bekele, "Crop-Livestock Interactions," Appendix 7, drawn from the Ada Agricultural Development Office. Percentage of change calculations are my own. Like many survey figures in Ethiopia these should be understood as comparative data.

Note: Percentage of change calculations are my own.

The decline in forest, though modest compared with declines in fallow and pasture, has also had an effect on agriculture. The bulk of Ada forest now consists of protected mountain slopes, where the forested area has probably increased over the past half century. The few wood fuel resources remaining on the plain are insufficient for household needs. By 1980 the average household fuel use consisted of 41 kilograms of dung cakes and only 10 kilograms of wood.[91] The dependence on dung as fuel—typical also of northern Shawa—has meant that manure has not been available to the farm as fertilizer. The local market for dried dung cakes has, however, provided some income for women. Stall feeding of livestock also means manure has been easier to collect and dry.

Intensification to Specialization

The circumstances which fostered the closure of Ada's crop-livestock system in the second half of the twentieth century and brought specialization in crop production also created opportunities for livestock systems. Forage markets created by the decline of pasture in Ada's closed economy also stimulated experimenta-

91. Gryseels and Anderson, *Research,* 10.

tion in using agroindustrial by-products to support a small-scale peri-urban livestock economy. In the postwar period a number of agroindustrial plants appeared
to serve the urban market: large flour mills at Dabra Zayt and Akaki, the Wonji
sugar factory, cotton processing in the Awash Valley, and small-scale oilseed mills
throughout the country.

By the late 1970s the market for forage in Addis Ababa's hinterland prompted
development agencies, small-scale entrepreneurs, and even the Ministry of Agriculture to begin using the discarded by-products of these industries—oilseed cake,
molasses, cottonseed cake, and *farushka* [wheat chaff]—as livestock feed concentrate to supplement the teff straw already available in local markets. These feed supplements, available initially either at no cost or at very low prices provided the basis
for the burgeoning peri-urban economy, which served growing urban markets for
meat and milk. Ada was ideally situated for such a market, and small-scale livestock
stall-feeding operations began to buy southern cattle, bring them to Ada and spend
three months fattening them for festival markets at Christmas and Easter.

Dasta Haile Mariam, a former tenant and migrant from Ankober who became one
of Ada's most successful fattening specialists, started shortly after the revolution:

> I started fattening cattle with chid [teff straw] and *gwaya* [vetch]. I bought an
> ox and sold it for E$1000 after fattening it. After that I bought sheep for
> E$290 and sold the sheep for E$790. I got a profit of E$1,790 in a single
> turnover. It was in 1983. Then in 1985 I had 10 oxen. . . . I bought them for
> E$5600. I used boiled teff and some straw and *gwaya*. Later some people
> from the Ministry of Agriculture heard about my work and supplied me with
> feed [wheat chaff from the Dabra Zayt flour mill]. Again I brought oxen from
> Bale for E$4500 and finally sold them for E$10,000. In 1986 the villagiza-
> tion program began [people sold off property]. Using the chance I bought 18
> cattle for less than E$7000, and I sold the fattened cattle for about E$18,000.
> With the E$18,000 I bought 31 oxen from Bale. I sold them all and brought
> 41 oxen from Ankober.
>
> Q: How many times do you fatten oxen during a year?
>
> Twice, for Christmas and Easter. If properly done three months is enough. . . .
> In the past I used to take oxen to Addis Ababa to sell. Today the buyers come
> to my house to buy them and pay by check.[92]

92. Interview with Dasta Haile Mariam (Godino), 11 June 1992. Though farmers claim
oxen fattening is a new industry in Ada, one commercial concern existed in 1969, though it
seems not to have survived to the revolution. Borton et al., *Development Program,* 75-76. It
was owned by an Armenian, Savajian, who had 2200 head and sold 45-50 head a week to
Addis Ababa.

A peri-urban milk and butter economy has also emerged in the environs of the town of Dabra Zayt, managed by women using cross-bred cows and wheat chaff concentrate. In the milk and butter market women have obtained Fresian-Boran cross-bred cows from a livestock research station or the Ministry of Agriculture. The industry, like oxen fattening, is seasonal, thriving primarily in the rainy season, when forage is more widely available and households have the least disposable income.[93]

In crop production the conjunctural elements of Ada's closed agronomic economy and investment of international development efforts in the late 1980s put in place a new effort to intensify and diversify Ada's black soil agriculture. The effort came not from farmer initiative but from a joint research extension project of ILCA, the Ministry of Agriculture, and the Institute of Agricultural Research, and required farmer adoption of an adaptation of the ox plow and a green revolution "package" consisting of three related elements designed to introduce a complex set of crop and soil management practices to the highland ox-plow farming system. These include: 1) the broadbed maker, an implement which combines two existing plow beams, handles, and a yoke, 2) fertilizer (DAP and Urea), and 3) improved wheat varieties of long-maturing bread and durum wheat which require sowing in early July.

The early planting of improved varieties of wheat on broadbeds promises to deliver both improved yields, to provide an October crop for the hungry season, which precedes the harvest of cereal crops in December–January, and to intensify the cultivation of black soils by overcoming the waterlogging problem. Yields vary from region to region but the results of ILCA field trials and reports from farmers themselves indicate a range of straw yields of 20–37 quintals per hectare.[94]

This seemingly straightforward package of green revolution inputs, however, involves a series of complex changes in the agricultural system practiced with a wide range of local variations across highland Ethiopia. First, it requires the use of broadbeds, an agronomic feature new to most areas of the highlands, but practiced historically on the extensive vertisol plain of Jirru in northern Shawa, where management of the waterlogging of vertisols has been in place since perhaps the sixteenth century. The use of an agronomic strategy which already exists within part of the highland farming lexicon is an important feature of the project design. Second, the use of the broadbed maker implement and improved, long-maturing wheat varieties requires early planting and thus a shift in the seasonal allocation of human and draft-animal labor for seedbed preparation, weeding, and harvest.

The expansion of a new system of cropping and soil management involving new inputs also marks a substantial change in the relationship of ox-plow agriculture to

93. This peri-urban industry is heavily dependent on markets for forage, given farm (and male) priorities for oxen forage and the high feed requirements of cross-bred cows. See Irene Whalen, untitled report to ILCA, 1985.

94. International Livestock Centre for Africa, *Outreach Sub-Project Deneba/Inewari (Shewa Province): Progress Report, Work Plan, and Budget, 1988* (Addis Ababa, 1988), 17.

external forces, especially the changing terms of political and market means through which both improved seed and fertilizer reach the individual farm. While the state has been a critical force in shaping the political climate for agricultural prices, access to external sources of seed and use of fertilizer are relatively recent phenomena in smallholder agriculture. The wide adoption of the package across the vertisol areas of the highlands necessarily means a major expansion of smallholder dependence on the availability and timely arrival of inputs and on policy decisions made in Addis Ababa about issues of credit, debt, and the privatization of input markets.

It remains to be seen whether the increased linkage of local farm production with external inputs and new technology is compatible with Ada's ox-plow system. While Ada's farmers have been willing to accept new resources from development agencies and the state, their willingness in the long term to appropriate new agronomic forms introduced from the tradition of empiricism remains to be seen.

Conclusion

The movement of Ada's agricultural system through time and its response to a particular set of historical contexts demonstrate the capacities of a closed ox-plow agronomic economy to respond, intensify the use of farm and external inputs, and specialize. This response stands in marked contrast with the areas of the ox-plow complex which have moved toward maize as a dominant crop, a specialization based on subsistence imperatives and a different set of historical conjunctures (see chapter 5).

Ox-plow cultivation in Ada has both survived and adjusted to intensive market conditions and scarce livestock resources of its closed agronomic economy. If technology of the plow has continued to be resistant to change, not withstanding the evidence of the broadbed maker, crop-livestock integration has continued apace.

7

Conclusion:
People of the Plow,
People of the City

In the preceding six chapters of this book the primary locus of action has been the rural landscape: farms, fields, and storehouses where farmers managed their resources to feed themselves, their communities, and a largely rural aristocracy. Over the course of the twentieth century Ethiopia's new towns and cities have emerged as the modern medium by which farms and farmers relate to the world economy and the state. New urban populations not only dominate the modern state apparatus but simultaneously have become a new source of demand for both food and tax revenue. Towns, small cities, and a new capital city mushroomed over formerly rural landscapes. The intersection of urban growth, consumption needs, and elite monopoly on state power created irresistible tensions by the 1970s and resulted in Ethiopia's 1974 revolution, which toppled the remnants of a landed aristocracy in favor of new, more purely urban, interests. By 1991, decline and state intervention in the agricultural economy culminated in a new upheaval, this time a rural response which brought down Addis Ababa's rule and challenged, if only temporarily, the urban basis of governance.

This chapter concerns the ways in which directions of agricultural change in Ethiopia have more often reflected new urban imperatives and less often the conditions on farms and region. The relationship between the state, agriculture, and the city has changed dramatically over the course of the twentieth century, the pace of change accelerating in the postwar period. Historically, state-agrarian relations depended on tribute relationships and labor obligations rather than rents. Towns, when they existed, served as marketplaces for middle- and long-distance trade and often as ecclesiastical centers but were too few to transform themselves into centers of market demand for food.[1] If the imperial court and rural elite fed themselves and their functionaries on products of the agrarian economy, they exercised little, if any, direct influence on agriculture itself, except in their immediate vicinity. At Ankober, as we have seen, the royal presence intensified and focused demand but did not fundamentally alter the technology or cropping patterns of small farmers; though it may well have increased the incidence of specialized agriculture. Before the late nineteenth century Ethiopia's "roving capitals" moved periodically because of the need to assert imperial authority, adjust to seasonal comforts, and capture trade routes, rather than because of any environmental necessity.

In the first third of the twentieth century the nature of urban-rural relations changed along with the Ethiopian region's engagement with the world economy. When the state began penetrating the rural economy by increasing its direct control over customs and land taxation, it also effected a weakening of the rural elite's economic base. This pattern created provincial towns in the north and strategic urban settlements *(katama)* in the newly incorporated regions of the south as loci for state power and markets for food.[2] Addis Ababa's foundation and growth in the late nineteenth century were an expression of Ethiopia's modern political economy; by the opening of the twentieth century, economic and political processes fixed it as the political economy's permanent center.

In the post-war era, penetration of the rural economy by the state and by urban interests increased virtually in direct proportion to the growth of urban populations in Addis Ababa, Asmara, and provincial centers. By the mid-1960s Ethiopia sported some 200 towns of various sizes, which concentrated state power, held almost 10 percent of the population, and grew at an annual rate of 7 percent. These towns and cities housed landowners, state functionaries, and merchants, who dominated trade and the marketing of food, though urban interests still exercised little direct control over small-farm agriculture. In small towns it was now possible for travelers to buy food—often impossible in the nineteenth century—usually prepared by women in hotels and tea houses who sought to escape rural drudgery. Urban demand for food, however, had begun to transform cropping strategies in certain areas which had access to urban markets. In the postwar period for Ada, for example, Addis Ababa's market and new transport

1. Traditional historiography has argued that Ethiopia's political economy functioned without urban centers. Towns such as Ankober, Adwa, and Saqota, however, existed as both market and administrative centers. For differing perspectives see Gamst, "Peasantries and Elites without Urbanism," 373-92; and Crummey, "Some Precursors of Addis Ababa."

2. McCann, *Poverty to Famine*, 127-43. For roving capitals see Horvath, "The Wandering Capitals." See chapter 4 for Ankober evidence of a political rather than an environmental rationale for the capital's shift to Addis Ababa.

links prompted attempts at mechanization and a general concentration in teff (see chapter 6). Food which had historically moved to elite households as tribute, now arrived in towns as rent, where urban-based landlords consumed it or marketed it to salaried urbanites. As the cash economy expanded, farmers also marketed their own grain directly to pay taxes, to acquire inputs, or to buy soap, medicine, razor blades, and sugar available in small shops, which sprang up around the urban marketplaces.[3]

In these circumstances, rural-urban terms of trade emerged as a critical issue and made a more direct state intervention in agriculture a necessity. These new towns changed the look of the countryside: their corrugated iron roofs stood out, as did the blue-green hue of eucalyptus trees planted in their residents' compounds. The eucalyptus, imported by Menilek in the late nineteenth century, came to be less a sign of nature than of urban investment. Ironically, in the intensively cultivated highland areas, trees visible in the distance were sure evidence to travelers of the presence of an urban, and not a rural, landscape.

The Urban Voice

Urban visions of Ethiopia's agriculture included metropolitan colonial policy and a new tradition of empiricism. Italian studies during the 1935–41 occupation, and two decades earlier in their Eritrean colony, were the first to attempt a systematic quantification and classification of agriculture.[4] Their purpose was not rural development in a contemporary sense, but an explicit attempt to increase wheat production for urban and metropolitan consumption. Italian schemes of demographic colonization envisioned rural communities of transplanted Italian peasants, each settlement drawn from specific villages in Puglia or Romagna. On a broader basis, Italian policy foresaw new technology as a means to introduce increased production to feed colonial cities and perhaps even the metropolitan population in Italy. Toward this end, by 1940 they had imported 338 tractors, established or planned experiment stations in each province, and recruited a "forest police." Though they quickly abandoned serious attempts to transform the

3. For an insightful description of this process, see Taye Mengistae, "Urban-Rural Relations in Agrarian Change: An Historical Overview," in Siegfried Pausewang et al., eds., *Ethiopia: Rural Development Options* (London, 1990), 30-37.

4. The best examples of Italian agricultural empiricism are the crop analysis by Raffaele Ciferri and his colleagues at the then Regio Istituto Agronomico per L'Africa Italiana. Nineteenth-century travelers sometimes attempted estimates of yield, though largely as part of a naturalist perspective rather than on the basis of a systematic survey. For an effective review of this literature, see Crummey, "Plow Agriculture." 13-14.

rural economy, their vision of modern agriculture's role in feeding cities fore-shadowed postwar trends in state policy.[5]

The Prologue has discussed the changing narrative voice of agricultural description in the twentieth century. An important part of that change has been the addition of an urban focus. If nineteenth-century narratives described a rural economy which was an organic part of society, politics, and moral life, twentieth-century accounts have observed rural conditions through the eyes of urban travelers based in towns. Addis Ababa, Asmara, and especially the railway town of Dire Dawa grew as economic centers as much as political centers, transforming their immediate landscape and economic hinterlands in their roles as markets and food-consuming centers. Shortly after its foundation, Addis Ababa became the stepping off point and/or destination of virtually all scientific observers and the measure against which they judged rural life.

In 1907, when Menilek recognized the state's new status as manager of the national political economy, he announced a European-style cabinet, including a minister of agriculture and labor. Under its founding mandate, the emperor enjoined the ministry to "improve the land, produce much grain and change agricultural work . . . in the European method." The mandate's main thrust, however, lay in the rationalization of state revenues drawn from the rural economy, that is, to enforce the agricultural tithe which supported the imperial army and to measure and to assess the lands of the nobility in the provinces. The ministry had neither the duty nor the means to attempt any direct role in cultivation and farm management until the postwar era, when in 1943 the Ministry of Agriculture became a ministry in its own right.[6]

State commitment to the transformation of its agricultural base waited until over a decade after liberation. It began humbly as an offshoot of the United States' expansion of its interests in the Red Sea area. In fact, however, the 1 May 1952 signing of the Technical Aid agreement between the Agricultural and Mechanical Arts College of Oklahoma (later Oklahoma State University) and the Imperial Govern-

---

5. For a survey of these plans, see Istituto Agricolo Coloniale, *Main Features*. In 1937, for example, Italian authorities decided to reinstate the *asrat* (tithe) collected through local Ethiopian officials and to abandon their attempts to tax agriculture more directly. See *Bolletino Ufficiale del Governo dell'Amara*, 27 December 1937.

6. Mahetma Sellase, *Zikra Nagar* (Addis Ababa, 1950), 318-71, quote from 318. Eritrea, after its 1952 federation with Ethiopia, retained its own Directorate of Agriculture and Forestry. The Agricultural Secondary School, established in 1933 but abandoned two years later, was a private venture, not an indicator of ministry plans. The Ministry of Agriculture and Labor was later named the Ministry of Agriculture and Commerce. For 1943 reorganization see Huffnagel, *Agriculture*, 443-46. See Huffnagel, *Agriculture*, 451; Sylvia Pankhurst, "Ambo Agricultural School," *Ethiopia Observer* 10 (1957), 309-10. The school reopened in 1952 under German supervision.

ment of Ethiopia under the auspices of the U.S. Point Four program preceded the much larger military assistance agreement of the following year.[7] The Oklahoma State contract provided the foundation, training, and philosophical directions of Ethiopia's tiny agricultural development infrastructure for the next three decades. It called for Ethiopia and the U.S. to share the cost of developing an agricultural secondary school and college, as well as establishing agricultural research stations, demonstration centers, and an agricultural extension service. The Jimma Agricultural School opened in late 1952 and the Bishoftu Agricultural Research Station opened two years later (see chapter 6). The agricultural university at Alamaya, near Harar, accepted its first students in October 1956 as the Imperial Ethiopian College of Agricultural and Mechanical Arts.

Observers of these innovations infused their descriptions of this new approach to changing the rural economy with optimism for Ethiopian agriculture and the vision of "Emperor Haile Sellassie, who has ever consistently planned for the development of Ethiopia's vast agricultural potential." Sylvia Pankhurst's 1957 visit to the agricultural college at Alamaya offered a symptomatic view of the expected triumph of empiricism in agriculture. The primary theme of Mrs. Pankhurst's 1956–57 articles on Ethiopia's fledgling agricultural research and training facilities was one of confidence that the imported technology and techniques of management would lead to a "brightly promising future."[8] At the college, visitors found the new buildings "delightfully furnished," including a gleaming stainless steel kitchen "equipped with the most efficient modernity from the U.S.A.," which included, among its innovations, window screens. Dr. L. A. Parcher, dean of the college, quipped that the flies were "the only old fashioned things here." On the grounds of the college and the other new centers, foreign advisors and their students raised Rhode Island red and leghorn chickens, Friesian bulls, and merino sheep, which they expected would be the progenitors of future generations of Ethiopia's livestock. Adjacent test plots held orderly experiments with imported U.S. cultivars of maize, sorghum, and wheat; the campus also displayed corrugated silos and John Deere disk harrows.[9]

Mrs. Pankhurst, however, also astutely cautioned her audience about the potential irrelevance of such empiricism and modernity to Ethiopian rural needs:

Many visitors to the College of Agriculture who admire the electric equipment of kitchen and cafeteria, the light well ventilated spacious buildings, the

7. *Ethiopian Herald,* 17 May 1952. For military assistance agreement, see Harold G. Marcus, *Ethiopia, Great Britain, and the United States, 1941-1974: The Politics of Empire* (Berkeley, 1983), 89.

8. Sylvia Pankhurst, "Ethiopian Agriculture in Retrospect and Prospect," *Ethiopia Observer* 9 (October 1957), 278-79.

9. Sylvia Pankhurst, "Imperial College of Agriculture and Mechanical Arts," *Ethiopia Observer* 10 (1957), 312-17.

efficient workshop and all the modern furnishings and paraphernalia ask doubtfully how it will be possible for young Ethiopians educated under these conditions to bury themselves in some primitive village? Will they remain contented, healthy and productive while living in a thatched tukul with an earth floor, and earthy plaster walls, unglazed windows, a total absence of water laid on and of all the modern conveniences, books, music, and educated companionship, to which they have grown accustomed during eight years of secondary school and university studies?[10]

In fact, Mrs. Pankhurst's questions were prophetic. The success of the college and programs at Jimma, Ambo, and Holeta in training a generation of Ethiopian agricultural specialists, who continued graduate study abroad or who worked in provincial ministry offices, helped fix the cultural and conceptual distance between urban managers and rural producers in the fabric of Ethiopia's modern political culture.

The American land-grant model, which permeated Oklahoma State's strategies in developing Ethiopia's agricultural infrastructure, envisioned that extension services would be the essential link between agricultural science at the research stations and agricultural practice on smallholder farms. In 1954 the Ministry of Agriculture assigned two Ambo graduates to initiate sheep improvement projects at Fiche (Shawa) and Asela (Arussi). Three years later the ministry had a total of 29 agents, covering only 6 of Ethiopia's 12 provinces, with 25 more 12th-grade graduates set to join their ranks in 1957. Each of these agents received the tools to effect an empirical revolution on the rural landscape:

> Each agent is furnished with simple farm equipment, an office, living quarters, privy of simple rural type, a water filter, seed store, barn, and meeting hall. . . . In the effort to advance Ethiopian agriculture an important stage further, every Extension Agent will next season [1958] also be supplied with a steel plough, a machine for digging, a machine for planting, and a mule to draw them.[11]

Extension efforts began haltingly, since the Ethiopian government bore the entire posttraining cost. When the Ministry of Agriculture reorganized into eight technical divisions in 1958, extension services received a small part of the total allocation for agriculture which was 1 percent of the budget for the Imperial Government. Excepting training, external aid contributed nothing to extension agents' placement or resource base for rural work.

The state revealed its lack of interest in smallholder agriculture in its successive five-year plans. In the first five year plan (1957–62) the government not surpris-

10. S. Pankhurst, "Imperial College," 317.

11. S. Pankhurst, "Bishoftu Agricultural Research Station," 332.

ingly chose to enhance its slow economic growth by investments in urban infrastructure: electric power, manufacturing, and housing for the growing urban bureaucratic class. Investment in agriculture was less than 6 percent of the total budget and consisted of surveys of large-scale irrigation potential for export production, developing the Wonji sugar factory, and improving productivity on commercial farms (i.e., land owned by the royal family and the newly urbanized aristocracy). Among the activities was a United Nations survey of the Awash Valley, which advocated an investment of E$300 million to develop 150,000 hectares of irrigable land for cotton and livestock exports. The plan also had some economic successes. In the 1957–61 period Ethiopia's electric power production increased from 73 million to 145 million kilowatt-hours, providing light and power to urban homes and factories. The plan notwithstanding, the general economic growth, laissez faire rural policy, and stable rainfall of the 1950s and early 1960s had trickled down to farmers—at least to some of them—whose per-capita income rose some 43 percent in the 1954–61 period.[12]

In the second Ethiopian development plan (1963–68), investment in agriculture grew threefold to E$363 million, 21 percent of the total agriculture budget, providing for increased investment in data collection, research stations, veterinary services, and expansion of extension services. This plan's primary focus, however, was to stimulate exports and provide import substitution. It projected increases in cereal and food production of a little over 100 percent, but provided for increases in cotton, cotton yarn, and livestock from commercial farms of 700, 900, and 700 percent, respectively.[13]

Emphasis on extension continued to be minimal. By the mid-1960s there were only 120 extension workers located in Ethiopia's 14 provinces. Most agents found themselves in provincial or district market towns, from where they staged "field days," tried to organize youth clubs, distributed model farmer awards, and built a few model poultry houses. The plan also responded to pressure from students and foreign economic advisors by establishing in 1966 the Ministry of Land Reform, charged with preparing new legislation on agrarian reform. The state's financial commitment to transforming small-scale agriculture amounted to next to nothing.[14]

Ethiopia's third development plan (1968–73) shifted its interest with some earnestness to agriculture, reflecting the market for international aid, that is, wider international, multilateral interests in agriculture as a means to economic growth in

12. *Ethiopia Information Bulletin* 19 (1965), 9-10. This percentage of income growth may reflect commoditization of the rural economy and the rise of urban markets for food rather than an expansion of productivity or economic welfare.

13. *Ethiopia Information Bulletin,* 9

14. Huffnagel, *Agriculture,* 455-56. Also see Ministry of Agriculture, *EPID Project Areas by Province* (Addis Ababa, 1974), 2.

developing countries.[15] The urbanized development planners, however, targeted mechanization, processing, and exports of agricultural production rather than work with smallholders. Of the 10 percent of the total national budget invested in agriculture, the lion's share went to tax exemptions for tractors and plant protection chemicals. Moreover, agricultural schemes of over U. S. $200,000 were tax exempt for five years. Overall, the plan allocated only 1 percent of the budget to small-farm agriculture, which still made up over 95 percent of total production.

The decade before the 1974 revolution nevertheless brought significant changes in state policies toward agriculture and the application of empiricism to Ethiopia's ox-plow farms. In 1966 Ethiopia signed an agreement with the newly formed Swedish International Development Authority (SIDA) to apply an approach developed by projects in Bangladesh and India which came to be known as integrated rural development. The focus on smallholder farms in this strategy reflected the dominant ideology in international development of the late 1960s and 1970s. The approach prescribed a package to enhance production on small farms, including green revolution inputs, rural infrastructure development, small-farm credit, local marketing cooperatives, and extension services based on a "visiting and training" method. From five sites offered by the imperial government, the SIDA group chose to work in the Chilalo area of Arussi Province, a half day's drive south of Addis Ababa. The project evolved under the acronym CADU (Chilalo Agricultural Development Unit).

Arussi's highlands, though long occupied by Arsi Oromo pastoralists, were ideal for producing highland cereals, especially wheat and barley, which complemented the teff-specialized areas around Addis Ababa and found ready and accessible urban markets along Ethiopia's best road networks. The economic results on small farms in the project were impressive. Between 1968 and 1984 total fertilizer use in the project area increased from 2822 to 83,091 quintals; from 1966 to 1981 wheat yield increased from 5.0–9.0 to 12.8–16.3 per hectare. Use of herbicides and improved seeds also expanded dramatically, suggesting important changes in ox-plow farm agronomy. These results appear not only to have been a triumph of scientific agriculture applied to small farms, but also to have reflected a remarkable capacity to gather data on the agricultural economy. The elasticity of production in Chilalo resulted from new attention to small farms but also from massive—by Ethiopian standards—investment of foreign aid. Between 1967 and 1975 the Swedes contributed E$37 million for inputs (fertilizer and improved seeds) and infrastructure.[16]

The social results of CADU's strategy to put high input investment into small ox-plow farms appear less certain. In its first seven years CADU's strategy

15. John Cohen argues persuasively that the shift also reflected the influence of better rural data collection and the presence of foreign fieldwork-based research. Cohen, *Integrated Rural Development*, 45.

16. Cohen, *Integrated Rural Development*, 72, 196-97.

improved small-farm productivity, but also encouraged urban landowners to evict ox-plow tenants in favor of consolidating holdings either directly under tractor cultivation or indirectly through mechanized contract farms using graduates from the new agricultural schools as managers to produce wheat for Addis Ababa markets. Between 1967 and 1974 average farm size increased from 2.5 to almost 3 hectares per household, a reflection of evictions and mechanization of larger farms rather than success in transforming production on small ox-plow farms.

After the 1975 land reform which nationalized farms over 10 hectares, tenant claims and new in-migrations to the project area reduced farm size to less than half a hectare. Moreover the evidence of the effects of green-revolution inputs on ox-plow farm income at CADU indicates that the high investment did not yield commensurate increases in income, largely because of decreasing farm size and the strict price control mechanisms imposed by economic managers in Addis Ababa after 1974.[17]

Despite its mixed social and economic results, the CADU model had profound effects on Ethiopian state policy regarding agriculture after the mid-1960s. In 1970–71 the WADU (Wollamo Agricultural Development Unit) project applied integrated rural development strategies in Sidamo, and in 1971–72 a similar project, AADU, began in Ada (see chapter 6), both with external funding. Each of these projects revolved around central premises of small-farm development: use of foreign donors and expertise, use of high-cost imported inputs, and concentration on high potential areas (usually in the south). The Ministry of Agriculture's Setit-Humera project on the Begamder-Eritrea border also fit this mold, though the state's role there followed rather than led private capital and local dynamics within the regional Eritrean economy.

A by-product of integrated development schemes was the 1971 formation of a new state strategy which linked management of high-cost, integrated development schemes to broader based policies of placing minimum package inputs in ox-plow farmers' hands. The Ministry of Agriculture's new Extension and Project Implementation Department combined the work of extension with the CADU experience of applied empirical research on small ox-plow farms. This strategy took advantage of international funding for two "Minimum Package" programs which sought to expand the successes of CADU more widely to ox-plow farms in high-potential areas by providing only fertilizer and improved seeds without the costly investments in infrastructure.[18] It was under this policy that Ada farmers began receiving fertilizer in the years preceding the revolution. The policy provided, for the first time, state investments in the small-farm sector combined with an expansion of in the number of extension workers prepared to work in the rural setting. It made inputs and trained

17. Cohen, *Integrated Development*, 125-26.

18. Interview with Demessie Gebre Mikael, project head, Ministry of Agriculture (Rome), 16 November 1991.

extension agents, even if in limited measure, available to small farms in high-potential areas without restrictions on prices or controls over management.

Under these programs farmers began to experiment with chemical fertilizers and improved varieties of wheat and maize, though in fact these inputs reached only a small fraction of ox-plow farmers. By 1975 only 10 percent of Ethiopia's farmers had used fertilizer or improved seeds. The comparative level of the investments in agriculture, however, was still small (c. 10 percent of the total annual budget) compared to the pressures of urban growth on food supply and temptations to control agricultural prices and change agricultural taxation. In the last case, a 1968 farmers' revolt in Gojjam against an agricultural income tax warned the government against too heavy-handed a change in taxation policy.[19]

The 1974 Revolution:
New Urban Imperatives

Though rural imagery—"land to the tiller" and the 1972–74 famine—dominates much of the literature on the 1974 revolution, that upheaval was fundamentally an urban phenomenon. The unrest which brought Haile Selassie's government down was the cumulative disaffection of teachers, taxi drivers, bus drivers, high school students, and, ultimately, young military officers. The rural metaphor "land to the tiller," championed by university students, disguised the wider, more entrenched, urban-class interests involved in the revolution. If many of the revolutionaries had come from rural areas, their career goals and expectations for their families and children planted them firmly in the city. The appearance of the Wallo famine victims at the outskirts of Addis Ababa in 1973 and 1974, horrifying the urban intelligentsia and spurring them to action against the government, does not alter the fundamental urban nature of the revolution itself and who dominated the new state apparatus. The military regime and the urban bureaucracy which replaced Haile Selassie's government learned quickly that urbanites' sense of well-being derived from stable food prices more than from perceptions of social justice. Consequently the new rulers placed agricultural price control high on their list of priorities.

The land reform proclamation of 1975 was the revolution's most profound single act. It outlawed private ownership of land and land holdings over 10 hectares (the maximum a single ox-plow household could manage), abolished rural wage labor, and guaranteed access to land to all claimants. Moreover, the socialist policies which allowed the military and urban bureaucrats to eliminate the economic base of the landed aristocratic class by nationalizing land also provided the means and desire to exercise direct control over agriculture as well. Commercial farms in

19. Hoben, *Land Tenure*, 216-26. For 1975 fertilizer use, see Gryseels and Anderson, *Research*, 10.

TABLE 7.1. Per-capita cereal production for Ethiopia, 1975–86.

	75–76	79–80	80–81	81–82	82–83	83–84	84–85	85–86
Kg/capita	154	179	153	143	170	135	119	104
Index	100	114	100	92	111	88	78	67

Source: Adapted from Stefan Brüne, "The Agrarian Sector: Structure, Performance and Issues (1974–1988)," in Pausewang et. al. *Ethiopia: Rural Development Options,* 17.

the Awash Valley and the south became state farms managed as part of a national economic plan. The land reform satisfied student demands to eliminate what they thought of as Ethiopia's "feudal" economy by severing the income flow from rural land to urban landlords. The terms of land reform, drafted by young bureaucrats and intellectuals for the military junta, reflected, however, a political agenda far more than a careful analysis of the ox-plow complex itself. Land reform had, in effect, placed land and agriculture within the reach of urban managers. Within the land reform regulations drafted by Addis Ababa's activist intelligentsia for the military government was a subtle but powerful preconception of a specific model of the ox-plow complex. While it attacked the abuses of tenancy associated with the south, it also assumed an annual cropping model which made no special provisions for perennial crops or the security of tenure needed for managing natural resources. In its view, land was a partible commodity over which only seasonal rights mattered.[20]

The terms of the land reform created no new wealth, but leveled existing resources in land. As was evident in Ankober, for example, stratification within the agricultural community continued on the basis of unequal household labor and ownership of oxen. While land reform forbade the accumulation of large farms, it encouraged the subdivision of existing ones, reducing farm size in many cases to less than half a hectare and requiring an intensification of production which the ox-plow complex had never demonstrated. Land reform nevertheless brought a major social and economic transformation. The immediate effect on production in the 1975–80 period was positive. The abolition of rents, increased access to production inputs, high world coffee prices, and consistent rainfall in those years raised rural consumption, total cereal production, and coffee revenues. By 1980, however, the conditions which had favored small farm production shifted, bringing a steady decrease in per-capita grain production (see table 7.1).

The stagnant or even declining per capita production rates of Ethiopian farms in the 1970s and 1980s have been conjunctural, reflecting erratic rainfall and the Malthusian scissors of demography, but also reflecting historically unprecedented state intervention in the production process itself. State policy in the 1980s derived

20. For a comprehensive view of the effects of land reform, see Dessalegn Rahmato, *Agrarian Reform in Ethiopia* (Trenton, 1984).

from a simultaneous desire to fuel the industrialized wage-labor sector by extracting profits from agriculture, the need to keep urban food prices low, and a political imperative to keep rural populations under its control. For their part, farmers had lost their traditional networks—petitions, local elite intermediaries, patron-client relations—for communicating displeasure with their rulers. Peasant associations set up to administer land claims in fairly quick order ceased to be populist organs for expressing rural needs and emerged as instruments through which the central government in the name of democratic centralism could implement its own choices in managing the rural economy. More subtly, intervention also reflected a dominant ethos among the urban bureaucracy that change in the agricultural sector could come about only by managing production and controlling rural dwellers.

In the wake of the revolution, the state quickly reorganized its relationship with the agricultural economy. The new socialist military government abolished the extension service of the Extension and Project Implementation Department (EPID) and ended the minimum package program, which had offered inputs directly to small farms without attempting to control them. The military government's action also recognized the populist sentiments of extension workers; EPID's young and educated field staff had largely deserted their posts to join the rural resistance organized by the Ethiopian People's Revolutionary Party (EPRP). In fact, extension agents proved to be among the few bureaucratic workers who had regular contact with small farms and farmers, and as a consequence have been consistently populist and thus anti-government in their orientation.[21]

Instead of implementing the broadly based distribution of inputs in the minimum package program, the state preferred to provide fertilizers, tractors, pesticides, and improved seed to state farms or under easy credit terms to independent farmers as an incentive for them to join producer cooperatives. Producer cooperatives constituted a more directly invasive intervention in small-farm management. In 1979, the state initiated a plan for creating producer cooperatives out of ox-plow farms, envisioning over half of Ethiopia's farmers in such structures by 1994. Toward this end, the Ministry of Agriculture actively encouraged and often coerced farmers to form cooperatives by pooling livestock, land, and labor. Local agents surveyed lands and provided cooperatives with preferred locations and access to inputs. By 1991 these programs of easy credit and Potemkin villages resulted in huge debts owed by the defunct producer cooperatives.[22]

21. This point was argued to me by Demessie Gebre Mikael, ex-head of the Ministry of Agriculture's Project Department, who suggested that it accounted for the state's reluctance to rebuild a genuine extension service in the years after the revolution. Interview (Rome), 16 November 1991.

22. By early 1992 one peasant association in Dejen Awraja (Gojjam) had an accumulated debt of E$200,000. Ada had a total of E$9,377,378. (data from the Ada and Dejen Ministry of Agriculture offices).

State farms in their initial concept were to provide the means for the state to gain direct control over production for export and urban food supply, an end run around peasant production. State farms were the consummate products of socialist planning: they employed wage labor, relied upon state investments in capital, and managed their crops by rigid, centrally mandated plans for sowing, harvest, and cultivation. Between 1977 and 1990 state farms expanded their holdings from 55,000 to 210,000 hectares, absorbed 64 percent of all state expenditure on agriculture, and accumulated a net loss of approximately $300 million. By 1985–86 state farms contributed only 1 percent of the country's food requirements and as the major consumers of imported agricultural inputs, created a net loss of foreign exchange earnings.[23]

Despite these advantages, by 1985–86 producer cooperatives cultivated only 2 percent of Ethiopia's arable land compared with the 94 percent still under individual holdings.[24] Among the cooperatives which existed, very few used their pooled resources or borrowed capital to change their ox-plow orientation. Far more than the Soviet tractors offered by the Ministry of Agriculture, farmers, especially women, preferred flour mills and covered water supplies—both investments which affected women's labor but were less favored by urban planners.[25] Thus, while these state attempts to reshape the nature of Ethiopia's agricultural economy absorbed the lion's share of state resources for agriculture, they did little to alter the shape and economics of the ox-plow complex or to transform the highland agricultural landscape.

The most pervasive state invasion of the agricultural economy was control over farm-gate grain prices through the Agricultural Marketing Corporation (AMC), which imposed fixed prices and, after 1979, sales quotas to provide food for urban markets and profit margins for state coffers. AMC farm-gate prices were "in all cases less than 50 percent of the Addis Ababa open market price."[26] For Gojjam's red teff, for example, the AMC offered farmers E$37, sold it in rationed urban gov-

23. Dessalegn Rahmato, "Cooperatives, State Farms, and Smallholder Production," in Pausewang et al., *Ethiopia,* 106-7. Planners in the Ministry of State Farms sheepishly admitted to me in 1990 that state farms could not utilize drought early-warning data on each farm because the ministry set planting dates a year in advance.

24. World Bank, *Ethiopia: Public Investment Program Review* (Washington, D.C., 1988), 46, quoted in Stefan Brüne, "The Agricultural Sector: Structure, Performance, and Issues (1974-1988)," in Pausewang et al., *Ethiopia,* 23.

25. Farmer refusal to accept the Nazret tractors assembled in Ethiopia from Soviet parts brought the collapse of that joint venture. In other cases tractor and oxen credit offered from the Aid Bank created a serious debt burden which few cooperatives could pay. See McCann, "An Evaluation of Oxfam Hararge Projects."

26. Eshetu Chole, "Agriculture and Surplus Extraction: The Ethiopian Experience," in Pausewang, et. al. *Rural Development,* 95.

ernment shops for E$57, and on the open market for $E94. Similar price differen-
tials affected wheat from Arsi and the famous *manya* teff of Ada. For other areas,
the long arm of the urban state did not have sufficient reach. Remote areas like Gera
and the Denki lowlands held little interest for the AMC. Those regions' maize and
sorghum had little appeal for urban dwellers, nor could the AMC manage to transport grain from remote rural markets.

Though practice varied over time and across different regions, the AMC's pricing and quota policies dramatically affected farm strategies. Farmers adjusted their
cropping schemes to avoid controls. Since farmers were aware that the AMC preferred cereals with urban demand (teff, wheat, and barley), they sowed oats, vetch
(gwaya), maize, and sorghum when they could. In some areas AMC quotas forced
farmers who grew only maize to purchase teff or wheat on the open market to fill
their quotas. While some farmers in remote areas could resist, the AMC retained
tight control over the areas of Gojjam, Shawa, Arsi, which constituted Addis
Ababa's food supply. These areas depended on the trucks and major roads controlled by the state to move their produce. Because grain is a good with a low value
to volume it was not amenable to smuggling. Though distant from the capital, Gojjam was especially easy prey to an AMC monopoly because of the Blue Nile gorge
which circumscribed the region, forming a single choke point—the Abbay
bridge—through which grain had to move to reach national markets. While Gojjam farmers could not avoid AMC price quotas directly, they responded by shifting
their crop mix away from teff and decreasing fertilizer consumption in direct proportion to the increasing AMC presence.[27] Areas near the capital, such as Ada and
Becho, by contrast, increased their concentration in teff. Administered from Addis
Ababa, the AMC set national prices, which tied local farmers to urban decisions in
a fashion unprecedented in highland agricultural history.

The changing rural-urban terms of trade in the Ethiopian economy best demonstrate the urban priorities of the state. As intended, the 1975 land reform initially
reversed the transfer of rural agricultural wealth to urbanized landowners and urban
consumers. In the 1960s rural cereal consumption had fallen somewhere between
75 and 91 percent of urban levels. In the wake of the 1974 revolution, the abolition
of rents, lack of marketing controls, and increases in per-capita production appear
to have narrowed this gap. By 1981–82, however, the World Bank estimated rural
cereal consumption in Ethiopia at only 69 percent of urban levels in a climate of

27. This phenomenon was reported to me by my colleague Dr. Ian Watt of the Addis
Ababa University geography department, whose research specialized in the economics of
fertilizer use.

28. Ian Watt, "Regional Patterns of Cereal Production and Consumption," in Zein Ahmed
Zein and Helmut Kloos, eds., *The Ecology of Health and Disease in Ethiopia* (Addis Ababa,
1988), 121-22; World Bank, *Ethiopia,* 46, quoted in Eshetu Chole, "Agriculture and Surplus
Extraction," 96-97.

overall decline in total consumption. The decline is also evident in farm prices. By 1985–86 farmers received prices for their primary crops, which in real terms had declined relative to the rising consumer price index by 32 percent for teff, 9 percent for wheat, 48 percent for maize, and 15 percent for sorghum.[28] These conditions formed the context for the famines of the 1984–85 season and also for the political climate during the 1985–90 period in which farmers absorbed the state's direct assault on farm life.

In July 1985 in an internal memorandum the Ethiopian government ordered the beginning of a program of villagization to move rural households from the historically dominant dispersed settlement patterns to compact "villages" of new houses arranged in grid patterns. In its subsequent publications and media programs the government argued that villagization would speed rural development by making farm households easier to reach by public health, water, and electricity. In July 1987 in the eastern Harerge town of Babile, I watched an evening Amharic-language television documentary called "Mendera Misrata Lemindenew?" (Why Villagization?), presented on the town square's new generator-powered television. The audience was not the surrounding Oromo—and Somali-speaking farmers, but townsfolk, mainly northern migrants with little or no connection to agriculture and under no jeopardy of being villagized.

Critics—mainly expatriate Ethiopians and foreigners—noted as well that the compact settlements enhanced tax collection and security and would be the basis for forced cooperatives. Political insecurity spared Tigray, northern Gonder, and northern Wallo from villagization, but Shawa, Harerge, Kaffa, Arsi, and Wallaga bore the brunt of the program. By March 1987 Chairman Mengistu Haile Mariam boasted that 15.4 percent of Ethiopia's rural population (5.7 million people) lived in new villages; by mid-1988 government policy had moved one-third of Ethiopia's rural population to compact sites, a level far beyond that achieved in Tanzania's ill-fated Ujamaa program of the 1970s.[29]

The villages transformed the rural landscape. Where single farm compounds or small hamlets had dotted the hillsides, a pattern in place since at least the sixteenth century, denuded rectangular villages now appeared laid out along a rigid

29. While official villagization figures may have been optimistic, they nevertheless indicate the widespread rural impact of the policy. The national villagization program had its origins in a 1977 program in the Bale region, which villagized farmers for the purposes of security and control. For data and discussion on villagization, see Mengistu Haile Mariam's speech to the Central Committee of the Workers Party of Ethiopia, 23 March 1987; Alemayehu Lirenso, "Villagization: Policies and Prospects," in Pausewang et al., *Ethiopia*, 135-43; John Cohen and Nils-Ivar Isaksson, *Villagization in the Arsi Region of Ethiopia* (Uppsala, 1987); Concern, "An Overview—Resettlement and Villagization," unpublished paper, May 1987; and Angela Raven-Roberts, "Report of Villagization in Oxfam America Assisted Project Areas in Hararge Province, Ethiopia," unpublished report to Oxfam America, June 1986.

grid. Farmers themselves built their new houses out of the remnants of their former houses on sites selected by teams of planning "experts" who arrived from the nearest town or from Addis Ababa itself. The plan called for not only building the new houses but also demolishing farmers' former dwellings, a policy which both provided building materials for the new villages and dissuaded farmers from returning to their former sites. Nineteenth-century travelers had noted that farmers almost always avoided roadside housing sites. The new treeless villages, by contrast, were quite visible along roadsides, where they were new sights for motorists accustomed to small distinctive compounds and rural hamlets. The square, regimented housing style imposed by villagization homogenized the diversity of domestic architecture, which had historically distinguished Gojjam from Wallo and Shawa from Hararge.

To travelers on footpaths or on muleback who ventured off the all-weather roads, the remains of former homestead sites provided the physical evidence of the economic shock that government policy imposed on agriculture. The old compounds bore the dead look of an archaeological site. Farmers took most of the wood, iron, and roof thatch to build their new village houses; they left behind objects too fragile or heavy to move: clay storage pots, empty granaries, and thorn fencing. Villagization called for building new rural villages in the "spirit" of rational planning, but also mandated the destruction of small market towns, which had grown spontaneously in the 1950s and 1960s. In Gera, for example, villagization planners from Addis Ababa ordered the demolition of "obsolete" market centers at Oba and Bera Dedo, which had housed shops, hotels, and homes.[30]

If urban-based bureaucratic planners had changed the human landscape with villagization, they had a more devastating, though less visible, effect on the economies of small farms. Farmers in the ox-plow complex had carefully chosen homestead sites for their proximity to fields, and water supplies for their location and off the rich bottom lands ideal for cultivation. Villagization planners observed no such nuances. In December 1990 I flew in the Relief and Rehabilitation Commission's Twin Otter at low altitude from Addis Ababa to Wallo, over heavily villagized areas of northern Shawa and southern Wallo. A consistent pattern of village formation presented itself: former homestead sites still visible from the air in northern Shawa and southern Wallo were perched atop hilly outcroppings above their fields. For the new villages, planners had moved the site grids off the slopes and onto the flatlands, that is, onto the best fields.

Dispersed sites were also useful for poultry and human health, since distance prevented the spread of disease. The villagization committees which arrived to select village sites had accessibility to roads, conformity with planning guidelines, and level construction sites in mind. In regard to the new village sites, they did not

30. Interview with residents of Bera Dedo (Gera) on site of the former town (Gera), January 1990.

and probably could not consider on an individual basis the new distances individual farmers had to travel to reach their fragmented fields and threshing floors.[31]

Site teams were invariably made up of males from towns, and they brought with them a firm and gendered image of annual crop, cereal agriculture derived from northern ox-plow traditions. Moving to villages in areas like southwestern Gera meant farmers abandoned perennial crops like ensete, *chat,* coffee, fruit trees, and root crops (see chapter 5). But in most areas planners paid little attention to the size and placement of compounds, the critical domain in which women's production took place. Restricted in size by topography or concern for covering cropland, compounds in Harerge were 1000 square meters or less. These cramped quarters prevented or curtailed the replanting of ensete and root crops, and also severely limited space for women to dry dung for fuel, process grain, and raise poultry.[32] Predictably, it was the support which rebel troops found in these areas that which formed the prelude to Addis Ababa's fall in 1991.

The Mengistu government's 1991 collapse took place quickly. Rebel forces which had maintained an ethnically based resistance in Tigray had moved easily south through the Amhara and Oromo areas of Gondar, Gojjam, Wallo, and Shawa by 1990. Government soldiers deserted in large numbers, and rural populations in Gojjam, Wallo, Wallaga, and Shawa overlooked their long-standing suspicion of Tigrayans and allowed them free passage to their final destination, Addis Ababa. Those who retreated before the young rebel soldiers were not farmers fearing the advancing army but townsfolk seeking ultimate refuge in the biggest town of all. Addis Ababa's population, to the end, never challenged Mengistu's rule and strongly resented the presence of Ethiopian People's Revolutionary Democratic Front (EPRDF) forces, which the rural population had either tolerated or encouraged.

If Ethiopia's 1974 revolution had been an expression of urban power against an archaic landed elite, the 1991 revolution was a final and resounding rejection by farmers of the domination of their lives by urban imperatives in the prices paid for their crops, the allocation of their labor, and, finally, where they could live and work. Villagization, which had faced little opposition from even the government's harshest critics among the intelligentsia, had proved a final straw. Neither the military nor the economic planners which implemented it had seen or understood its real effects on farms or farmers. Urban management under the guise of socialism and centralism had devastated farm-level productivity and, with villagization, had physically changed the human landscape to a point beyond toleration.

31. Interview with Redd Barna staff, Gera, January 1990; Alemayehu Lirenso, "Villagization," 139. In one district the average distance between fields and the village increased from 1.6 to 2 kilometers.

32. Alemayehu Lirenso, "Villagization," 140. In Bera Dedo, Gera farmers told me they had shifted almost entirely to annual crops because of the small size of compounds. Indeed, they had failed to replant ensete even after three years.

Rural Voices: Crop Change and Modernization
in the Ox-Plow Complex, 1938–90

On most issues of how they understand changing political and economic climates, farm voices in history are frustratingly mute. There are, however, some windows of understanding in regard to what actions they take on their beliefs. Cropping patterns over time provide an aggregate response by farmers to conditions of climate, politics, markets, demography, and new economic conditions. Such patterns in Ethiopia's ox-plow complex have remained remarkably stable over time in terms of the total crop repertoire and in patterns of local variation based on elevation and soil type. Change is evident, however, in the percentage shares of those crops, reflecting the aggregate effects of farm-level, seasonal decisions over which crops to plant and how to adjust the farm labor and livestock resources to cope with new external conditions.

In 1938 an Italian team from the Regio Istituto Agronomico per l'Africa Italiana in Florence compiled data on cereal production in Ethiopia, providing the first empirically based measure of the relative balance of cereal crops across the highlands.[33] The survey figures reflect a fuller geographic spread than the bulk of nineteenth-century travel accounts which concentrated on the northern Shawa, Tigray, and Gonder regions, recognizing the role of sorghum and maize in the production of all ox-plow zones. Nonetheless, barley clearly dominated aggregate production, reflecting its concentration of cultivation at particular elevations, its subsistence value, its role in rural diets, and its cost of production.[34]

The cropping evidence from the Italian occupation provides an important point of comparison with data from the last three decades. In that period, the presence of a modern bureaucratic state structure and multi-lateral international organizations such as the Food and Agriculture Organization and the World Bank has brought an empirical foundation for assessing Ethiopia's agricultural production. For many years, however, the conventional wisdom of both expatriate academics and agricultural specialists in Ethiopia has held that Ethiopia's highland farms concentrated

33. Tables have been adapted from Ciferri and Bartolozzi, *La produzione cerealicola*, 5-6, 13. Regions in Italian East Africa were based roughly on the principle of consolidating precolonial provinces. Eritrea was equivalent to Tigray plus Eritrea, Amara consisted of Gojjam, Begamder, and Wallo; Harar included Bale but excluded the Ogaden (placed within Somalia), Shawa was truncated slightly in the south and west; Galla and Sidamo included Wollega, Kaffa, Gamo Gofa, Ilubabor, and Sidamo.

34. Between November 1938 and April 1939 Emilio Conforti, a staff member of the Servizi dell'Agricoltura dell'Africa Italiana conducted a field survey of crops in the three central regions of Italian Ethiopia (Shawa, Harar, and Galla and Sidama). His field-level measurements, though based on road travel and limited seasonal exposure, nonetheless offer important corroboration and local nuances to the sweep of Ciferri's figures. Conforti, *Impressioni.*

TABLE 7.2. Cereal production in Ethiopia
by grain, 1938.

Grain	Metric tons	% of total
Sorghum	165,000	23.0
Teff	162,000	22.6
Barley	192,000	26.7
Wheat	75,000	10.5
Maize	97,000	13.5
Millet	25,000	3.4

TABLE 7.3. Regional percentage distribution of cereal production, by grain, 1938.

Region	Sorghum	Teff	Barley	Wheat	Maize	Millet
Eritrea	18.1	24.7	33.9	46.4	6.2	48.0
Amara	30.3	49.4	23.4	8.0	15.5	20.0
Shawa	6.1	9.2	20.8	26.5	10.3	8.0
Harar	33.3	4.3	18.2	16.4	16.5	0.0
Galla and Sidamo	12.2	12.4	3.7	2.7	51.5	24.0
	100.0	100.0	100.0	100.0	100.0	100.0

TABLE 7.4. Percentage distribution of cereal production, by region, 1938.

Grain	Eritrea	Amara	Shawa	Harar	Galla and Sidamo
Sorghum	15.9	24.9	10.3	43.9	19.1
Teff	21.3	39.9	15.5	5.5	19.0
Barley	34.6	22.4	41.3	27.9	6.7
Wheat	18.6	3.0	20.6	9.9	1.9
Maize	3.2	7.3	10.3	12.8	47.6
Millet	6.4	2.5	2.0	—	5.7
	100.0	100.0	100.0	100.0	100.0

Note: In tables 7.2, 7.3, 7.4, 7.5 I have recalculated Ciferri's data to account for the addition of Somalia to the Italian figures. The Somali region of Italian East Africa also included the Ogaden. Though Rafaele Ciferri was perhaps the most knowledgeable international scholar of Ethiopia's cereals, he reveals little of his methods for this survey.

TABLE 7.5. Total hectarage of cereal
production in Ethiopia, by grain, 1938.

Grain	Total hectares	% of total
Sorghum	110,000	14.4
Teff	147,270	19.3
Barley	274,290	36.0
Wheat	109,975	14.4
Maize	97,000	12.7
Millet	22,740	3.0

TABLE 7.6. Percentage distribution of rural cereal production in Ethiopia, by grain, 1961–84.

Year	Barley	Teff	Sorghum	Maize	Wheat	Millet
1961–62	26.51	25.29	20.99	14.18	13.10	—
1966–67	26.55	24.02	20.62	15.65	13.17	—
1970–71	26.38	23.04	22.33	15.31	12.94	—
1974–75	15.33	23.38	15.59	23.71	18.43	3.56
1975–76	10.51	22.77	19.12	30.62	11.24	5.75
1976–77	19.25	24.48	16.76	22.31	10.84	4.62
1977–78	17.03	27.37	16.77	23.38	10.84	4.62
1978–79	16.62	28.34	15.65	24.15	11.20	4.04
1979–80	15.56	23.38	27.19	23.39	7.81	2.66
1980–81	19.05	25.58	26.48	15.85	9.32	2.87
1981–82	17.42	21.97	23.35	21.67	11.53	2.85
1982–83	17.06	22.01	20.89	23.75	12.40	2.99
1983–84	15.52	20.57	21.75	28.30	10.73	2.68

Source: Watt, "Regional Patterns," 121.

Note: Because of rounding, some rows of percentages do not add to 100.0.

TABLE 7.7. Annual cereal consumption per capita (in kilograms), by grain, 1966–83.

Grain	1966–67	1966–68	1979–80	1982–83
Sorghum	29.2	19.5	32.1	16.8
Teff	34.6	27.4	28.9	24.8
Barley	31.4	22.3	70.1	13.4
Wheat	15.9	8.2	9.5	10.3
Maize	27.2	27.0	31.8	25.0
Millet	—	3.8	4.5	4.6
Total	138.3	108.2	176.9	94.9

Source: Watt, "Regional Patterns," 121.

primarily on teff, barley, and wheat as the basis of production and local consumption.[35] This same assumption regarded other cereals, such as sorghum, millet, and maize (i.e., crops suited to lower altitudes and marginal lands), as secondary.

The data in table 7.6 indicate that the most significant long-term trends evident nationally over the past three decades have been the steady decline in barley, a highland crop, which moved from the dominant cereal to third rank, and the increasing role of maize, a mid- to low-altitude crop which has come to dominate southern Ethiopia, though still not as thoroughly as it has the cropping systems of eastern and southern Africa.

Teff expanded its strong position and surpassed barley as the major crop in the mid-1960s, probably reflecting both urban tastes and specialization in districts such as Ada, Bacho, and much of Gojjam.[36]

The consumption data in table 7.7 indicate a specific "climate of production" in which the percentage changes of crops took place, that climate being the declining rural consumption of cereals.[37] The decline in cereal consumption from the 1960s through the early 1980s spans the revolution and socialist policies but is ambiguous. It may represent either declining production or decreasing rural-urban terms of trade, or some combination of both.

Overall, these figures reveal clear trends in national cereal production through the 1960s, the 1970s, and into the 1980s. The most obvious national trends evident in these data are: 1) the overall decline in production and consumption of cereals, that is, crop choices within overall economic decline; 2) the persistent decline of barley as the major cereal crop between 1961 and 1984; 3) the smaller, but steady decline of teff and wheat; 4) the consistent rise of maize to a position of predominance among staple cereals in the national food supply.

Though the aggregate data can never be precise, the nineteenth- and twentieth-century data presented here indicate a clear historical trajectory in cropping patterns. The relative balance of cereals within the cropping repertoire between 1938 and 1963 shows remarkable consistency. The major shift in the balance of crops occurred in the early to mid-1970s, likely as a result of the 1974 revolution and the 1975 land reform proclamation. Perhaps as important, the overall shift of maize from a crop of

35. Crummey, "Plow Agriculture," 7, cites Simoons, *Northwest Ethiopia*, 87. Huffnagel's ordering of cereals implies that teff, barley, and wheat are the three most important crops in the ox-plow systems. Huffnagel, *Agriculture*, 181-96.

36. Watt, "Regional Patterns," 111-15. For comparative figures for Eritrea, see Tomaso Silliani, *L'Africa Orientale Italiana (Eritrea e Somalia)* (Roma, 1933); F. Cappeleti and G. Mainardi, Documentazione statistica sulle attività agricole e zootecniche durante occupazione militare Brittanica in Eritrea al 30/6/1947 Istituto Agronomico per l'Oltremare, 1947, fasc. 788; and G. Lodi, Rapport annuale sull'Eritrea, Istituto Agronomico per l'Oltremare, Centro di documentazione, 1955, fasc. 2466.

37. Watt, "Regional Patterns," 109, 112-15.

minor importance to its position as a dominant national food crop reflects two critical shifts in the structure of ox-plow cultivation. First, the shift appears to parallel the changing political economy in the wake of the revolution, that is, the changes in the political environment. From the foundation of modern agricultural policy right through the 1980s state, policy favored the incorporation of maize as a means of expanding farm productivity. Maize's high yields to labor attracted both agronomists from the U.S. prairie and Mengistu Haile Mariam, who viewed teff as an elite crop which drained rural labor. While comparatively little crop breeding in teff took place in experimental stations, the Ministry of Agriculture introduced 150 varieties of improved maize, usually provided at no charge to farmers.[38]

A major surge of maize cultivation in southern Ethiopia appears to have taken place within the first third of the twentieth century, resulting in maize's relatively strong position in the cropping pattern summarized in Italian field surveys and crop estimates.[39] The rise of maize cultivation in the southwest may well reflect the arrival of new, more attractive cultivars in that period as well as after 1952. As chapter 5 argues, it may also reflect the expansion of coffee cultivation and the ideal mix of maize with the coffee labor cycle, which coincided with the arrival of open-pollinated improved seeds that raised yields to above 15 quintals per hectare even on unfertilized fields.[40]

The Eritrean, Ankober, and Italian survey evidence suggests that the rise of maize appears to have been less the wholesale adoption of it by older northern farms than the expanded role of southern farms and marginal lands in national food production. Yet, the dominant position of maize in regions like Kaffa and Harerge is also a fairly recent phenomenon (see chapter 5). In southern systems the ox plow had adjusted itself to an annual cycle that synchronized annual cropping with cultivation of ensete, coffee, *chat,* or tubers, which either supplemented diet or, after the mid-nineteenth century, provided access to an international market. Maize and its peculiar characteristics of early maturity, low labor needs, and relatively high yield fitted ideally into the mix of perennial and annual cereal crops that characterized southern systems. Those southern farms which continue to integrate annual with perennial crops may, however, reflect less an equilibrium between systems than a balancing act, that is, systems in flux in which annual cropping may eventually

38. Huffnagel, *Agriculture,* 447. For information on maize seed distribution methods and cultivars I am grateful to Ato Tassew Gabayehu of the Institute of Agricultural Research, Jimma, interviewed November 1991.

39. Conforti, *Impressioni,* 61.

40. Imported open-pollenated varieties may well have effected a dramatic increase in maize yield. In 1960 Huffnagel argued that 8-9 quintals per hectare was an average yield, whereas more recent surveys by the Institute of Agricultural Research place average on-farm yields at 17 quintals per hectare. See Huffnagel, *Agriculture,* 190 and interview with Tassew Gabayehu (Jimma), November 1991.

dominate, as in northern systems.[41] Just as forests and mixed farming with perennial crops have given way to the need for annual cropping in northern areas of longest settlement under the plow, southern systems may well be in transition also, albeit slowly.

Maize's most recent surge has taken place in a somewhat different context. Table 7.6 indicates that, while maize's share of the national cereal production has fluctuated, its overall role has rapidly expanded. The importance of this trend, however, is not an indicator of the success of modern agriculture. On the contrary, the historical context suggests that it is more likely the most visible sign of a crisis within the ox-plow complex which still dominates Ethiopia's rural economy, food supply, and source of potential capital accumulation. For farm families working less than a hectare of land, maize offers a quick maturing food supply which grows well on newly deforested land. It is also suited to the marginal lowlands brought increasingly into production below the escarpment in Wallo, Tigray, and Shawa. Yet, its high yield provides only subsistence, since maize cannot be stored long enough to reach the next hungry season; and it will not yield at all if the rains fall short. The increasing dominance of maize in a farm economy which is producing fewer perennial crops on smaller plots with less diversity is thus a clear message from Ethiopia's farmers that the basis for the ox-plow complex's historical success has eroded rapidly.

41. Simoons considers evidence that ensete was historically much more common in the north than it was to become in the nineteenth and twentieth centuries. Simoons, *Northwest Ethiopia*, 96-99.

EPILOGUE:
The Trajectory of Ox-Plow Agriculture

The long-term perspective of change offered by this book suggests at least two major trends in agriculture in the second half of the twentieth century, an epoch when the conjunctural elements of an activist state, population pressure, the emergence of large urban markets for food, and, more recently, political instability have set the climate of agricultural production. First, the growth and then collapse of the coffee-maize complex in many areas and the need to sow early-maturing, non-market, subsistence crops pushed maize into a dominant position in the national food supply. Farms in this emerging maize monoculture have moved away from economic diversity toward a single crop, away from agronomic innovation, and have become more vulnerable to environmental and economic shocks.

A movement toward specialization in areas like Ada, however, offers a second, alternative, vision of ox-plow cultivation's future. Given the effects of urban growth on the vast majority of Ethiopia's people of the plow outlined in chapter 7, the irony is that the intensification of production in units of labor and land may well be achieved by those close enough to the city to let market forces diminish the diversity in crops and livestock which has marked the ox-plow complex's historical resilience. The evidence of intensification of farm economies in Ada and of other "closed" economies suggests that the population's pressure to increase cropped land has forced more efficient use of on-farm and industrial by-products to sustain livestock and the domestic domain (fuel, building materials, milk). The integration of small ox-plow farms with urban markets, the agroindustrial sector, and urban sources for inputs (improved seeds and fertilizers) has, however, led these farms away from self-reliance toward further dependence on events and policies of the national political economy. The emergence of new agronomic applications of the basic ox-plow technology associated with this intensification affirms that system's resilience but expands farmers' vulnerability to new conjunctures of politics and the environment.

Appendices

Bibliography

Index

APPENDIX A   *Crop Names*

English	Latin	Amharic	Oromo
teff	*Eragrostis teff*	tef	tafı
barley	*Hordeum* (many varieties)	gabs	gerbo
wheat	*Triticum* (many varieties)	sinde	qamadi
eleusine	*Eleusine coracana*	dagusa	dagusa
red sorghum	*Sorghum vulgare*	zangada	masinga dima
white sorghum	*Sorghum vulgare*	mashila	masinga hadi
lentils	*Lens esculenta*	misser	mishira
field peas	*Pisum sativum*	ater	atara
horsebeans	*Vicia faba*	baqela	beqila
chick-peas	*Cicer arietinum*	shimbra	shumbura
maize	*Zea mays*	boqolo/ yabahar mashila	boqolo
rough pea (vetch)	*Lathyrus sativus*	gwaya	gayo
oats	*Avena fatua*	senar	
niger seed	*Guizotia abyssinica*	nug	nugi
Irish potato	*Solanum tuberosum*	dinnich	dinicha/ifate
castor beans	*Ricinus communis*	gulo	qobbo
soy beans	*Glicine hispida*	adangware	
haricot beans	*Phaseolus vulgaris*	boloqe	boloqe
linseed	*Linum usitsatissimum*	talba	talba
rape seed	*Brasica napus*	goman zar	sagni rafu
chat	*Catha edulis*	chat	jumaa
ensete	*Ensete ventricosum*	ensat/qocho	warqe/qocho

English	Latin	Amharic	Oromo
sweet potato	*Ipomoea batatas*	sukwar dinnich	
emmer wheat	*Triticum dicoccum*	ajja	ajja
taro	*Colocasia antiquorum*	godare	
sunflower	*Helianthus annus*	suf	suffi
fenugreek	*Tigonella fenum*	abish	abishi

APPENDIX B *Glossary of Agricultural Terms*

amrach	agricultural cooperative promoted by socialist government policy after 1974.
balegn	the exchange of human labor for the use of oxen.
belg	the short rainy season, usually March–May.
butta	Oromo celebration marking a rite of passage in the gada system.
chid	teff straw used for forage and construction.
dabo	agricultural cooperative labor, for example, for harvest or threshing.
edari	fallow land.
erbo	share-cropping system based on one-fourth share for the cultivator.
fachasa	land unit used in southwestern Ethiopia, about one acre.
farushka	wheat middlings used as forage, usually the by-product of a flour mill.
gabbar	a tribute-paying farmer, typically in a non-Amhara region.
gada	Oromo social-political system which marks changing generational status.
gasha	a land measure which varied widely over region and time; commonly 40 hectares or 100 acres.
gasha maret	literally, "land of the shield"; used to describe land taken by conquest where rights to income accrue directly to the emperor.
gaye	soil burning to decrease acidity and increase short-term fertility in long fallow agricultural system.
gert	grain remaining around threshing floor after tax collection.
gombisa (or gumbi)	Oromo word for outside granary.
gotera	Amharic word for outside granary.
gult	the "bundle" of income and labor rights granted to prominent officials or soldiers; implies control over income but not agricultural access to land.

269

gwaro	kitchen garden land around the homestead.
hudad	corvée labor or name for land controlled directly by state and cultivated with corvée labor.
keremt	main rainy season between June and October.
koticha	local term for black cotton vertisol.
magazo	land use practice in which landholder exchanged the seasonal use of a plot of land for a percentage of the crop, labor, or use of oxen.
madebet	literally, "kitchen"; land whose income and taxes in foodstuffs flowed directly to the palace.
maqanajo	Shawan term for the agreement between farmers to exchange their individual oxen on alternate days to yoke them for plowing one another's land. Also known as mallafagn and kendi.
malkanya	local elite who controlled income and labor rights over land cultivated by local farmers.
marasha	literally "plowshare"; also refers to the ox plow as a whole.
meher	main rainy season between June and October, also called keremt.
qalad	leather thong used for land measurement; also a term used for imperial land measurement schemes.
qanbar	yoke.
qollo	agreement for seasonal rental of an ox.
rest	partible land-use rights which guaranteed access to agricultural land.
ribi	livestock rearing agreement for owner and herder to share products and progeny of livestock in exchange for care outside the city or in lowland pastoral zones.
samon maret	land whose income accrues to a local church.
siso	share cropping agreement based on one-third share for the cultivator.
timad	a yoke of oxen; also an estimate of land. One timad is the amount of land a pair of oxen can plow in a day. Four timad equal roughly one hectare.

BIBLIOGRAPHY

Archival Materials

Istituto Agronomico per l'Oltremare (IAO) (Florence)

L'agricoltura dell'Abissinia, 1938. Fasc. 534; fasc. 1505.

"Autonomia Alimentare del Harar." N.d. IAO. Fasc. 1809.

Bonetti, S. and Gandussio. G. Relatzione agrologica e zootecnica della piana di Calamino, Macalle, Eritrea, 1939. Fasc. 584.

Boninsegni, G. Le coluture cerealicole, leguminose e oleifere nel territorio di Macalle, Eritrea, 1940. Fasc. 650.

Brizioli, F. Il caffé nella zona della pendici orientali in Eritrea, 1932. Fasc. 1952.

Cappelletti, F., and Mainardi, G. Documentazione statistica sulle attività agricole e zootecniche durante l'occupazione militare britannica in Eritrea al 30/6/1947. 1947. Fasc. 788.

Commissariato di Governo di Biscioftu. "La chiesa etiopica nel R. Commissariato di Governo Biscioftu." 1939. IAO Fasc. 791.

Di Martino, F. Profilo agrario del Seraé e considerazioni sulle possibilité di un miglioramento dell'agricoltura indigena e di sviluppo nazionale, 1940. Fasc. 1906.

Gubellini, M. Documentazione sull'agricoltura Eritrea durante l'Amministrazione Militare Britannica. 1946. Fasc 826.

Ispettorato Agrario dell'Eritrea. Sperimentazione agraria in Eritrea dal 1932 al 1938. Fasc. 1505.

Istituto Agronomico Africa Italiana. Appunti sui metodi sguiti dall'Italia nell'Africa Orientale Italiana per l'evoluzione dell'economica agraria e sui risulti raggiunti, 1949. fasc. 4567.

Istituto Agronomico Africa Italiana. Appunti sull'avvaloramento agrario dell'Impero. 1937. Fasc. 1926.

Istituto Agronomico per l'Oltremare. Attivita agricola essistenti nell'Harar al 1939. Fasc. 1985.

Istituto Agronomico per l'Oltremare. Attivita Esistenti nell Regione dello Scioa al 1939. Fasc. 1985.

Istituto Agronomico per l'Oltremare. Attivita Esistenti nella Regione dell'Amara al 1939. Fasc. 1985.

Lodi, G. Rapporto annuale sull'Eritrea. 1955. Fasc. 2466.

Massa, L. "Bibliografia." N.d. Fasc. 1762.

Morgagni, E. Nuovi indirizzi agricoli delle pendici orientali eritree, 1940. Fasc. 1888.

L'andamento dell'agricoltura nell'anno 1934. Fasc. 1903.

Attività agricola in Eritrea dal 1923 al 1931. 1931. Fasc. 1956.

"Progamma per l'incremento e il miglioramento della produzione avicola della Scioa." 1940. IAO Fasc. 1803.

Provvedimenti a favore dell'agricoltura indigena in Eritrea. 1932. Fasc. 1969.

Ufficio Agrario della Colonia Eritrea. Relazione annuale dell'agricoltura e pastorizia in Eritrea nel 1939. 1940. Fasc. 1300.

Verdecchia, Giovanni. "L'Agricoltura Indigena nello Scioa." I.A.O. Fasc. 525 (Scioa).

Istituto Italo-Africano

Chiomio, P. Giovanni. "Note di viaggio nel Sud Etiopico (1927-1928)." Unpublished typescript, 3 vols. N.d., Istituto Italo-Africano, b-5-0-10/1-3.

Periodicals

L'Agricoltura Coloniale, vol. 1 (1907-8); vol. 2 (1909-10).
Bollettino dell"Istituto Coloniale Fascista (1933-34).
Bollettino Ufficiale del Governo dell'Amara (1936-39)
Ethiopia Information Bulletin (1965).

Other Sources

Abdul Mejid Hussein, ed. *Rehab: Drought and Famine in Ethiopia.* London: International Africa Institute, 1976.

Acton, Roger. *The Abyssinian Expedition and the Life and Reign of King Theodore, with One Hundred Illustrations Engraved from Original Sketches by the Special Artists and Correspondents of the Illustrated London News.* London: Illustrated London News, 1868.

Addis Hiwet. *Ethiopia: From Autocracy to Revolution.* London: Review of African Political Economy, 1975.

Agostini, Augusto. "L'opera della milizia forestale nell'A.O.I." *Gli Annali dell'Africa Italiana* 4, 3 (1941), 851-55.

Ahmed Hassen Omer. "Aspects of the History of Efrata-Jille Wareda (Shoa Region) with Particular Reference to Twentieth Century." B.A. thesis, Addis Ababa University, 1987.

Alberro, M., and S. Haile Mariam. "The Indigenous Cattle of Ethiopia—Part I." *World Animal Review* 41 (1982), 2-10.

Alberro, M., and S. Haile Mariam. "The Indigenous Cattle of Ethiopia—Part II," *World Animal Review* 42 (1982), 27-34.

Alemayehu Lirenso, "Socio-economic Constraints to the Production of Belg Crops in Ethiopia." Social Science Research Council, Working Paper No. 4, New York, 1991.

Alemayehu Lirenso. "Villagization: Policies and Prospects," in Siegfried Pausewang, Fantu Cheru, Stefan Brune, and Eshetu Chole, eds., *Ethiopia: Rural Development Options.* 135-43. London: Zed Books Ltd., 1990.

Almeida, Manoel de. *Some Records of Ethiopia, 1593-1646.* Trans. and ed. C. F. Buckingham and G. W. B. Huntingford. London: Hakluyut Society, 1954.

Alvares, Francisco. *The Prester John of the Indies: A True Relation of the Lands of the Prester John, being a Narrative of the Portuguese Embassy to Ethiopia in 1520.* Trans.

and ed. C. F. Buckingham and G. W. B. Huntingford, 2 vols., Hakluyt Society, vols 114-15. Cambridge: Cambridge University Press, 1961.

Amborn, Hermann. "Agricultural Intensification in the Burji-Konso Cluster of South-western Ethiopia." *Azania* 24 (1989), 71-83.

"Animal traction and Vertisol Cropping," *ILCA Newsletter* 6 (1987), 1.

Annaratone, Carlo. *In Abissinia.* Roma: Voghera, 1914.

*Annuario statistico italiano. Statistica delle movimento marittimo.*Roma, 1886-1934.

Antinori, Orazio. "Lettera del M. O. Antinori a S.E. il comm. Correnti presidente dell società," *Bollettino della Società Geografica Italiana* 16 (1879), 374-83.

Antinori, Orazio. "Relazione del M.O. Antinori," *Bollettino delle Società Geografica Italiana* 16 (1879), 388-403.

Antinori, Orazio. "Sul vitto e sul modo di aggiustare i cibi presso il popolo di Scioa." *Bollettino della Società Geografica Italiana* 16 (1879), 388-403.

Antonelli, Pietro. Rapporti sullo Scioa del Conte Pietro Antonelli al R. Ministero degli Affari Esteri (dal 22 maggio 1883 al 19 giungno 1888) "Zemeccia: Ossia spedizione dell'esecito scioano," annex to Pietro Antonelli to Ministero degli Affari Esteri, Antoto, 19 September 1887, annex to letter. Roma: n.p., 1889.

Antonelli, Pietro. "Lettera del Conte Pietro Antonelli alla sua familglia." *Bollettino della Società Geografica Italiana* 19 (1882), 80.

Asmarom Legesse. *Gada: Three Approaches to the Study of African Society.* New York: Free Press, 1973.

Asnakew Woldeab. "Physical Properties of Ethiopian Vertisols." In S.C. Jutzi, I. Haque, John McIntire, and J. E. S. Stares, eds., *Management of Vertisols in Sub-Saharan Africa,* 111-23. Addis Ababa: International Livestock Centre for Africa, 1988.

Assefa Bekele, Yitateku Negge, and Tewolde Berhan Gebre Egziabher. *Zobel: An Experiment in Relief and Rehabilitation.* Addis Ababa: University Famine Relief and Rehabilitation Committee, 1974.

Aubry, Alphonse. *Une mission au royaume du Choa et dans le pays Gallas.* Tome XIV *Archives des Missions Scientifique et Lettéraires.* Paris: Ernest Leroux, 1888.

Aubry, Alphonse. "Mission au Royaume de Choa et den les pays Galla." *Achives des Mission Scientifique et literairus* 14 (1888), 296.

Baccarini, P. *La patria d'origine dell piante coltivate in Eritrea.* Firenze: Stabilimento tipografico pei Minuri Corrigendi, 1909.

Bahru Zewde. "A Historical Outline of Famine in Ethiopia." In Abdul Mejid Hussein, ed., *Rehab: Drought and Famine in Ethiopia,* 52-58. London: International African Institute, 1976.

Baldratri, Isaia. *Mostra agricola della colonia Eritrea: Catalogo illustrativo.* Firenze: Tipografia Luigi Niccolai, 1903.

Barrett, Vincent, Gregory Lassiter, David Wilcock, Doyle Baker, and Eric Crawford. "Animal Traction in Eastern Upper Volta: A Technical, Economnic, and Institutional Analysis." Michigan State University. International Development Paper, No. 4, 1982.

Bauer, Dan F. "For Want of an Ox. . : Land, Capital, and Social Stratification in Tigre." In Harold G. Marcus, ed., *Proceedings of the First United States Conference on Ethiopian Studies,* 235-48. East Lansing: African Studies Center, Michigan State University, 1975.

Dan F. Bauer. *Household and Society in Ethiopia.* 2nd Ed. East Lansing: African Studies Center, Michigan State University, 1985.

Dan F. Bauer "Land, Leadership, and Legitimacy in Enderta, Tigre." Ph.D. dissertation, University of Rochester, 1972.

Beke, Charles. "Abyssinia-Being a Continuation of Routes in That Country." *Journal of the Royal Geographic Society* 14 (1844), 3.

Bekele Shiferaw. "Crop Livestock Interactions in the Ethiopian Highlands and Effects on Sustainability of Mixed Farming: A Case Study from Ada District." M.S. thesis, Agricultural University of Norway, 1991.

Benedictis, A. de "L'autarchia alimentare dell'impero, problemi e prime realizzazione." *L'Agricoltura Coloniale* 32 (1938), 1-12.

Bent, J. Theodore. T*he Sacred City of the Ethiopians, Being a Record of Travel and Research in Abyssinia in 1893.* London: Longmans, Green, and Co. 1893.

Bernatz, John Martin. *Scenes in Ethiopia.* Vol. 2: *The Highlands of Shoa.* Munich and London: Bradbury and Evans, 1852.

Bianchi, Gustavo. *Alla terra dei Galla.* Milano: Società Felsinea, 1882.

Borelli, Jules. *Ethiopie méridionale: Journal de mon voyage aux pays Amhara, Oromo, et Sidama.* Paris: Librairies-Imprimeries Reunies, 1890.

Borton, Raymond, Mammo Bahte, Almaz Wondimu, and John Asfaw, *A Development Program for the Ada District based on a Socio-economic Survey.* Stanford: Stanford Research Institute, 1969.

Boserup, Ester. *The Conditions of Agricultural Growth.* London: Allen and Unwin, 1965.

Brandt, Steven A. "New Perspectives on the Origins of Food Production in Ethiopia." In J. D. Clark and Steven Brandt, eds., *From Hunters to Farmers*, 173-90. Berkeley: University of California Press, 1984.

Bruce, James. *Travels to Discover the Source of the Nile; in the Years 1768, 1769, 1770, 1771, 1772, and 1773.* 5 vols. Edinburgh, 1790; 7 vols. Edinburgh, 1804.

Brüne, Stefan. "The Agricultural Sector: Structure, Performance and Issues (1974-1988)." In Siegfried Pausewaung, Fantu Cheru, Stefan Brüne, and Eshetu Chole, eds.,. *Ethiopia: Rural Development Options.* 15-37. London: Zed Books Ltd., 1990.

Brunialti, Attilio. "La spedizione Italiana nello Scioa." in *Nuova antologia miscellanea africana, anni 1873-1889,* .288-333. Roma: Direzione Centrale Affari Coloniali, 1938.

Bryson, Reid, and Christine Paddock."On the Climates of History." In Robert Rotberg and Theordore Rabb, eds., *Climate and History in Interdisciplinary History*, 1-18. Princeton: Princeton University Press, 1981.

Buxton, David. "The Shoan Plateau and Its People: An Essay in Local Geography." *Geographical Journal* 114 (1949), 157-72.

Camera dei Deputati. *Il Missione Antonelli in Etiopia.* Roma: n.p.,1890.

Campbell, Mary B. *The Witness and the Other World: Exotic European Travel Writing, 400-1600.* Ithaca: Cornell University Press, 1988.

Capucci, Luigi. "Condizioni dell'agricultura nello Scioa." *Bolletino della Società Africana d'Italia* 6 (1887), 30-35.

Caroselli, F .S. "Aspetti economici dell'avvaloramento agrario nell'impero." *L'Agricoltura Coloniale* 35 (1941), 47-54.

Castro, Lincoln de. N*ella terra dei negus: Pagine raccolte in Abissinia.* 2 vols. Milano: Fratelli Treves, Editori, 1915.

Caulk, Richard. "Armies as Predators: Soldiers and Peasants in Ethiopia c. 1850-1935." *International Journal of African Historical Studies* 11 (1979), 457-93.

Cecchi, Antonio. *Da Zeila alle frontiere del Caffa*. 2 vols. Roma: Ermanno Loescher, 1886.

Cecchi, Antonio."Dallo Scioa al Ghera, la morte di Chiarini" In *Nuova Antologia Miscellanea Africana, Anni 1873-1889*. 122-36. Roma: Direzione Centrale Affari Colniali, 1938.

Cecchi, Antonio."Le mie vicende durante la prigiona nel Ghera." In *Nuova sntologia miscellanea africana, anni 1873-1889*. 257-79.

Cecchi, Antonio, and Giovanni Chiarini. "Relazione alla presidenza della Società Geografica." *Bollettino della Società Geografica Italiana* 16 (1879), 410-23.

Cecchi, Antonio, and Giovanni Chiarini. "Relazione dei signori G. Chiarini and A. Cecchi sui mercati principali dello Scioa." *Bollettino della Società Geografica Italiana* 16 (1879), 445-55.

Cerruli, Enrico. "Canti amarici dei Musulmani di Abissinia." *Reale Accademia dei Lincei. Rendiconti*. Series 6, 2 (1926), 433-47.

Cerruli, Enrico. "Canti amarici delle corti e delle campagne." In *Mélanges Marcel Cohen*, 415-23. La Haye: 1970.

Cerruli, Enrico. "Canti amarici di ieri." In A. Francesco Gabrieli, *Studi orientalistici*, 61-69. Roma: Università di Roma, .

Cerruli, Enrico. "Canti burleschi di studenti delle scuole abissinie." *Revista degli Studi Orientale* 13 (1931-32), 342-50.

Cerruli, Enrico. "Canti populare amarici," *Reale Accademia dei Lincei. Rendiconti*. Series 5, 25 (1916), 563-658.

Cerruli, Enrico. "La colonizzazione del Harar." *Gli Annali dell'Africa Italiana* 5, 1 (1943), 63-79.

Cerruli, Enrico. *Etiopia Occidentale (dallo Scioa alla frontiera del Sudan): Note del viaggio 1927-1928*. Roma: Sindicato Italiano Arti Grafiche, 1933.

Cerruli, Enrico. "The Folk Literature of the Galla of Southern Abyssinia," *Harvard African Studies* 3 (1922), 11-228.

Cerruli, Enrico. "La Missione Cattolica della Consolata e la coltivazione del tè al Kaffa." *Oriente moderno* 13 (1933), 483-85.

Cerruli, Enrico. "Poesie di guerra e di amore dei Galla dell'A.O.I." in Gli Annali dell'Africa Italiana 5, 1 (1942), 117-25.

Cerruli, Enrico. "La poesia populare amarica." *L'Africa Italiana. Bollettino della Società Africana d'Italia* 35 (1916), 172-78.

Cerruli, Enrico. "Una raccolta amarica di canti funebri," *Rivista degli Studi Orientale* 10 (1923-25), 265-80.

Chambers, Robert. "Climate, Seasonality, and the Tropics," in Chambers, Robert, Richard Longhurst, and Arnold Pacey, eds. *Seasonal Dimensions to Rural Poverty*. London: Frances Pinter Ltd., 1981.

Chambers, Robert. *Rural Development: Putting the Last First*. New York: John Wiley and Sons, Inc., 1983.

Chambers, Robert, Richard Longhurst, and Arnold Pacey, eds. *Seasonal Dimensions to Rural Poverty*. London: Frances Pinter Ltd., 1981.

Chiaromonte, A. "Il problema delle cavallette nell'A.O.I.." In Sindacato Nazionale Fascista Techici Agricoli. *Agricoltura e impero*, 589-620. Roma: Tipografica Editrice Sallustiana, 1937.

Chiovenda, Emilio. *Osservazioni botaniche agrarie ed industriali fatte nell'Abissinia settentrionale nell'anno 1909*. Monografiche e rapporti coloniali, No. 24. Roma: Ministero dell Colonie, Direzione Centrale degli Affari Coloniali, Ufficio di Studi Coloniali, 1912.

Ciferri, Raffaele. *Frumenti e granicoltura indigena in Etiopia.* Firenze: Tipografia Mariano Ricci, 1939.

Ciferri, Raffaele. *L'Africa Orientale Italiana: Centro d'evoluzione di piante coltivate.* Roma: Istituto Poligrafico dello Stato, 1939.

Ciferri, Raffaele. *La cerealicoltura in A.O.I.* Vol. 8: *Generalita botanico-agrarie sui sorghi.* Roma: Ramo Editoriale Agricoltori, 1941.

Ciferri, Raffaele. "Primo rapporto sul caffé nell'Africa Orientale Italiana." *Agricolo Coloniale* 34 (1940), 135-44.

Ciferri, Raffaele, and Enrico Bartoluzzi. *La produzione cerealicola dell'Africa Orientale Italiana nel 1938.* Firenze: Regio Istituto Agronomico per l'Africa Italiana, 1940.

Ciferri, Raffaele, and Renzo Giglioli. *I cereali dell'Africa Italiana.* Vol. 1: *I frumenti dell'Africa Orientale Italiana, studiati su materiali originali.* Firenze: Regio Istituto Agronomico per l'Africa Italiana, 1939.

Ciferri, Raffaele, and Renzo Giglioli. *La cerealicoltura in A.O.I.* Vol. 2: *I frumenti piramidali, turgidi, polacchi e dicocchi.* Roma: Ramo Editoriale degli Agricoltori, 1939.

Ciferri, Raffaele, and Renzo Giglioli. *La cerealicoltura in A.O.I.* Vol. 3: *I frumenti volgari e compatti.* Roma: Ramo Editoriale degli Agricoltori, 1939.

Ciferri, Raffaele, and Renzo Giglioli. *La cerealicoltura in A.O.I.* Vol. 4: *Caratteristiche dei gruppi minori di frumenti etiopici e loro "formulazione."* Roma: Ramo Editoriale degli Agricoltori S.A., 1940.

Ciferri, Raffaele, and Renzo Giglioli. "La cerealcoltura in A.O.I.: I frumenti duri." *L'Italia Agricola* 76 (1939), 247-57.

Ciferri, Raffaele, Isaia Baldrati. *La cerealicoltura in A.O.I.* Vol. 6: *Il "teff" (Eragrostis Teff).* Roma: Ramo Editoriale degli Agricoltori, 1940.

Cillis, Emanuele de. "I problemi della cerealicoltura nell'A.O.I." Sindacato Nazionale Fascista Techici Agricoli, *Agricoltura e Impero*, 177-89. Roma: Tipografica Editrice Sallustiana, 1937.

Clapham, Christopher. *Transformation and Continuity in Revolutionary Ethiopia.* Cambridge: Cambridge University Press, 1988.

Clark, J.D. ,and Steven Brandt, eds., *From Hunters to Farmers.* Berkeley: University of California Press, 1984.

Clay, Jason W. and Bonnie K. Holcomb. *Politics and the Ethiopian Famine 1984-1985.* Cambridge, Mass: Cultural Survival, 1985.

Cohen, John, and Nils-Ivar Isaksson. *Villagization in the Arsi Region of Ethiopia.* Uppsala: Scandinavian Institute of African Studies, 1987.

Cohen, John, and Dov Weintraub. *Land and Peasants in Imperial Ethiopia: The Social Background to a Revolution.* Assen: Van Gorcum and Co., 1975.

Cohen, John M. *Integrated Rural Development: The Ethiopian Experience and the Debate.* Uppsala: Scandinavian Institute of African Studies, 1987.

Colaci, Francesco. "Commerico e agricoltura in Abissinia." *Cosmos* 8 (1884), 51-58.

Combes, E. D. and M. Tamisier. *Voyage en Abyssinie, dans le pays des Gallas, de Choa et d'Ifat.* 4 vols. Paris: Louis Desessart, 1838.

Concern. "An Overview—Resettlement and Villagization." Unpublished paper, May 1987.

Conforti, Emilio. *Impressioni agrarie su alcuni itinerari dell'altopiano etiopico.* Firenze: Regio Istituto Agronomico per l'Africa Italiana, 1941.

Consociazione Turistica Italiana, *Guida dell'Africa Orientale Italiana.* Milano: Consociazione Turistica Italiana, 1938.

Contenson, Henri, de. "Les fouilles de Haoulti en 1959." *Annales d'Etiopie* 5 (1963), 41-86.

Conti Rossini, Carlo. *Poverbi tradizioni e canzoni Tigrine.* Roma: Ufficio Studi del Ministero dell'Africa Italiana, 1942.

Cordell, Dennis, Joel Gregory and Victor Piché. "African Historical Demography: The Search for a Theoretical Framework." In Dennis Cordell and Joel Gregory, eds., *African Population and Capitalism,* 14-34. Boulder: Westview, 1987.

Cossins, Noel. "Day of the Poor Man," Unpublished mimeo, Ethiopian Relief and Rehabilitation Commission, 1975.

Cossins, Noel. "Still Sleep the Highlands: A Study of Farm and Livestock Systems in the Central Highlands of Ethiopia." Unpublished report prepared for the Provisional Military Government Livestock and Meat Board, 1974.

Cronon, William. "A Place for Stories: Nature, History, and Narrative." *Journal of American History* (March 1982), 1347-76.

Cronon, William. *Changes in the Land: Indians, Colonists, and the Ecology of New England.* New York: Hill and Wang, 1983.

Crummey, Donald. "Abyssinian Feudalism." *Past and Present* 89 (1980), 115-38.

Crummey, Donald. "Ethiopian Plow Agriculture in the Nineteenth Century." *Journal of Ethiopian Studies* 16 (1983), 1-24.

Crummey, Donald. "Gondarine Rim Land Sales: An Introductory Description and Analysis." In *Proceedings of the Fifth International Conference on Ethiopian Studies.* Chicago: University of Illinois at Chicago Circle, 1978.

Crummey, Donald. "Some Precursors of Addis Ababa: Towns in Christian Ethiopia in the Eighteenth and Nineteenth Centuries." In *Proceedings of the International Symposium on the Centenary of Addis Ababa.* Addis Ababa: Institute of Ethiopian Studies, 1987.

Crummey, Donald. "Three Amharic Documents of Marriage and Inheritance from the Eighteenth and Nineteenth Centuries." In *Proceedings of the Eighth International Conference of Ethiopian Studies,* 315-28. Addis Ababa: Addis Ababa University Press, 1988.

Cutler, Peter. "Forecasting Famine: Prices and Peasant Behavior in Northern Ethiopia." *Disasters* (August 1984), 48-56.

d'Abbadie, Antoine. *Géographie de L'Éthiopie: Ce que jai entendu, faisànt suite a ce que jai vu.* Paris: Gustave Mesnil, 1890.

d'Abbadie, Arnauld *Douze ans de séjour dans la Haute-Ethiopie (Abyssinie).* Vatican City: Biblioteca Apostolica Vaticana, 1980.

Dainelli, Giotto. "Del commercio tra l'Eritrea e l'Etiopia nell'anno 1905." *Bollettino della Societá Africana Italiana* 25 (1906), 137-46.

Damon, G. *The Cultivated Sorghums of Ethiopia.* Dire Dawa Imperial Ethiopian College of Agriculture, 1962.

Daniel Gamachu. *Environment and Development in Ethiopia.* Geneva: International Institute for Relief and Development, 1988.

Darkwah, R. Kofi. *Shewa, Menilek and the Ethiopian Empire 1813-1889.* London: Heinemann, 1975.

Dessalegn Rahmato. *Agrarian Reform in Ethiopia.* Trenton, N.J.: Red Sea Press, 1984.

Dessalegn Rahmato. "Cooperatives, State Farms, and Smallholder Production." In Siegfried Pausewaung, Fantu Cheru, Stefan Brüne, and Eshetu Chole, eds.,. *Ethiopia: Rural Development Options.* 100-10.. London: Zed Books Ltd., 1990.

Dessalegn Rahmato. *Famine and Survival Strategies: A Case Study from Northern Ethiopia.* Uppsala: Scandanavian Institute of African Studies, 1991.

Dessalegn Rahmato. "Moral Crusaders and Incipient Capitalists: Mechanized Agriculture and Its Critics in Ethiopia." In *Proceedings of the Third Annual Seminar of the Department of History* . Addis Ababa: 1986, 80.

Dessalegn Rahmato. "The Peasants and the Comrades: Problems and Prospects of Socialist Transition in Rural Ethiopia." Paper presented to the African Studies Association, Denver, 1987.

DeVries, Jan. "Measuring the Impact of Climate on History: The Search for Appropriate Methodologies." In Robert Rotberg and Theodore Rabb, eds., *Climate and History in Interdisciplinary History*, 69-90. Princeton: Princeton University Press, 1981.

Dias, Jill. "Famine and Disease in the History of Angola c. 1830-1930." *Journal of African History* 22 (1981), 349.

Donahue, Roy L. *Ethiopia: Taxonomy, Cartography, and Ecology of Soils.* East Lansing: African Studies Center, Michigan State University, 1972.

Donham, Donald. "Old Abyssinia and the New Ethiopian Empire: Themes in Social History." In Donald Donham and Wendy James, eds., *The Southern Marches of Imperial Ethiopia: Essays in History and Social Anthropology.*, 3-48. Cambridge: Cambridge University Press, 1986.

Donham, Donald and Wendy James,eds. *The Southern Marches of Imperial Ethiopia: Essays in History and Social Anthropology.* Cambridge: Cambridge University Press, 1986.

Duchesne-Fournet, Jean. *Mission en Ethiopie (1901-1903).* 2 vols. Paris: Masson et Co., 1908-9.

Edwards, Sue. *Some Wild Flowering Plants of Ethiopia.* Addis Ababa: Addis Ababa University Press, 1976.

Ege, Svein. "Chiefs and Peasants: The socio-political structure of the Kingdom of Shoa about 1840." Unpublished Hovedoppgave dissertation, University of Bergen, 1978.

Ehret, Christopher. "On the Antiquity of Agriculture in Ethiopia," *Journal of African History* 20 (1979), 161-78.

Eichberger, Willis G. "A Study of Traditional Farming in Four Areas of the Ethiopian Highlands." Unpublished typescript, Institute of Ethiopian Studies, 1968.

Ellero, Giovanni. "Il Uolcait." *Rassegna di Studi Etiopici* 6-7 (1948), 89-112.

Ellis, Gene. "Man or Machine, Beast or Burden: A Case Study of the Economics of Agricultural Mechanization in Ada District, Ethiopia." unpub. Ph.D. dissertation, University. of Tennessee, 1972.

Ellis, Gene. "The Feudal Paradigm as a Hindrance to Understanding Ethiopia." *Journal of Modern African Studies* 14 (1976), 275-95.

Eshetu Chole. "Agriculture and Surplus Extraction: The Ethiopian Experience." In Siegfried Pausewaung, Fantu Cheru, Stefan Brüne, and Eshetu Chole, eds.,. *Ethiopia: Rural Development Options.* 89-99. London: Zed Books Ltd., 1990.

Fassil G. Kiros. "Agricultural Land Fragmentation: A Problem of Land Distribution Observed in Some Ethiopian Peasant Associations." *Ethiopian Journal of Development Research* 4 (1980), 1-12.

Feierman, Steven. *Peasant Intellectuals: Anthropology and History in Tanzania.* Madison: University of Wisconsin Press, 1990.

Felcourt, J. E. de. *L'Abyssinie: Agriculture chemin de fer.* Paris: Emile Larose, 1911.

Forbes, Rosita. *From Red Sea to Blue Nile: Abyssinian Adventure.* New York: Macaulay, 1925.

Fortunati, G., and P. Cascianti. *L'Agave e Sisal in Eritrea*. Roma: Stabilimento Capaccini, 1909.

Franchetti, Leopoldo. *Sulle colonizzazione agricola dell'altipiano etiopico*. Roma: Tipografia di Gabinetto del Ministero degli affari esteri, 1890.

Gabra Sellase. *Tarika Zaman Dagmawi Menilek Negusa Nagast Zaltyopya*. Addis Ababa: Artistic, 1959 E.C.

Gamst, Frederick. "Peasantries and Elites without Urbanism: The Civilization of Ethiopia." *Comparative Studies in Society and History* 12 (1970), 373-92.

Gebru Tareke. "Peasant Resistance in Ethiopia: The Case of Wayane." *Journal of African History* 25 (1984), 77-92.

Gennari, Giulio. *L'agricoltura nell'Africa Orientale Italiana*. Roma: Istituto Poligrafico dello Stato, 1938.

Getachew Asamenew and Mohammed Saleem, "The Concept and Procedure of Improved Vertisol Technology Development in Ethiopia." Paper presented to the IBSRAM's Workshop on the Management of Vertisols in Africa, Nairobi, March 1991.

Getachew Asamenaw and Mohammed Saleem, "Farmer Participatory Roles in Technology Development and Transfer: A Case Study from On-farm Vertisol Technology Experience in Ethiopia." Paper presented to the Institute of Agricultural Research Workshop on Farmers; Participatory Research, Addis Ababa, February 1992.

Getachew Asamenew, S. C. Jutzi, Abate Tedla, J. McIntire, and E. Ikavalko. "Improved Draught Power for Increased Food and Feed Production on Ethiopian Highland Vertisols." Paper presented to the Second National Livestock Improvement Conference, Addis Ababa, 1988.

Getachew Tecle Medhin and Telahun Makonnen. "Socio-economic Characteristics of the Central Highlands of Ethiopia—Ada Woreda." unpublished report, Ministry of Agriculture, 1974.

Ghezzi, Carla. "Fonti di documentazione e di ricerca per la conoscenza dell'Africa: Dall'Istituto coloniale italiano all'Istituto italo-africano." *Studi Piacentini* 7 (1990), 167-92.

Giordano G. "Utilizzazione dell foreste dell'Etiopia." *Rivista di Agricoltura Subtropicale e Tropicale* 1-3 (1949), 29-35.

Giudotti, Rolando. *L'Agricoltura in Eritrea nel 1936*. Firenze: Istituto Agricolo Coloniale Italiano, 1937.

Giudotti, Rolando, and Gioacchino Dallari. *Bassopiano occidentale oltre il Setit e territorio del Tana*. Florence: Istituto Agricolo Coloniale Italiano, 1937.

Glantz, Michael. "Drought, Famine, and the Seasons in Sub-Saharan Africa," in R Huss-Ashmore and S. Katz, eds. *Anthropological Perspectives on the African Famine*. New York: Gordon and Breech Science Publishers, 1987.

Glantz, Michael. ed. *Drought and Hunger in Africa: Denying Famine a Future*. Cambridge: Cambridge University Press, 1987.

Glantz, Michael, and Richard Katz. "When is a Drought a Drought?" *Nature* 267 (1977), 192-93.

Goe, Michael. "Current Status of Research on Animal Traction," *World Animal Review* January-March (1983), 2-17.

Goe, Michael. "The Ethiopian Maresha: Clarifying Design and Development," *Northeast African Studies* 11 (1989), 71-112.

Goe, Michael. "Tillage with the Traditional Maresha in the Ethiopian Highlands." *Tools and Tillage* 6 (1990), 127-56.

Goe, Michael, and Robert McDowell. *Animal Traction: Guidelines for Utilization.* Ithaca: Cornell Agricultural Mimeograph 81, 1980.

Goody, Jack. "Class and Marriage in Africa and Eurasia," *American Journal of Sociology* 76 (1971), 585-603.

Goody, Jack. "Future of the Family in Africa." Unpublished paper, n.d.

Goody, Jack. *Production and Reproduction: A Comparative Study of the Domestic Domain.* Cambridge: Cambridge University Press, 1976.

Goody, Jack. *Technology, Tradition, and the State in Africa.* London: Oxford University Press, 1971.

Governo del Harar (Sezione Studi). *Tre anni di occupazione.* Harar: Governo del Harar, 1939.

Graham, Douglas. "Report on the Agricultural and Land Produce of Shoa." *Journal of the Asiatic Society of Bengal* 13 (1844), 253-96.

Green, David A. G. *Ethiopia: An Economic Analysis of Technological Change in Four Agricultural Production Systems.* East Lansing: African Studies Center, Michigan State University, 1974.

Griaule, Marcel. "Labour in Abyssinia." *International Labour Review* 23 (1931), 181-202.

Griaule, Marcel. "Le travail sur l'aire au Wollo (Abyssinie)." *Journal de la Société des Africanistes* 12 (1942), 81-86.

Grottenelli, Vinigi. *Richerche geografiche ed economiche sulle populazione.* Vol. 2. *Missione de Studio al Lago Tana.* Roma: Reale Accademia d'Italia, 1939.

Gryseels, Guido. "Role of Livestock on Mixed Smallholder Farms in the Ethiopian Highlands: A Case Study from the Baso and Worena Wereda near Debre Berhan." Ph.D. dissertation, University of Wageningen, 1988.

Gryseels, Guido, and Frank Anderson. *Research on Farm and Livestock Productivity in the Central Ethiopian Highlands: Initial Results, 1977-1980.* Addis Ababa: International Livestock Centre for Africa, 1983.

Gryseels, Guido, and Samuel Jutzi. *Regenerating Farming Systems after Drought: ILCA's Ox/Seed Project, 1985 Results.* Addis Ababa: International Livestock Centre for Africa, 1986.

Gryseels, Guido, Abiye Astatke, Frank Anderson, and Getachew Asemenew. "The Use of Single Oxen for Crop Cultivation in Ethiopia," *ILCA Bulletin* 18 (1984), 20-25.

Gryseels, Guido, Frank Anderson, Jeffrey Durkin, and Getachew Asemenew. *Draught Power and Smallholder Grain Production in the Ethiopian Highlands.* Addis Ababa: nternational Livestock Centre for Africa, n.d.

Guluma Gemeda. "Some Notes on Food Crop and Coffee Cultivation in Jimma and Limmu Awrajas, Kafa Administrative Region (1950-1970s)." In *Proceedings of the Third Annual Seminar of the Department of History.* 91-102. Addis Ababa: Addis Ababa University, 1986.

Guyer, Jane. "Synchronizing Seasonalities: From Seasonal Income to Daily Diet in a Partially Commercialized Rural Economy (Southern Cameroon)." In David Sahn, ed., *Causes and Implications of Seasonal Variation in Household Food Security.* 137-50. Washington D.C.: International Food Policy Research Institute, 1987.

Haile M. Larebo. "The Myth and Reality of Empire Building: Italian Land Policy and Practice in Ethiopia (1935-41)." Ph.D. thesis, University of London, 1990.

Hakansson, Thomas. "Social and Political Aspects of Intensive Agriculture in East Africa," *Azania* 24 (1989), 12-20.

Hallpike, C. R. "Konso Agriculture." *Journal of Ethiopian Studies* 8 (1970), 31-43.

Hancock, Graham. *Ethiopia: The Challenge of Hunger.* London: Victor Gollancz Ltd., 1985.

Haque, I., Desta Beyene, and Marcos Sahlu. *Bibliography on Soils, Fertilizers, Plant Nutrition, and General Agronomy in Ethiopia.* Addis Ababa: International Livestock Centre for Africa, 1985.

Harms, Robert. *Games against Nature.* New York: Cambridge University Press, 1987.

Harris, W. Cornwallis. *The Highlands of Ethiopia.* 2nd. ed., 3 vols. London: Longman, Brown, Green, and Longmans, 1844.

Heuglin, M. Theodore von. *Reise nach Abessinien: Den Gala-Landern, Ost-Sudan und Chartúm in den Jahsen 1861 und 1862.* Bem: Hermann Coftenoble, 1868.

Heuglin, M. Theodore von. *Reisen in Nord-Ost-Afrika.* Gotha: Verlag von Fuftus Berthes, 1857.

Hobbs, Wesley, and Leonard F. Miller. *A Proposed Grain Storage Programme for Ethiopia.* Dire Dawa: n.p., 1965.

Hoben, Allan. *Land Tenure among the Amhara of Ethiopia.* Berkeley: University of California Press, 1973.

Hoben, Allan "Family, Land, and Class in Northwest Europe and Northern Ethiopia." In Harold G. Marcus, ed., *Proceedings of the First United States Conference on Ethiopian Studies.* 151-70. East Lansing, African Studies Center, Michigan State University, 1975.

Horvath, Ronald. "The Wandering Capitals of Ethiopia," *Journal of African History* 10 (1969), 205-19.

Huffnagel, H. P. *Agriculture in Ethiopia.* Rome: FAO, 1961.

Humphreys, Charles. "An Empirical Investigation of Factors Affecting Peasant Crop Production (Based on a Survey of Ada Wereda, Ethiopia)." Ph.D. dissertation, Fletcher School of Law and Diplomacy, Tufts University, 1975.

Hurni, Hans. "Degradation and Conservation of the Soil Resource in the Ethiopian Highlands." Paper presented at the First International Workshop on African Mountains and Highlands, Addis Ababa, 1986.

Iliffe, John. "The Origins of African Population Growth." *Journal of African History* 30 (1989), 165-69.

International Livestock Centre for Africa (ILCA). *Animal Power for Improved Management of Deep Black Clay Soils (Vertisols) in the Ethiopian Highlands, On-Farm Technology Evaluation on the High Elevation Plateau of Were Ilu/Degolo (Southern Wello Province).* Addis Ababa: ILCA, 1988.

International Livestock Centre for Africa. *Animal Power for Improved Management of Vertisols in High Rainfall Areas of Highland Ethiopia: Report of Phase 1 and Programme for Phase 2 (1989-1991).* Addis Ababa: ILCA, 1989.

International Livestock Centre for Africa. *Animal Power for Improved Management of Deep Black Clay Soils (Vertisols) in the Ethiopioan Highlands: Outreach Sub-Project Wereta (Gonder Province). Final Report.* Addis Ababa: ILCA, 1988.

International Livestock Centre for Africa. *Outreach Sub-Project Deneba/Inewari (Shewa Province): Progress Report, Work Plan, and Budget, 1988.* Addis Ababa: ILCA, 1988.

Isenberg C. W. and J. L. Krapf. *The Journals of Rev. Mssrs. Isenberg and Krapf, Detailing Their Proceedings in the Kingdom of Shoa and Journeys in other Parts of Abyssinia.* London: Frank Cass, 1968.

Istituto Agricolo Coloniale. *Main Features of Italy's Action in Ethiopia 1936-1941*. Florence: Istituto Agricolo Coloniale, 1946.

Istituto Agricolo Coloniale. *Relazione di una missione di agricoltura in A.O.I.* Relazione e monografiche agrario-coloniale, No. 43. Firenze: Istituto Agricolo Coloniale Italiano, 1937.

Johnston, Charles. *Travels in Southern Abyssinia through the Country of the Adal to the Kindom of Shoa during the Years 1842-43*. 2 vols. London: J. Madden, 1844.

Jutzi, Samuel C. "Contribution of Vertisols to National Food Self-Sufficiency: Actuals and Potentials." Paper presented to the Tenth Anniversary Conference of the Ethiopian Journal of Agricultural Sciences, Addis Ababa, October 1988.

Kaplan, Steven. "Kifu-Qen: The Great Famine of 1888-1892 and the Beta Israel (Falasha)." *Paideuma* 36 (1990), 68-77

Kaplan, Steven. *The Monastic Holy Man in Ethiopia*. Wiesbaden: Franz Steiner Verlag, 1984.

Kassahun Seyoum, Steven Franze, and Tesfaye Kumsa. "Initial Results of Informal Survey of Coffee Producing Areas of Manna and Gomma Woredas, Kefa Region." Working Paper No. 4/88. Department of Agricultural Economics and Farming Systems Research, Institute of Agricultural Research, Addis Ababa, 1988.

Kassahun Seyoum, Hailu Tafesse, and Steven Franzel. *The Profitability of Coffee and Maize among Smallholders: A Case Study of Limu Awraja, Ilubabor Region*. Research Report No. 10. Addis Ababa: Institute of Agricultural Research, 1990.

Kjekshus, Helge. *Ecology Control and Economic Development in East African History: The Case of Tanganyika, 1850-1950* . Berkeley: University of California Press, 1977.

Kloos, Helmut. "Development, Drought, and Famine in the Awash Valley of Ethiopia." *African Studies Review* 25 (1982), 21-48.

Kobischanov, Yuri. *Axum*. University Park: Pennsylvania State University Press, 1979.

Lauro, Raphael di. "Panorama politico-economico dei Galla e Sidama." *Rassengna Economica dell'Africa Italiana* (1938), 1084-87.

Lefebvre, Théophile, A. Petit, Vignaud Quartin-Dillon.. *Voyage en Abyssinie exécuté pendant les années 1839, 1840, 1841, 1842, 1843*. 6 vols. Paris: Arthus Bertrand, 1845-48.

Lenin, Vladamir. *The Development of Capitalism in Russia*. Moscow: Progress Press, 1977.

LeRoy Ladurie, Emmanuel *Times of Feast, Times of Famine: A History of Climate since the Year 1000*. Garden City: Doubleday and Co., 1971.

LeRoy Ladurie, Emmanuel *The Peasants of Languedoc*. Urbana: University of Illinois Press, 1976.

Levine, Donald. *Wax and Gold: Tradition and Innovation in Ethiopian Culture*. Chicago: University of Chicago Press, 1965.

Lockhart, Donald, trans. *The Itinerário of Jerónimo Lobo*. London: Hakluyt Society, 1984.

Logan, W. E. M. *An Introduction to the Forests of Central and Southern Ethiopia*. Oxford: Oxford University Press, 1964.

Ludolph, J. *Ad suam Historiam Aethiopicam antehac editam Commentarius*. Frankfort: 1691.

McCann, James C. "Agriculture and African History." *Journal of African History* 31 (1991), 507-13.

McCann, James C. "Britain, Ethiopia, and the Lake Tana Dam Project, 1922-35." *International Journal of African Historical Studies* 14 (1981), 667-99.

McCann, James C. "Children of the House." In Suzanne Miers and Richard Roberts, eds., *The End of Slavery in Africa.* 332-56. Madison: University of Wisconsin Press, 1988.

McCann, James C. "Climate and Class in Ethiopia," Paper presented to the African Studies Association, Denver, 1987.

McCann, James C. "A Dura Revolution and Frontier Agriculture in Northwest Ethiopia, 1898-1920." *Journal of African History* 31 (1990), 121-34.

McCann, James C. "An Evaluation of Oxfam Hararge Projects: A Report Submitted to Oxfam (U.K.)." Unpublished report, Oxfam, August 1987.

McCann, James C. *From Poverty to Famine in Northeast Ethiopia: A Rural History.* Philadelphia: University of Pennsylvania Press, 1987.

McCann, James C. "A Great Agrarian Cycle? Productivity in Highland Ethiopia, 1900 to 1987." *Journal of Interdisciplinary History* 20 (Winter 1990), 389-416.

McCann, James C. "Households, Peasants, and the Push Factor in Northern Ethiopian History." *Review* 9 (1986), 369-411.

McCann, James C. "Households, Peasants, and Rural History in Lasta, Northern Ethiopia 1900-35." Ph.D. dissertation, Michigan State University, 1984.

McCann, James C. "The Political Economy of Dura: Frontier Agriculture on Ethiopia's Mazega." African Studies Center Working Paper, African Studies Center, Boston University, 1988.

McCann, James C. "Preliminary Report on Oxfam/ILCA Oxen Seed Project." Unpublished report to Oxfam America, July 1985.

McCann, James C. "Report to Oxfam America on Evaluation of Ox/Seed Project." Unpublished report, Boston, 1986.

McCann, James C. "The Social Impact of Famine in Ethiopia." In Michael Glantz, ed., *Drought and Hunger in Africa: Denying Famine a Future.* 245-68. Cambridge: Cambridge University Press, 1987

McClellan, Charles. "Perspectives on the Neftenya-Gabbar System: The Darasa Ethiopia." *Africa* (Rome) 33 (1978), 426-40.

McClellan, Charles. *State Transformation and National Integration: Gedeo and the Ethiopian Empire, 1895-1935.* East Lansing: African Studies Center, Michigan State University, 1988.

McIntire, John, Danile Bourzat, and Prabhu Pingal. *Crop Livestock Interactions in Sub-Saharan Africa.* Washington D.C.: World Bank, 1990.

Maddox, Gregory. "Njaa: Food Shortages and Famine in Tanzania between the Wars." *International Journal of African Historical Studies* 19 (1986), 17-34.

Mahetma-Sellasse Walda Masqal. *Zikre Nagar.* Addis Ababa: Artistic Printers, 1950.

Mandala, Elias. *Work and Control in a Peasant Economy: A History of the Lower Tchiri Valley in Malawi 1859-1960.* Madison: University of Wisconsin Press, 1990.

Manetti, Carlo. Guida panoramica dell'agricoltura in Africa Orientale. Roma: Edizioni Geometra Italiano, 1936.

Mangano, Guido. "Il problema cotoniero nell'A.O.I." In Sindacato Nazionale Fascista Techici Agricoli, *Agricoltura e impero,* 193-212. Roma: Tipogrfica Editrice Sallustiana, 1937.

Mantel-Niecko,J oanna. *The Role of Land Tenure in the System of the Ethiopian Imperial Government in Modern Times.* Warsaw: Wydawnictwa Universytetu Warszawkiego, 1980.

Marchini, Ascanio. "Piano di colonizzazione demografico nel territorio di Harrar." In Sindacato Nazionale Fascista Techici Agricoli, *Agricoltura e Impero,* 655-79. Roma: Tipogrfica Editrice Sallustiana, 1937.

Marcus, Harold G. *Ethiopia, Great Britain, and the United States, 1941-1974: The Politics of Empire*. Berkeley: University of California Press, 1983.

Marcus, Harold G. "The Infrastructure of the Italo-Ethiopian Crisis: Haile Sellassie, the Solomonic Empire and the World Economy, 1916-1936." In *Proceedings of the Fifth International Conference on Ethiopian Studies*, 559-68. Chicago: University of Illinois Press, 1979.

Marcus, Harold G. *The Life and Times of Menilek II: Ethiopia, 1844-1913*. Oxford: Clarendon Press, 1975.

Marcus, Harold G., ed.. *The Modern History of Ethiopia and the Horn of Africa: A Selectand Annotated Bibliography*. Stanford: Hoover Institution, 1972.

Markakis, John. *Ethiopia: Anatomy of a Traditional Polity*. Oxford: Clarendon Press, 1974.

Martini, Ferdinando. *Il diario Eritreo di Ferdinando Martini*. 3 vols. Firenze: Vallecchi,1946.

Martini, Ferdinando. "L'Italia e l'Eritrea." In *L'Eritrea economica*. Roma: Istituto Geografico de Agostini, 1913.

Martini, Ferdinando. *Nell'Affrica Italiana*. Milano: Fratelli Treves, 1895.

Martini, Ferdinando. *In Abissinia e fra i galla*. Florence: Tipografica di Enrico Ariani, 1895.

Martini, Ferdinando. *I miei trentacinque anni di missione nell'alta Etiopia*. 12 vols. Milan: Tipografica S. Giuseppe, 1886.

Maugini, Armando. "Agricoltura indigena e colonizzazione nell'A.O.I." Sindacato Nazionale Fascista Techici Agricoli, *Agricoltura e impero*. 61-81. Roma: Tipogrfica Editrice Sallustiana, 1937.

Maugini, Armando. "Appunti sulle prospettive agricole dell'Impero." *L'Agricoltura coloniale* 35 (1941), 409-14, 447-51.

-Maugini, Armando. "Preface." In Raffaele Ciferri, and Renzo Giglioli. *I cereali dell'Africa Italiana* Vol. 1: *I frumenti dell'Africa Orientale Italiana, studiati su materiali originali*. Firenze: Regio Istituto Agronomico per l'Africa Italiana, 1939.

Mazzoni, Guglielmo. "Le colture stagionale dell tradizione rurale scioana." L' Agricoltura coloniale 9-10 (1943), 225-31.

Memmo, G., F. Martoglio, and C. Aldini. "La cura della peste bovina nella Colonia Eritrea." *Bollettino della Società Africana Italiana* 23 (1904), 130-32.

Mengistu Lamma. *Masehafa tezeta zalaqa Lamma Haylu*. Addis Ababa: Berhanna Selam Press, 1959 E.C.

Merchant, Carolyn. *Ecological Revolutions: Nature, Gender, and Science in New England*. Chapel Hill: University of North Carolina Press, 1989.

Merid Wolde Aregay, "Land Tenure and Agricultural Productivity, 1500-1850." In *Proceedings of the Third Annual Seminar of the Department of History*, 115-30. Addis Ababa: Addis Ababa University, 1986.

Mesfin Wolde Mariam. *An Introductory Geography of Ethiopia*. Addis Ababa: Haile Sellassie I University Press, 1972.

Mesfin Wolde Mariam. *Rural Vulnerability to Famine in Ethiopia, 1958-77*. Addis Ababa: Vikas Publishing House and Addis Ababa University, 1984.

Mesfin Wolde Mariam. *Suffering under God's Environment: A Vertical Study of the Predicament of Peasants in North-Central Ethiopia*. Bern: African Mountains Association and Geographica Bernensia, 1991.

Meseret Showamare, *Climatic Records for ILCA research sites, 1985*. Addis Ababa: International Livestock Centre for Africa, 1986.

Messerli, B. and K. Aerni, eds., *Simen Mountains, Ethiopia*. Vol. 1: *Cartography and Its Application for Geographical and Ecological Problems*. Bern: Geographische Institut der Universitat Bern, 1978

Michels, Joseph. "The Axumite Kingdom: A Settlement Archaeology Perspective." In *Proceedings of the Ninth International Congress of Ethiopian Studies*, vol. 6, 173-83. Moscow: Nauka Publishers, 1988.

Miller, Joseph. "The Significance of Drought, Disease, and Famine in the Agriculturally Marginal Zones of West-Central Africa." *Journal of African History* 23 (1982), 20.

Ministero dell'Africa Italiana, *La legislazione agrario dell'Africa Italiana*. Roma Libreria dello Stato, 1941.

Ministry of Agriculture. *EPID Project Areas by Province*. Addis Ababa: Ministry of Agriculture, 1974.

Ministry of Agriculture. *General Agricultural Survey: Preliminary Report 1983/84 (1976 E.C.)*. vol. 2: *Kaffa*. Addis Ababa: Ministry of Agriculture, 1984.

Ministry of Agriculture. "Humera Agricultural Development Project (HADP)." Final draft report, Ministry of Agriculture, Addis Ababa, May 1972.

Mohammed Hassen. *The Oromo of Ethiopia: A History 1570-1860*. Cambridge: Cambridge University Press, 1990.

Montandon, George. *Au pays Ghimirra: Récit de mon voyage à travers le massif éthiopien (1909-1911)*. Neuchatel: Attinger Frères, 1913.

Mordini, Antonio. "Un riparo sotto roccia con pitture rupestri nell'Ambà Focadà." *Rassegna di Studi Etiopici* 1 (1941), 54-60.

Muhammad Hassan. "The Relationship between Harar and Its Surrounding Oromo between 1800-1887." B.A. thesis, Haile Sellassie I University, 1973.

Murphy, H. F. *A Report on the Fertility Status and Other Data on Some Soils of Ethiopia*. Alemaya: Haile Sellassie I College of Agriculture and Mechanical Arts, 1968.

Nastrucci, Mario. "Molitura dei cereali nell'Hararino." *L'Agricoltura Coloniale* 33 (1939), 495-504.

Nastrucci, Mario. "Raccolta, trebbiatura e conservazione dei cereali nell'Hararino." *L'Agricoltura Coloniale* 33 (1939), 370-74.

National Meteorological Services Agency. "Climate and Drought Conditions in Ethiopia." Unpublished paper prepared for the Scientific Roundtable on the Climate Situation and Drought in Africa. Addis Ababa, February 1984.

Neate, Paul. "Animal Traction and Vertisol Cropping," *ILCA Newsletter* 6, 1 (1987), 1-2.

Nesbitt, L. M. *Desert and Forest: The Exploration of the Abyssinian Danakil*. London: Jonathan Cape, 1937.

Nicholson, Sharon. "The Methodology of Historical Climate Reconstruction and Its Application to Africa," *Journal of African History* 20 (1979), 31-49.

Nistri, Pier Francesco. "Itinerari e studi agrari nella regione del Lago Tana." In *Missione di studio Lago Tana*. vol 1: *Relazione preliminari*, 161-74. Roma: Reale Accademia d'Italia, 1938.

Oklahoma State University. *The Agriculture of Ethiopia*. Stillwater: Oklahoma State University, 1964.

Oklahoma State University. *Ethiopian Farm Tools*. Alamaya: n.p., 1960.

Oklahoma State University. *Oklahoma State University in Ethiopia: A Decade of Progress*. Stillwater: OSU, n.d.

L'opera dell'Italia nei suoi territori Africana. Roma: Arte della Stampa, 1940.

Orent, Amon. "From the Hoe to the Plow: A Study in Ecological Adaptation" in *Proceedings of the Fifth International Conference on Ethiopian Studies.* 187-94. Chicago, 1979.

Oxfam America. *Ethiopia Relief and Development.* Boston: Oxfam America, 1988.

Palma, Silvana. "La fototeca dell'Istituto-Africano: Appunti e prolemi di un lavoro di riordino." *Africa* 44 (1989), 596-609.

Pankhurst, Helen. "The Value of Dung." In *Ethiopia: Problems of Sustainable Development: A Conference Report.* 75-88. Trondheim: University of Trondheim, 1989.

Pankhurst, Richard. *Economic History of Ethiopia, 1800-1935.* Addis Ababa: Addis Ababa University Press, 1968.

Pankhurst, Richard. "Ethiopian Agriculture." *Ethiopia Observer* (1957), 306-8.

Pankhurst, Richard. "The Great Ethiopian Famine of 1888-92: A New Assessment" *Journal of the History of Medicine and Allied Sciences* 21 (1966), 95-124, 271-94.

Pankhurst, Richard. *History of Ethiopian Towns from the Middle Ages to the Early Nineteenth Century.* Weisbaden: Franz Steiner Verlag, 1982.

Pankhurst, Richard. *History of Ethiopian Towns from the Mid-Nineteenth Century to 1935.* Wiesbaden: Steiner Verlag, 1985.

Pankhurst, Richard. "YaHedar Besheta," *Journal of Ethiopian Studies* 12 (1975), 103-31.

Pankhurst, Rita and Richard Pankhurst. "A Select Annotated Bibliography of Travel Books on Ethiopia," *Africana Journal* 9, 2 (1978), 113-32; 9, 3 (1978), 101-33.

Pankhurst, Sylvia. "Ambo Agricultural School." *Ethiopia Observer* 10 (1957), 309-10.

Pankhurst, Sylvia. "Bishoftu Agricultural Research Station." *Ethiopia Observer* 10 (1957), 330-33.

Pankhurst, Sylvia. "Coffee Cultivation, and Processing." *Ethiopia Observer* 10 (1957), 324-29.

Pankhurst, Sylvia. "Ethiopian Agriculture in Retrospect and Prospect." *Ethiopia Observer* 9 (October. 1957), 278-82.

Pankhurst, Sylvia. "Imperial College of Agriculture and Mechanical Arts." *Ethiopia Observer* 10 (1957), 312-17.

Pankhurst, Sylvia. "Jimma Agricultural School." *Ethiopia Observer* 10 (1957), 318-26.

Parkyns, Mansfield. *Life in Abyssinia.* London: John Murray and Sons, 1853.

Patterson, K. David. "The Demographic Impact of the 1918-19 Pandemic in Sub-Saharan Africa: A Preliminary Assessment." In *African Historical Demography,* Vol. 2, 402-29. Edinburgh: University of Edinburgh, 1981.

Pausewang, Siegfried. *Peasants, Land and Society: A Social History of Land Reform in Ethiopia.* London: Weltforum Verlag, 1983.

Pausewang, Siegfried, Fantu Cheru, Stefan Brüne, and Eshetu Chole, eds. *Ethiopia: Rural Development Options.* London: Zed Books Ltd., 1990.

Paviolo, Italo. "La coltivazione del caffe nell'A.O.I.." In Sindacato Nazionale Fascista Techici Agricoli, *Agricoltura e Impero,* 237-62. Roma: Tipografica Editrice Sallustiana, 1937.

Pearce, Nathaniel. *The Life and Adventures of Nathaniel Pearce.* 2 vols. London: Henry Colburn and Richard Bentley, 1831.

Pellegrineschi, A. V. *Etiopia: Aspetti economici.* Milano: Casa Editrice Giuseppe Principato, 1936.

Perham, Margery. *The Government of Ethiopia.* Oxford: Oxford University Press, 1959.

Piani, Giovanni. "L'agricoltura indigena nel Governo dell'Harar e i mezzi per farla progredire." *L'Agricoltura Coloniale* 38 (1939), 401-8.

Plowden, Walter. *Travels in Abyssinia and the Galla Country*. London: Gregg International Publisher, 1868; reprinted, 1972.

Pochen-Eiche, Peter. "The Application of Farming Systems Research to Community Forestry: A Case Study of the Hararge Highlands, Eastern Ethiopia." Ph.D. dissertation, Albert-Ludwigs Universitat, 1986.

Pollera, Alberto. *L'Abissinia di ieri: osservazioni e ricordi*. Roma: Scuola Tipografica Pio X, 1940.

La produzione dell'Etiopia. Roma: L'Economia Italiana, 1938.

Prouty, Chris. *Empress Taytu and Menilek II*. London: Raven Educational Press, 1986.

Quaranta, Ferdinand. *Ethiopia: An Empire in the Making*. London: P. S. King and Son Ltd., 1939.

Raffray, Achille. *Africa Orientale: Abyssinie*. Paris: E. Plon and Co., 1876.

Rasmusson, Eugene. "Global Climatic Change and Variability: Effects of Drought and Desertification in Africa."In Michael Glantz, ed., *Drought and Hunger in Africa: Denying Famine a Future*, 3-22. Cambridge: Cambridge University Press, 1987.

Raven-Roberts, Angela. "Report of Villagization in Oxfam America Assisted Project Areas in Hararge Province, Ethiopia." Unpublished report to Oxfam America, June 1986.

Reale Società Geografica Italiana, Roma. *L'Africa Orientale*. Bologna: Zanichelli, 1936.

Relazione di una missione di agricoltori in A.O.I. Firenze: Istituto Agricolo Coloniale Italiano, 1937.

Redd Barna-Ethiopia. "Food Availability Survey in Gera: Report to Redd Barna-Ethiopia." Gera Integrated Rural Development Project P.-4008, November 1989.

Relief and Rehabilitation Commission. *1987 Belg Production: Early Warning System Belg Synoptic Report*. Addis Ababa: Relief and Rehabilitation Commission, 1987.

Relief and Rehabilitation Commission. Combatting the Effects of Cyclical Drought in Ethiopia. Addis Ababa: Relief and Rehabilitation Commission, 1985.

Relief and Rehabilitation Commission. *Food Shortage Survey Report on Tigrai*. Addis Ababa: Relief and Rehabilitation Commission, 1979.

Relief and Rehabilitation Commission. "Food Shortages Survey Report on Wello (Lasta, Wadla Delanta, and Werehimeno Awrajas)." Unpublished typescript, Addis Ababa, 1978.

Relief and Rehabilitation Commission. *Food Shortages Survey Report on Wello*. Addis Ababa: Relief and Rehabilitation Commission, 1979.

Ricci, Lanfranco. "Enrico Cerulli." *Rassegna di Studi Etiopici* 32 (1988), 5-44.

Richards, Audrey. *Land, Labour, and Diet in Northern Rhodesia*. London: Oxford University Press, 1939.

Richards, Paul. "The Environmental Factor in African Studies." *Progress in Human Geography* 4 (1980), 589-600.

Robinson, Warren and Yamazaki, Famika."Agriculture, Population, and Economic Planning." *Journal of Developing Areas* 20(1986), 327-38.

Rochet d'Héricourt, Charles. *Voyage sur la côte orientale de la mer Rouge dan le pays d'Adel et le royaume de Choa*. Paris: Arthus Bertrand, 1841.

Rochet d'Héricourt, Charles. *Second voyage sur les deux rives de la Mer Rouge dan la pays des Adels et la Royaume de Choa*. Paris: Arthus Bertrand, 1846.

Roorda, T. M. M. "Soil Burning in Ethiopia: Some Effects on Soil Fertility and Physics." In S.C. Jutzi, J. McIntire, and J. E. S. Stares, *Management of Vertisols in Sub-Saharan Africa*. 124-25. Addis Ababa: International Livestock Centre for Africa, 1988.

Rosenfeld, Chris Prouty. *A Chronology of Menilek II of Ethiopia*. East Lansing: African Studies Center, Michigan State University, 1976.

Rotberg, Robert, and Theodore Rabb, eds. *Climate and History in Interdisciplinary History.* Princeton: Princeton University Press, 1981.

Rubenson, Sven. *The Survival of Ethiopian Independence*. London: Heinneman, 1976.

Rubenson, Sven. ed. *Acta Aethiopica.* Vol. 1: *Correspondence and Treaties, 1800-1854.* Evanston and Addis Ababa: Northwestern University Press and Addis Ababa University Press, 1987.

Rüppell, Eduard. *Reise in Abyssinien*. 2 vols. Frankfurt: Sigmund Schmerber, 1838-40.

Salt, Henry. *Twenty-Four Views in St. Helena, the Cape, India, Ceylon, the Red Sea, Abyssinia, and Egypt.*. London, 1809.

Salt, Henry. *A Voyage to Abyssinia and Travels in the Interior of that Country*. Philadelphia: M.Carey, 1816.

Samatar, Abdi Ismail. *The State and Rural Transformation in Northern Somalia, 1884-1986.* Madison: University of Wisconsin Press, 1989.

San Marzano, Roberto di. *Dalla piana somala all'altipiano Etiopico*. Roma: Edit. L'Azione Coloniale, 1935.

Santagata, F. *La colonia Eritrea nel Mar Rosso davanti all'Abissinia*. Naples: n.p., 1935.

Scarin, Emilio. *Hararino: Ricerche e studi geografici*. Firenze: G. C. Samsoni Editore, 1942.

Scott, James. *The Moral Economy of the Peasant: Rebellion and Subsistence in Southeast Asia*. New Haven: Yale University Press, 1976.

Sen, Amartya. *Poverty and Famine: An Essay in Entitlements and Deprivation.* Oxford: Oxford University Press, 1981.

Senni, Lorenzo. "Il problema forestale in A.O.I." In Sindacàto Nazionale Fascista Techici Agricoli, *Agricoltura e impero*, 401-418. Roma Tipografica Editrice Sallustiana, 1937.

Sillani, Tomaso. *L'Africa Orientale Italiana (Eritrea e Somalia)*. Roma: n.p., 1933.

Silver, Timothy. *A New Face on the Countryside: Indians, Colonists, and Slaves in South Atlantic Forests, 1500-1800*. New York: Cambridge University Press, 1990.

Simoons, Frederick. "The Agricultural Implements and Cutting Tools of Begemder and Semyen, Ethiopia." *Southwestern Journal of Anthropology* 14 (1958), 386-406.

Simoons, Frederick. *Northwest Ethiopia: Peoples and Economy*. Madison: University of Wisconsin Press, 1960.

Simoons, Frederick. "Some Questions on the Economic Prehistory of Ethiopia." in J.D. Fage and Roland Oliver, eds, *Papers in African Prehistory*, 124-29. Cambridge: Cambridge University Press, 1970.

Sindacato Nazionale Fascista Techici Agricoli. *Agricoltura e impero*. Roma: Tipografica Editrice Sallustiana, 1937.

Soleillet, Paul. *Voyages en Ethiopie (jan 1882-oct 1884): Notes, lettres et documents divers.* Rouen: Cagniard, 1886.

"La spedizione scientifica italiana in Etiopia." *Bollettino della Società Africana Italiana* 27 (1908), 226-28.

Stitz, Volker. "Distribution and Foundation of Churches in Ethiopia." *Journal of Ethiopian Studies* 13 (1975), 1, 17.

Stitz, Volker. "The Amhara Resettlement of Northern Shoa during the 18th and 19th Centuries." *Rural Africana* 11 (1970), 70-78.

Sukert, Ezio. "Il Caffa dal punto di vista agrologico." *L'Agricoltura Coloniale* 37 (1943), 234-39.

Sutton, John. "Towards a History of Cultivating the Fields." *Azania* 24 (1989), 98-103.

Taddesse Tamrat. Church and State in Ethiopia. Oxford: Clarendon Press, 1972.

Taddesse Tamrat. "Feudalism in Heaven and Earth: Ideology and Political Structure in Medieval Ethiopia." In Sven Rubenson, ed. Proceedings of the Seventh International Conference on Ethiopian Studies. Addis Ababa, Lund, East Lansing: Addis Ababa University Press, Scandanavian Institute of African Studies, and African Studies Center, Michigan State University, 1984.

Taddia, Irma. *Eritrea-Colonia: Paesaggi, strutture, uomini del colonialismo 1890-1952.* Milano: Franco Angeli, 1986.

Tancredi, Alfonso. "La missione della Società Geografica Italiana in Etiopia settentrionale." *Bollettino della Società Geografica Italiana* 45 (1908), 1199-1250.

Tancredi, Alfonso. Nel piano del sale." *Bolletino della Società Geografica Italiana* 48 (1911), 57-84, 150-78.

Taye Mengistae. "Urban-Rural Relations in Agrarian Change. An Historical Overview." In Siegfried Pausewang, Fantu Cheru, Stefan Brüne, and Eshetu Chole, eds., *Ethiopia: Rural Development Options*, 30-37.London: Zed Press, 1990.

Tekeste Negash. *Italian Colonialism in Eritrea.* Uppsala: Scandanavian Institute of African Studies, 1987.

Teruzzi, Attilio. *L'Africa italiana nel secondo anno dell'impero.* Roma: Istituto Poligrafico dello Stato, 1938.

Tesema Ta'a. "The Basis for Political Contradictions in Wollega: The Land Apportionment Act of 1910 and Its Consequences." *Northeast African Studies* 6 (1989), 179-98.

Teshome Mulat. "Ada Baseline Survey. Part 4: Credit and Indebtedness in Rural Ada Woreda." IDR Research Document no, 16, 1974.

Tewolde Berhan Gebre Egziabher. "Land Management and Changes in Land Tenure in Ethiopian History." Unpublished typescript, 1989.

Tewolde Berhan Gebre Egziabher. "Historical and Socio-economic Basis of the Ethiopian Ecological Crisis." Unpublished typescript, Asmara University, March 1989.

Traversi, Leopoldo. "Da Entoto al Zuquala. Lettere del dott. Leopoldo Traversi al sig. conte Buturlin." *Bollettino della Società Geografica Italiana* 24 (1887), 583.

Traversi, Leopoldo. *Let Marefià: Prima stazione geografica italiana nello Scioa e le nostre relazioni con Etiopia (1876-1896).* Milano: Relae Società Geografica Italiana, 1931.

Trimingham, J. Spencer. *Islam in Ethiopia.* London: Frank Cass and Co., 1965.

Tsegaye Wodajo. *Agrometeorological Activities in Ethiopia: Prospects for Improving Agroclimatic/Crop Condition Assessment.* Columbia, Mo.: National Oceanic and Atmoshperic Administration, 1984.

Ufficio Studi del Ministero Africa Italiana. "Lineamenti della Legislazione per l'impero." *Gli Annali dell'Africa Italiana* 2, 3 (1939), 1-160.

Ufficio Studi del Ministero Africa Italiana. "La recognizione scientifica." *Gli Annali di Africa Italiana* 3 (1940), 933-72.

Ufficio Studi del Ministero Africa Italiana. "La valorizzazione agraria e la colonizzazione." *Gli Annali dell'Africa Italiana* 2, 3 (1939), 179-316.

United Nations Food and Agriculture Organization. *Vegetation and Natural Resources of Ethiopia and Their Significance for Land Use Planning.* Rome: United Nations Food and Agriculture Organization, 1982.

United Nations Development Programme and World Bank. *Ethiopia: Issues and Options in the Energy Sector.* Report no. 4741-ET.

Valentia, George Annesley, Viscount. *Voyages and Travels to India, Ceylon, and the Red Sea, Abyssinia, and Egypt, in the Years 1802, 1803, 1804, 1805, and 1806.* 3 vols. London: William Miller, 1809.

Vanderhym, J.G. *Une expédition avec le négous Ménélik.* Paris, 1896.

Varvikko, Tuomo. *Development of Appropriate Feeding Systems for Dairy Cattle in the Ethiopian Highlands. Research and Development Project in Collaboration between International Livestock Centre for Africa and the Ministry of Agriculture, Ethiopia.* Addis Ababa: International Livestock Centre for Africa, 1991.

Vavilov, N. L. The Origin, Variation, *Immunity, and Breeding of Cultivated Plants.* Waltham: Chronica Botanica, 1951.

Virgo, V. T., and R. N. Mamo. "Soil and Erosion Features of the Central Plateau Region of Tigre." *Geoderma,* 2 (1978), 131-58.

Voll, Sarah P. "Cotton in Kassala: The Other Scheme." *Journal of African Studies* 5, 2 (1978), 214-15.

Walsh, R.P.D. "The Nature of Climate Seasonality." In Robert Chambers, et. Richard Longhurst, and Arnold Pacey, eds., *Seasonal Dimensions to Rural Poverty,* 11-29. London: Fancis Pinter Lted., 1981.

Watt, Ian. "Regional Patterns of Cereal Production and Consumption." in Zein Ahmed Zein and Helmut Kloos, eds, *The Ecology of Health and Disease in Ethiopia* 94-135. Addis Ababa: Ministry of Health, 1988.

Watts, Michael "The Sociology of Seasonal Food Shortage in Hausaland." In Robert Chambers, et. Richard Longhurst, and Arnold Pacey, eds., *Seasonal Dimensions to Rural Poverty,* 201-6. London: Fancis Pinter Lted., 1981.

Webb, Patrick, Joachim von Braun, and Yisehac Yohannes. "Famine in Ethiopia: Policy Implications of Coping Failure at National and Household Levels." Draft MSS. Washington, D.C.: International Food Policy Research Institute, April 1992.

Webster, J. B. *Chronology, Migration, and Drought in Interlacustrine Africa.* New York: Africana, 1979.

Weissleder, Wolfgang. "The Political Ecology of Amhara Domination," Ph.D. dissertation, University of Chicago, 1965, 172-74.

Westphal, E. *Agricultural Systems in Ethiopia.* Wageningen: Center for Agricultural Publishing and Documentation, 1975.

Whalen, Irene. Untitled report to International Livestock Centre for Africa. 1985.

Wilks, Ivor. "Land, Labour, Capital and the Forest Kingdom of Asante: A Model of Early Change." In J. Friedman and M. J. Rowlands, eds. *The Evolution of Social Systems* , 487-534. Pittsburgh: University of Pittsburgh Press, 1978.

Wolde Michael Kelecha. *A Glossary of Ethiopian Plant Names.* 4th ed. Addis Ababa: Artistic Printers, 1987.

Woobeshet Shibeshi. "A Regional Study of Angolela." B.A. thesis, Haile Sellassie I University, 1970.

Wood, Adrian. "The Decline of Seasonal Labour Migration to the Coffee Forests of South-West Ethiopia," *Geography* 68 (1983), 53-56.

Wood, Adrian. "A Preliminary Chronology of Ethiopian Droughts." In David Dalby, R. J. Harrison, and F. Bezzasz, eds., *Drought in Africa-2,* 68-73. African Environment Special Report 6. London: n.p., 1977.

Workineh Degefu. "Some Aspects of Meteorological Drought in Ethiopia." In Michael Glantz, ed. *Drought and Hunger in Africa: Denying Famine a Future*, 23-36. Cambridge: Cambridge University Press, 1987.

World Bank. *Ethiopia: Public Investment Program Review*. Washington, D.C.: World Bank, 1988.

Worster, Donald. *Dust Bowl: The Southern Plains in the 1930s*. New York: Cambridge University Press, 1979.

Worster, Donald. *Rivers of Empire: Water Aridity, and the Growth of the American West*. New York: Cambridge University Press, 1985.

Wrigley, C. C. "Population in African History." *Journal of African History* 20 (1979), 129-31.

Wylde, Augustus. *Modern Abyssinia*. London: Methuen, 1901.

Yamane Kidane and Yilma Habteyes. "Food Losses in Traditional Storage Facilities in Three Areas of Ethiopia." In *Towards a Food and Nutritional Strategy for Ethiopia: The Proceedings of the National Workshop on Food Strategies for Ethiopia*, 407-30. Addis Ababa: Ethiopian Nutritional Institute, 1989.

Zanutto, Silvia. "Bibliografica dell'Africa Orientale Italiana." *Gli Annali de Africa Italiana* 4 (1941), 1335-405.

Zein Ahmed Zein and Helmut Kloos. *The Ecology of Health and Disease in Ethiopia*. Addis Ababa: Ministry of Health, 1988.

Zervos, Adrien. *L'Empire d'Ethiopie: le miroir de l'Ethiopie moderne, 1906-1935*. Athens: 1936.

INDEX

Abbay River. *See* Blue Nile river

Ada, 9; crater lakes of, 13; 1975 study of, 16; crop rotation in, 56; compared to Ankober, 193; geographic description, 194; as closed microagronomic economy, 194–97; church foundation in, 196, 198–99; climate, 196–97; settlement history, 197–99; on Shawan periphery, 197–204; as imperial pantry, 201; population increase, 216–22

Ada District Development Project (ADDP), 231–32

Addis Ababa, 9, 185, 186, 205; rainfall patterns, 31; growth of, 53, 219; foundation of, 113, 200, 201, 240; feeding of, 118; and rise of coffee-maize complex; growth of, 193; markets, 208; as center of policy decisions, 238; as market for food, 240–41, 242

Afallo (Gera): description of, 164; mission station, 164, 169; population decline in, 170–71; coffee production in, 184

Agaro, 177, 180–82

Agawmeder, 50

Agricultural history: methods for, 8–11; use of narrative in, 19

Agricultural Marketing Corporation (AMC): 185, 233, 251–52

Agriculture: as environmental history, 7; evolution of highland, 40

Agronomy: in Ankober, 129–37; Oromo seedbed flooding, 186–87. *See also* Microagronomic economies, Irrigation, Terracing

Aliyu Amba, 113, 128

Almeida, Manoel de, 12; on deforestation, 37; on livestock, 48; on ensete, 54–55; on teff, 55; on women's labor, 75

Alvarez, Francisco, abundance, 3, 12; reference to maize, 14; livestock exchange, 49; crops in Lasta, 51; on Tigray monastery garden, 52; on marriage and divorce, 73–74; on Axumite

irrigation, 97; on teff, 102

Amba Focada, 39 *fn1*, 48

Ambilineal descent, 72, 77

Amhara: expansion of Amhara-Tigrayan state, 40–44; as social category, 42; interaction with Oromo in Ada, 197–99

Amharic (language): use of in written sources, 8–9; dominance of, 182

Ankober, 249; farmers' response to 1985 tax, 7, 9; travelers to, 12; cotton in, 53; intensive agriculture in, 85; nineteenth-century population of, 89; marginal lands at, 137–43; loss of status as *madebet*, 202

Annual cropping, 71; arrival in southwest, 153; requirements of, 160–61; and population density in forest, 162, 189; gendered nature of, 255; dominant position in southern systems, 260

Antinori, Orazio, 115, 125–26

Antonelli, Pietro, 90, 125

Argobba, 116, 137–40, 142

Arsi, 18. *See also* Arussi

Arussi, 246. *See also* Arsi

Awash Valley, 18, 24, 137–38, 197, 249

Axum: relation of agriculture to, 4, 44; remains of irrigation at, 13; role of environment in decline of, 33; South Arabian influence on agriculture of, 39; dryland agriculture in, 39–40; intensive agriculture in, 40, 85; sources of food for, 50, 81; irrigation in, 61, 97–98

Babicheff plantation, 208, 213, 221

Badada Kilole, 197, 228–29

Bananas, 111. *See also* Fruit

Barley, 44, 51, 101–2, 135. *See also* Crop repertoire

Bauer, Dan F., 94

Belg, 15; effects on food production, 31. *See also* Climate

Bent, J. Theodore, 13, 97–98

Bernatz, Johan, 114, 125

Bianchi, Gustavo, 202

Bias in sources, 14–15

Bishoftu, 17, 207, 210, 211; Italian development of, 211–13; name change of, 213; Agricultural Research Station, 243. *See also* Dabra Zayt

Blue Nile River, 26, 32, 147